GIS AND REMOTE SENSING APPLICATIONS IN BIOGEOGRAPHY AND ECOLOGY

THE KLUWER INTERNATIONAL SERIES
IN ENGINEERING AND COMPUTER SCIENCE

UNIVERSITIES AT MEDWAY
DRILL HALL LIBRARY
SHORT LOAN

This book must be returned or renewed by the date printed on your self service receipt, or the last date stamped below if applicable. Fines will be charged as soon as it becomes overdue.

2 1 OCT 2009

UNIVERSITIES
at MEDWAY

WITHDRAWN
FROM
UNIVERSITIES
AT
MEDWAY
LIBRARY

GIS AND REMOTE SENSING APPLICATIONS IN BIOGEOGRAPHY AND ECOLOGY

edited by

Andrew C. Millington
*University of Leicester
United Kingdom*

Stephen J. Walsh
*University of North Carolina
Chapel Hill, U.S.A.*

Patrick E. Osborne
*University of Stirling
United Kingdom*

KLUWER ACADEMIC PUBLISHERS
Boston / Dordrecht / London

Distributors for North, Central and South America:
Kluwer Academic Publishers
101 Philip Drive
Assinippi Park
Norwell, Massachusetts 02061 USA
Telephone (781) 871-6600
Fax (781) 871-6528
E-Mail <kluwer@wkap.com>
Distributors for all other countries:
Kluwer Academic Publishers Group
Distribution Centre
Post Office Box 322
3300 AH Dordrecht, THE NETHERLANDS
Telephone 31 78 6576 000
Fax 31 78 6576 474
E-Mail <services@wkap.nl>
Electronic Services <http://www.wkap.nl>

 Library of Congress Cataloging-in-Publication Data

GIS and remote sensing applications in biogeography and ecology / edited by Andrew C. Millington, Stephen J. Walsh, Patrick E. Osborne.
    p.cm.
  Book arose from two joint symposia held in Leicester and Honolulu in 1999.
  Includes bibliographical references (p.).
  ISBN 0-7923-7454-1 (alk. paper)
   1. Biogeography—Remote sensing. 2. Ecology—Remote sensing. 3. Geographic information systems. I. Millington, A. C. II. Walsh, Stephen Joseph, 1951- III. Osborne, Patrick E.

QH84 .G47 2001
578'.09—dc21

2001038306

The cover is a composite image of Lee Ridge and vicinity, Glacier National Park, Montana, USA. The image is composed of a 1999 Landsat Thematic Mapper (TM) satellite image, an orthophoto quadrangle (b&w), and an ADAR 5500 digital aircraft image referenced to a 30 m Digital Elevation Model (DEM).

The Landsat TM image is visualized using a channel 4, 3, 2 color composite and represented using a RGB (Red-Green-Blue) color model. The ADAR 5500 image was collected at a spatial resolution of 1 m and processed using the Normalized Difference Vegetation Index (NDVI) to show the "green wave" across Lee Ridge. The image is used to assess climate change at the alpine treeline ecotone as part of research funded by the US Geological Survey, Biological Research Division. Cover art by Sean McKnight and Stephen J. Walsh, Landscape Characterization & Spatial Analysis Lab, Department of Geography, University of North Carolina - Chapel Hill, USA.

Copyright © 2001 by Kluwer Academic Publishers. Second Printing 2004.
All rights reserved. No part of this publication may be reproduced, stored in a retrieval system or transmitted in any form or by any means, mechanical, photo-copying, recording, or otherwise, without the prior written permission of the publisher, Kluwer Academic Publishers, 101 Philip Drive, Assinippi Park, Norwell, Massachusetts 02061
Printed on acid-free paper.
Printed in the United States of America

The Publisher offers discounts on this book for course use and bulk purchases.

TABLE OF CONTENTS

Co-Editors: Abbreviated Profiles ix

Acknowledgements: List of Reviewers x

1. **Introduction – Thinking Spatially**

    Andrew C. Millington, Stephen J. Walsh, and Patrick E. Osborne 1

2. **A Spectral Unmixing Approach to Leaf Area Index (LAI) Estimation at the Alpine Treeline Ecotone**

    Daniel G. Brown 7

3. **The Utilization of Airborne Digital Multispectral Image Dynamics and Kinematic Global Positioning Systems for Assessing and Monitoring Salt Marsh Habitats in Southern California**

    Douglas Stow, Alice E. Brewster, and Brian K. Bradshaw 23

4. **Spatial Variability in Satellite-Derived Seasonal Vegetation Dynamics**

    Simon D. Jones, Andrew C. Millington, and Barry K. Wyatt 47

5. **Documenting Land-Cover History of a Humid Tropical Environment in Northeastern Costa Rica Using Time-Series Remotely Sensed Data**

    Jane M. Read, Julie S. Denslow, and Sandra M. Guzman 69

6. **Patterns of Change in Land-Use and Land-Cover and Plant Biomass: Separating Intra- and Inter-Annual Signals in Monsoon-Driven Northeast Thailand**

    Stephen J. Walsh, Kelley A. Crews-Meyer, Thomas W. Crawford, William F. Welsh, Barbara Entwisle, and Ronald R. Rindfuss 91

7. **Barriers and Species Persistence in a Simulated Grassland Community**

   David M. Cairns 109

8. **Feedback and Pattern in Computer Simulations of the Alpine Treeline Ecotone**

   Matthew F. Becker, George P. Malanson, Kathryn J. Alftine, and David M. Cairns 123

9. **Spatial Pattern and Dynamics of an Annual Woodland Herb**

   Lucy Bastin and Chris D. Thomas 139

10. **Spatial Analysis of Micro-Environmental Change and Forest Composition in Belize**

    Peter A. Furley, Malcolm Penn, Neil M. Bird, and Malcolm R. Murray (with a contribution from Doug R. Lewis) 163

11. **The Radiate Capitulum Morph of Senecio Vulgaris L. within Sussex: the Use of GIS in Establishing Origins**

    Stephen Waite and Niall Burnside 179

12. **A Geographical Information Science (GISci) Approach to Exploring Variation in the Bush Cricket *Ephippiger Ephippiger***

    David M. Kidd and Michael G. Ritchie 193

13. **The GIS Representation of Wildlife Movements: A Framework**

    Ling Bian 213

14. **Stratified Sampling for Field Survey of Environmental Gradients in the Mojave Desert Ecoregion**

    Janet Franklin, Todd Keeler-Wolf, Kathryn A. Thomas, David A. Shaari, Peter A. Stine, Joel Michaelsen, and Jennifer Miller 229

15. **Development of Vegetation Pattern in Primary Successions on Glacier Forelands in Southern Norway**

    Günter A. Grimm 255

16. **Multi-Scale Analysis of Land-Cover Composition and Landscape Management of Public and Private Lands in Indiana**

    Tom P. Evans, Glen M. Green, and Laura A. Carlson 271

17. **Shifting Cultivation Without Deforestation: A Case Study in the Mountains of Northwestern Vietnam**

    Jefferson Fox, Stephen Leisz, Dao Minh Truong, A. Terry Rambo, Nghiem Phuong Tuyen, and Le Trong Cuc 289

18. **Linking Biogeography and Environmental Management in the Wetland Landscape of Coastal North Carolina**
    The difference between nationwide and individual wetland permits

    Nina M. Kelly 309

Index 329

CO-EDITORS: ABBREVIATED PROFILES

Andrew C. Millington is Professor of Geography at the University of Leicester, where he is currently head of department. He lectured previously at the Universities of Reading (UK) and Sierra Leone, and has been a Visiting Lecturer at the University of Gent and University College, Dublin. He is the inaugural chair of the IGU Biogeography Study Group (1996-2000) and honorary editor of *The Geographical Journal*. He has published over 50 scientific articles and edited five books, including *Vegetation Mapping* with R. Alexander (1999) and *Environmental Change in Dryland Environments* with K. Pye (1994). His research interests focus on the use of remotely-sensed data to map and monitor soils and vegetation in arid and humid tropical environments, modelling habitat and, more generally, in natural resource management. He has carried out research for the EU, UK and Dutch Governments, and the World Bank.

Stephen J. Walsh is a Professor of Geography, Director of the Landscape Characterisation & Spatial Analysis Lab (LC&SAL), and Research Fellow of the Carolina Population Centre at the University of North Carolina, Chapel Hill, USA. He is the former Amos H. Hawley Professor of Geography (1993-1996) and Director of the Spatial Analysis Unit at the Carolina Population Centre (1992-1997). He is the past Chair of the GIS (1997-1999) and the Remote Sensing (1995-1997) Specialty Groups of the Association of American Geographers (AAG). In 1997, Professor Walsh was the recipient of the Outstanding Contribution Award and Medal from the Remote Sensing Specialty Group of the AAG, in 1999 was award Research Honours from the Southeastern Division of the AAG, and in 2001 was awarded Research Scholarship Honours from the Association of American Geographers. He is on the editorial boards of *Plant Ecology* (1996-present), *Journal of Geography* (1995-present), *The Professional Geographer* (1997-2001), and the *Southeastern Geographer* (1992-2000). He also has recently co-authored special remote sensing and GIS application issues in the *Journal of Vegetation Science* (1994) and *Geomorphology* (1998). Professor Walsh's research interests are in GIS, remote sensing, spatial analysis, physical geography, and population-environment interactions. Specific research foci are in pattern and process at the alpine treeline ecotone, biodiversity and river dynamics, scale dependence and information scaling, and landuse/landcover change modelling. Major studies are underway in Ecuador, Thailand, and North Carolina and Montana, USA.

Patrick E. Osborne lectures in Environmental Management at the University of Stirling, Scotland, UK, and is co-director of the masters' programme. He was previously the Head of Ecology at the National Avian Research Center in Abu Dhabi where he directed research on conservation issues and compiled the book *The Desert Ecology of Abu Dhabi* (1996). He has travelled extensively and worked in Europe, Lesotho, Kenya, Mexico, the Middle East, Pakistan and Kazakhstan. Dr Osborne's research interests are in spatial ecology and landscape ecology, and the uses of new technologies in conservation biology, particularly birds. He is currently working on large-scale predictive models of bird distributions based on digital cartography and satellite imagery. His research students are studying the effects of forest fragmentation on birds, building simulation models of forests succession, comparing correlative mapping and Individual-Based Behaviour Models for predicting areas used by grazing geese, studying forest cover change in Nigeria, and devising conservation strategies for woodland grouse.

ACKNOWLEDGEMENTS -- LIST OF REVIEWERS

Thomas R. Allen, Old Dominion University
Stephen J. McGregor, University of North Carolina - Chapel Hill
Joseph P. Messina, Michigan State University
John Kupfer, University of Arizona
Duane Griffin, Bucknell University
Paul Mausel, Indiana State University
Stephen R. Yool, University of Arizona
David R. Butler, Southwest Texas State University
Ling Bian, State University of New York - Buffalo
Nina M. Kelly, University of California - Berkeley
Daniel G. Brown, University of Michigan
Stephen J. Walsh, University of North Carolina - Chapel Hill
Kelley A. Crews-Meyer, University of Texas - Austin
Steve Blackmore, Royal Botanic Gardens – Edinburgh
John Proctor, University of Stirling - Stirling
Godfrey Hewitt - University of East Anglia - Norwich
Dan Howard - New Mexico State University - Las Cruces
D. McCollin, University College Northampton - Northampton
Mike Hutchings, University of Sussex - Brighton
J.A. Matthews, University College of Swansea, Swansea
O.R. Vetaas, University of Bergen – Bergen, Norway

Chapter 1

INTRODUCTION – THINKING SPATIALLY

Andrew C. Millington
Department of Geography, University of Leicester, UK
acm4@leicester.ac.uk.

Stephen J. Walsh
Department of Geography, University of North Carolina, USA
swalsh@email.unc.edu

Patrick E. Osborne
Department of Environmental Management, University of Stirling, UK
p.e.osborne@stir.ac.uk.

Keywords:	geographic information science, remote sensing, biogeography, ecology.
Context	The idea for the book arose from two joint symposia that were organised by the Biogeography Study Group of the International Geographical Union; the Biogeography, Remote Sensing, and GIS Speciality Groups of the Association of American Geographers, and the Biogeography Research Group of the Royal Geographical Society-Institute of British Geographers and held in Leicester and Honolulu in 1999. These groups represent the majority of geographers who are conducting research in biogeography and ecology, and teaching this material to geographers.
	Many members of these groups have strong links with ecologists and biogeographers in biological sciences, but as geographers they have particular insights into biogeography and ecology, which arise from their training in spatial digital technologies that have been collectively referred to as Geographic Information Science (GISc). It is this issue – the application of spatial concepts and techniques of GISc to biogeography and ecology – that has formed the focus of the two symposia and this book. This material is increasingly being covered in a variety of disciplines and sub-disciplines (e.g., large-area ecology, landscape ecology, remote sensing and GIS), but many researchers in these fields lack the training in spatial concepts behind the techniques that they utilise. The spatial concepts that are covered in this book are richer than those found within landscape ecology at the present time, and we suspect that this book will promote the use of many of these concepts amongst landscape ecologists. We feel that it is timely to gather together a representative set of examples of the many varied spatial techniques and analytical approaches that are

being used by geographers, ecologists, and biogeographers to study plant and animal distributions, to assess processes affecting the observed patterns at selected spatial and temporal scales, and to discuss these examples within a strong conceptual spatial and/or temporal framework. Therefore, the aims of this volume are to:

- Identify the key spatial concepts that underpin GISc in biogeography and ecology;
- Review the development of these spatial concepts within geography and how they have been taken up in ecology and biogeography;
- Exemplify the use of the key spatial concepts underpinning GISc in biogeography and ecology using case studies from both vegetation science and animal ecology/biogeography that cover a wide range of spatial scales (from global to micro-scale) and different geographical regions (from arctic to humid tropical); and
- Develop an agenda for future research in GISc, which takes into account developments in biogeography and ecology, and their applications in GISc including remote sensing, geographic information systems, quantitative methods, spatial analysis, and data visualisation.

INTRODUCTION

There is no escaping the reality that the world is unavoidably spatial (Tilman and Kareiva 1997) and any technique or process that helps us to think spatially will bring us closer to understanding how nature works. The term Geographic Information Science (GISc) encompasses a whole range of quantitative, graphical and non-graphical, and spatial methods that enable us to capture, process, analyse, interpret, and display information about the arrangement of objects in space and their character and connectivity across spatial and temporal scales. In our own fields of biogeography and ecology, it is only relatively recent that advances in satellite technology, computer hardware and software, geographic information systems, and spatial analysis have brought GISc within reach of a wide audience of researchers. In this sense GISc is a young and rapidly developing area of endeavour that is still diffusing across the natural sciences. As questions, data, and methods are increasingly framed within a spatially-explicit context, where space and time scales are used to represent patterns and processes, new insights are being generated into familiar problems and new perspectives and technologies are being developed and applied to address new challenges. Synthesis of theory, data, and methods are now becoming the rule rather than the exception as a consequence of the broad availability of spatial digital technologies combined within the concept of GISc. The following chapters apply a host of these technologies to an array of biogeographical and ecological topics, and still, the areas of research where GISc can inform relative to these fields are too extensive to be solely reported here because of the broad applicability of GISc and the robustness of spatial tools and techniques that are contained. Others (e.g., Walsh et al. 1994; Butler and

Walsh 1998) have presented applications of GISc theory and practice within vegetation science and geomorphology, but new emergent technologies continue to make the application of GISc to landscape issues quite dynamic. Here, we describe a breath of scientific work where GISc serves as the unifying framework that ties the examples together within the study of biogeography and ecology. The chapters report on the application of new data and approaches to familiar problems as well as the use of GISc to explore new topics and areas of research.

It may be surprising that we introduce spatial analysis in this way and you would be forgiven for thinking that since geography and ecology are themselves intrinsically spatial, the subject would have a long history. Yet ecology, for example, has advanced during the last century largely by ignoring spatial processes because they complicate the development of theory and, equally, poses enormous problems for data collection to test models. This matters greatly in applied ecology (such as the conservation of biodiversity) because "each real world situation is unavoidably explicitly spatial, and land management aimed at habitat restoration or habitat preservation must be grounded in space" (Tilman and Kareiva 1997, page viii). Increasingly we are facing problems at larger spatial scales or extents, such as global climatic change and the introduction of new land-use policies. Yet because of a lack of adequate tools, ecologists have traditionally focused on scales or grains of only a few metres rather than tens of kilometres (May 1994), ill-preparing us to meet these challenges. Even where we have been able to gather adequate data at an appropriate spatial scale, data analysis has not been straightforward. Using the words of Legendre and Legendre (1998), we have been trained to believe that nature follows the assumptions of classical statistics; it does not. The spatial arrangements of many objects in nature (e.g., rock types, plants, animals) are neither random nor independent, often critical assumptions in classical statistics. Furthermore, the spatial structuring that they possess is functional and should not be regarded as a statistical nuisance. An aspatial view of the world is not only a gross simplification; without spatial structuring, ecosystems would simply cease to function (Legendre and Legendre 1998).

The availability of satellite and aerial photography time-series data combined with the plethora of sensor systems and reconnaissance platforms of varying spatial resolutions offer biogeographers and ecologists new opportunities to examine scale dependent relationships across time and space in a host of environments and geographic places. The integration of optical remote sensing systems (e.g., Landsat Thematic Mapper) with non-optical systems (e.g., Synthetic Aperture Radar systems) further enhances our ability to explore patterns and processes; GIS gives researchers the capability to integrate not only remote sensing images and landscape views but field-based measurements, map products, and existing digital data held in government repositories, marketed by commercial vendors, and/or distributed over the Internet by universities, individual researchers, and research stations. The power of information has been enhanced by the spatial-explicitness of it and the time-sensitive perspective that much of it captures. Research questions linked to complexity theory and hierarchy theory, for instance, are now being more aggressively explored because of the ability to capture landscape information over numerous grains and extents, for extensive time frames, and for an ever increasing set of thematic domains. Because of the convergence of theory, data, and methods that are appropriate for spatial inquiry, new challenges face the modern scientist but the potential for insight is substantial and the rewards significant.

In this book, we have tried to bring together examples of the work being carried out by biogeographers and ecologists who acknowledge the importance of thinking spatially. The mixture of contributions is somewhat eclectic, being driven by authors' own choices of topics to present to one of two related meetings convened on the theme of spatial analysis in biogeography and ecology. When we first conceived the idea for this book, we imagined individual contributions fitting into neat sections related to topics such as gradients and ecotones, spatial dynamics, or applications of remote sensing, often reflecting their origins in the traditional disciplines of ecology or geography. However, it soon became clear that the barriers between the disciplines have become so blurred that this was not possible. Indeed, we celebrate this as an advance that bodes well for the future because it promises rapid development. For example, whether a problem derives from geography or ecology, spatial data sets are often best handled with a Geographical Information System (GIS). Once using a GIS, many researchers will want to incorporate data layers from satellites because this is often the only way to acquire data of reliable and uniform quality across large areas. The result is that a biologist interested in understanding a plant's distribution, must also appreciate the qualities of satellite-derived data, how it can be processed for different purposes, and how it can be interpreted, for example, in terms of standing biomass. As Ling Bian's contribution shows (Chapter 13), an interest in a field problem such as animal movements can also highlight limitations in existing GIS technology. Perhaps more importantly, having appreciated what remotely sensed data can and cannot be used for, researchers are increasingly asking more appropriate spatial questions. We thus see a new breed of researcher asking functional, often ecological questions but armed with the tools of the geographer, spatial data analyst, or remote-sensing scientist.

We have thus organised this book in a simple way, trying to progress from chapters that focus largely on development of techniques to applications in management. We do not pretend in any way to have covered all the techniques being used nor to have focussed on the most appropriate applications. But we do believe that the range of examples from the alpine treeline ecotone to tropical deforestation, and from ecological settings as distinct as Thailand to Sussex, illustrates in an exciting way what can be done to analyse environmental data in a spatially explicit way. One advantage in not having sections in the book is the often unusual juxtaposition of subject matter, such as the GIS representation of wildlife movements (Chapter 13) next to the stratified sampling of environmental gradients in a desert ecoregion (Chapter 14), which we hope will encourage researchers to read beyond their own fields. It seems to us that advances in spatial analysis will be all the more rapid if techniques developed in one subject are tried elsewhere.

As a general overview of the approaches that are being followed in GISc for biogeography and ecology, and reported here through a set of examples, Figure 1 is presented. In this schematic, we emphasise the pronounced utility of remote sensing and GIS for characterising landuse/landcover (LULC) through measurements of state and conditions variables. Dependent variables might include LULC type, change trajectories of LULC, metrics of landscape organisation, plant biomass, and so on, whereas independent variables might be terrain descriptors from digital elevation models, climatic data collected at stations and/or output from GCMs, soil and geologic site information from field or map products, and disturbance types and patterns from site treatment records. Such variables can be characterised over time through satellite systems such as Landsat that gives historical snapshots beginning in 1972 and extending through the present, and aerial photography that offers landscape views often times back a half century or more. In addition to being time-sensitive, GISc can be space-sensitive in that data can be captured at a variety of grains and extents. In addition, spatial analytical techniques can be used, for example, through agglomeration techniques for the iterative recomputing of cell sizes used to partition space and summarise landscape features through spectral responses and/or environmental indications of site and situation. Approaches are also readily available to stratify the landscape into other spatial units such as watersheds, terrain units, ecological zones, and so on to conduct data collections, sampling approaches, or to generalise data in some hierarchical way. Social units also serve to bifurcate data for studies of human impacts on landscape change through, for example, deforestation and agricultural extensification. Finally, GISc is capable of creating numerous types of outputs to accommodate the diversity of data, researchers, and problems being addressed. The visualisation of derived or in some way value-added data is a common approach for examining graphically the spatial and/or temporal dynamics of parameters and systems. Plotting regression residuals to examine the spatial pattern of model fits is but one simple approach for using the spatial perspective to examine the results of non-spatial tests. But GISc techniques can also be used to account for spatial autocorrelation in data sampling and/or modelling routines by including autoregressive terms in the models and/or discerning the scale dependence of variables and landscapes so that distances between samples can be specified to minimise the influence of location on data values. Image animation through 3-D displays, image rotations, and the like further add power to the analyst for interpreting data and in generating subsequent hypotheses about system and variable behaviour over time and space. Landscape dynamics are being increasingly explored because of the availability of image time-series data, and change vector analysis is but one of many approaches for documenting the nature of change including the direction, timing, and intensity of landscape dynamics. In summary, Figure 1 suggests a system afforded through GISc technologies that is space-time sensitive, is graphical and non-graphical, and is integrative across thematic domains, scales, data structures, and analytical methods. Again, the challenges are many for integrating GISc into biogeographical and ecological research, but the opportunities for new and richer insights are substantial.

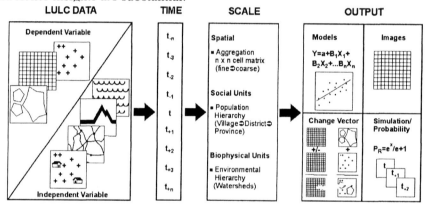

Figure 1. Schematic of the GISc analytical process in landscape studies.

REFERENCES

Butler DR, Walsh SJ. Applications of remote sensing and GIS in geomorphology. Geomorphology 1998; 3-4: 179-352.

Legendre P, Legendre L. Numerical Ecology. Developments in Environmental Modelling. Amsterdam: Elsevier, 1998.

May R.M. "The Effects of Spatial Scale on Ecological Questions and Answers." In *Large-scale Ecology and Conservation Biology*, P.J. Edwards, R.M. May and N.R. Webb, eds. Oxford, UK: Blackwell Scientific Publications, 1994.

Tilman D, Kareiva P, (eds.), *Spatial Ecology: Monographs in Population Biology*. Princeton, New Jersey: Princeton University Press, 1997.

Walsh SJ, Davis FW, Peet RK. Applications of remote sensing and geographic information systems in vegetation science. Journal of Vegetation Science 1994; 5:609-756.

Chapter 2

A SPECTRAL UNMIXING APPROACH TO LEAF AREA INDEX (LAI) ESTIMATION AT THE ALPINE TREELINE ECOTONE

Daniel G. Brown
School of Natural Resources and Environment, The University of Michigan, Ann Arbor, Michigan, USA
danbrown@umich.edu

Keywords: leaf area index (LAI), treeline ecotone, spectral unmixing, vegetation index (VI) - LAI relationships.

Abstract The objective of this research was to develop methods for mapping arboreal leaf area index (LAI) at the alpine treeline ecotone in Glacier National Park, Montana using Landsat Thematic Mapper (TM) imagery. A three-stage approach was tested for addressing the problem of mixed pixels in biophysical value estimation. This paper illustrates a proof of concept for this method. First, spectral unmixing was used to obtain estimates of the percentage of each pixel that was composed of tree, tundra, bare rock and shadow. Spectral signatures obtained through image interpretation were used for mixture modeling. The second step involved adjusting the pixel vegetation index (VI) values so that they represented the VI of the tree-only portion of the pixel, assuming an average VI for background components. Finally, the adjusted VI was regressed against leaf area index (LAI) measured in the field using a LiCor LAI-2000 and spatially referenced through differential GPS. Results using the adjusted VI values were compared with unadjusted VI, as were the results obtained using the normalized difference vegetation index (NDVI) and the simple ratio (SR).
Maps of the LAI specific to conifer trees were generated by adjusting VI values to represent only the tree component of pixels. The results indicate that adjusted NDVI can be used to predict the LAI of trees within mixed pixels much better than does unadjusted NDVI, which overestimates the LAI because it includes non-arboreal vegetation. NDVI provided better results than did SR. A number of issues affect the accuracy of LAI estimates in practice: the accuracy of estimates of end-member proportions obtained through unmixing, the adequacy of the end-member average NDVI for removing the effects of non-arboreal NDVI contributions, the non-synchronous nature of the satellite flight and field

work, and the accuracy of field estimates of LAI made using the LAI-2000.

INTRODUCTION

The leaf area index (LAI) of a forest canopy is an indicator of forest structure and function. LAI is a dimensionless ratio of the leaf surface area of a plant to the amount of ground surface beneath it. The index is used to describe the geometric structure of a plant or community. LAI has functional effects on forest canopy light penetration, snow accumulation and melt, interception, evapotranspiration, and, therefore, productivity and carbon budget (Waring and Running 1998). Mapped LAI values provide a description of the spatial pattern of forest structure and have served as inputs to functional models of ecosystem biogeochemistry (Running and Gower 1991).

Red and near-infrared (NIR) spectral reflectance information acquired from orbiting satellites have shown sensitivity to variations in leaf area index measursments made in the field (Peterson et al. 1987; Spanner et al. 1990; Myneni et al. 1997; White et al. 1997). Reflectance from green vegetation is much higher in NIR than in red wavelengths, whereas the increase in reflectance from red to NIR in other cover types is less pronounced. A number of spectral vegetation indexes have been developed to exploit this relationship. The normalized difference vegetation index (NDVI) combines red and NIR reflectance in a single index. NDVI values increase when NIR reflectance increases. A curvilinear relationship has been demonstrated between NDVI and LAI, limiting the usefulness of NDVI for differentiating LAI values above about 4.0. A simple ratio (SR) between red and NIR reflectance results in a nearly linear relationship with LAI.

Because NDVI and other index values are sensitive to any vegetation within a given pixel, attempts to map the LAI of canopy vegetation alone are complicated by background signal. Where background reflectance is strong and/or highly variable, the predictability of LAI from spectral information is weakened. One approach to correcting for the influence of background soil and understory involves the use of spectral information in the middle infrared (~1.5-2.0 µm) wavelengths (Nemani et al. 1993). By subtracting out the background signal detected in the middle infrared wavelengths, corrected NDVI and simple ratio indexes are designed to be more sensitive to the canopy vegetation.

At the alpine treeline, where trees mix within a fine spatial matrix with tundra, meadows, and non-vegetated surfaces of rock and snow, the effects of background reflectance on the spectral signal is pronounced. The predictability of tree LAI from NDVI is, therefore, reduced. Because tree canopy cover at treeline tends to be low (e.g., under 50 percent), LAI at a given location is as much a function of the canopy cover as it is the structure of the canopy and understory at that location. Because the VI signal at treeline includes reflectance from meadow and tundra vegetation the middle infrared correction is not appropriate for isolating the tree VI.

This paper demonstrates an approach to estimating the LAI of trees at the alpine treeline ecotone that combines linear spectral mixture modeling (Gong et al. 1994; Adams et al. 1995) with regression of vegetation index and LAI at treeline in Glacier National Park, Montana, USA. The approach addresses the effect that the mixing of multiple vegetation cover types within each pixel has on LAI estimation by first estimating the composition of each pixel, expressed as end-member fractions. The end-member fractions are then used to adjust the vegetation index values to estimate the vegetation index contribution of the trees in each pixel. Then, the adjusted vegetation index values and tree LAI, measured in the field, are regressed to develop a predictive relationship for mapping the LAI of trees only, with background LAI removed. The curves are compared with those generated by White et al. (1997), which describe relationships between vegetation indexes and LAI across multiple vegetation communities in Glacier National Park.

STUDY AREA

The study is set at the alpine treeline ecotone in eastern Glacier National Park, Montana, USA (Figure 1). The Park lies on the U.S.-Canada border in the Rocky Mountain cordillera and is roughly bisected by the continental divide. The eastern side of the Park has a substantially more continental climate than the wetter and milder western side. The sub-alpine conifer forests reflect these climatic differences, with the western side consisting of species common in the forests of the Pacific Northwest. The subalpine forests on the eastern side of the Park are dominated by Lodgepole pine (*Pinus contorta*), Quaking aspen (*Populus tremuloides*), and Engelman spruce (*Picea engelmannii*) at lower elevations and Subalpine fir (*Abies lasiocarpa*). The higher elevation treeline sites investigated in this study tend to be dominated by Subalpine fir, but also contain Whitebark pine (*Pinus albicaulis*) and Limber pine (*Pinus flexilis*) locally.

Trees at treeline tend to grow in clumps of matted growth forms, called krummholz, which are capped by a dense layer of needles with very little needle growth below. The canopy often contains substantial quantities of dead needles and exposed wood from "flags," which protrude from the mat. Where trees grow erect they often contain a number of gaps and dead branches or are shaped directionally by prevailing winds. Tree heights at our treeline sites ranged from under 0.5 m to over 3 m, with the majority under 2 m. The trees in the treeline landscape are intermixed with a diverse assemblage of tundra and meadow vegetation and rocky or gravelly surfaces that are devoid of vegetation. The majority of the rock and gravel is derived from underlying limestone and argillite formations. Repeat photography at selected sites within the study area suggest that the landscape at treeline has been quite stable over the past century (Butler et al. 1994).

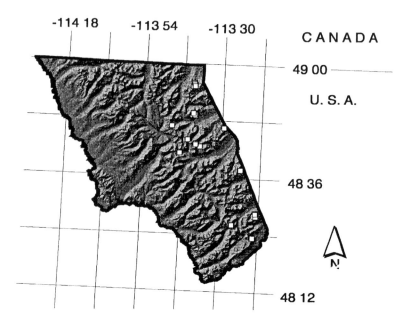

Figure 1. Glacier National Park, MT, USA with latitude and longitude lines. White boxes indicate quadrat locations and are exaggerated in size.

METHODS

Data Collection

Field Data

Data on tree cover and LAI were collected at a total of 38 sample quadrats during the summers of 1998 and 1999 (Figure 1). The square quadrats were selected at random from among known, undisturbed treeline sites. They were 50 m by 50 m in size and oriented along cardinal compass directions; though 10 sites were oriented 40° off true north due to an error in the compass declination setting. The center point for each quadrat was located through post-processing GPS measurements for differential correction. Most corrections were made using a base station in Missoula (about 200 km to the south and west), but infrequently those data were unavailable and a station in Helena (about 260 km to the south and east) was used.

Within each quadrat, a map of the site was sketched in field notebook and each individual tree or patch of trees (depending on growth form and clumping) was labeled and the species was noted. The size of each tree or tree-patch was estimated by recording the length of a primary and secondary axis of the tree or patch on the ground. The axis lengths were measured by tape measure or pacing. The axis lengths were then converted to area estimates by assuming the shape of each tree or patch was oval. Summing up the total area of trees and patches within a quadrat and dividing by the area of the quadrat gave the percentage of the quadrat that was tree-covered. The calculated values ranged from 2 to 50 percent, with an average of 17 percent.

The LiCor LAI-2000 Plant Canopy Analyzer was used to estimate the "effective" LAI of trees within the quadrat. The instrument measures light transmitted through the canopy using five concentrically nested sensors (Welles and Norman 1991). Readings taken above the canopy are compared with those taken below to determine the amount of area over which light is intercepted by the needles. Because of contamination by self-shading and branches the estimate is termed effective LAI. A minimum mask size of 30° was placed over the sensor to avoid contamination by the observer. Although the calculation of LAI from the LAI-2000 readings assumes that radiation is received from a diffuse source, as on a cloudy day, the field campaign was not sufficiently flexible to enable the work only on cloudy days. Therefore, readings were made on both sunny and cloudy days, introducing some variability into the measurements.

For each tree and patch within a quadrat the average effective LAI was estimated using one reading above the canopy and a minimum of five readings below the canopy, distributed throughout the patch or tree. Based on the individual tree and patch measurements, an area-weighted estimate of the average effective LAI of trees in the quadrat was calculated. Quadrat effective LAI, or LAI_q, was calculated as $\Sigma (A_p \cdot LAI_p) / A_q$, where A_p is the area of a given patch within the quadrat, LAI_p is the average LAI of that patch, and A_q is the area of the quadrat.

To compare the LAI calculations with the results presented by White and others (1997) the LAI values were adjusted twice. First, all LAI values were converted to projected LAI by multiplying by a combined clumping correction factor (CCCF), which adjusts for the non-random (i.e., clumped) distribution of needles and contamination by branches. Because of clumping and enhanced self-shading, the LAI tends to be underestimated by the LAI-2000 in coniferous forests (Gower and Norman 1991). However, woody tissue in the canopy can, even in the presence of clumping, cause an overestimation of projected LAI by the LAI-2000 (Deblonde et al. 1994). Ideally, the CCCF accounts for both of these major sources of bias. In the harsh treeline environment a variety of growth forms develop, including dense mats, flagged trees, and dead branches. Because of the small size of many of the trees and patches, many of the LAI readings were taken very near to the tree stems. All quadrat-scale LAI-2000 readings were, therefore, multiplied by 0.75, which is a lower CCCF than values reported for homogenous stands of large coniferous trees (White et al. 1997), but it emphasizes the influence of branch, stem, and dead needle contamination.

The second adjustment was to convert projected LAI values to total LAI by multiplying them by 2.0. Projected LAI represents one-sided leaf area, whereas total LAI accounts for the two-sided flat needles in the tree canopy. The resulting estimate is termed LAI_{tree}, because it includes only arboreal vegetation in the quadrats.

Remotely Sensed Data

A Landsat Thematic Mapper (TM) image of the park, Worldwide Reference System (WRS)-2 path 41 and row 26, was acquired on September 1, 1995 under clear sky conditions (Figure 2 - on CD-ROM). Although the imagery was flown three to four years before the field work was conducted, the relative stability of treeline vegetation (Butler et al. 1994) suggests that substantial changes were unlikely. Nonetheless, the inter-annual and seasonal variability of LAI may introduce uncertainty into this analysis (Deblonde et al. 1994). The image was georeferenced to zone 12 of the UTM coordinate system using orthographic rectification to remove the effect of terrain displacement. The geometric terrain correction made use of a mosaic of USGS 7.5-minute 30-meter digital elevation models (DEMs). A root-mean-squared (RMS) error of < 15 m was obtained using 21 control points, which were read from USGS 7.5-minute topographic quadrangles.

The radiometric calibration approach was identical to that of White et al. (1997) to ensure comparability of results. Raw digital numbers (DNs) in the TM image were converted, first, to radiance by (a) applying gain and offset values for sensor calibration provided in the image header file and (b) adjusting the visible channel values to correct for path radiance by subtracting the DN value obtained in clear lakes. Next, radiance values were converted to exoatmospheric reflectance using the method of Leprieur and others (1988; c.f. White et al. 1997). The reflectance calculation includes a correction for the effects of terrain on solar incidence using the DEMs.

Two vegetation indexes were computed using the red (TM3) and near-infrared (TM4) channels from the TM image. NDVI was computed as (TM4 − TM3) / (TM4 + TM3) and the simple ratio (SR) as TM4 / TM3.

Each of the index values was estimated for all of the field quadrats through image resampling. Bilinear interpolation and cubic convolution were both used for the resampling and found very little difference in the results. Nearest neighbor resampling was not considered because spatial averaging was required to compare the index values from 30 m pixels to field data collected in 50 m quadrats. The index values resampled through bilinear interpolation were subsequently used in the testing and analysis of unmixing and LAI estimation. The resampling process was the same for both correctly and incorrectly oriented quadrats. Although less than ideal, the difference in orientation had relatively little influence on the analysis because it affected the overlap of pixels and quadrats only at the corners, a relatively small portion of each quadrat.

Digital orthophoto quadrangles (DOQs) of the Park were used as a second source for validating tree cover estimates. The image files are orthographically rectified black-and-white aerial photographs that are stored at 1-meter resolution. The files were provided in a UTM-rectified digital image format and image subsets were extracted for our study sites. Tree cover within each quadrat was determined using a manually defined brightness value threshold in the DOQs. Once a brightness threshold was defined interactively, the threshold was applied such that pixels with brightness values below the specified threshold were labeled "trees" and pixels with higher brightness values were labeled "not trees."

Linear Mixture Modeling

Linear mixture modeling assumes that the pixel values, expressed as digital numbers (DNs), are linear combinations of reflectances from a limited set of constituent elements, called end-members. For a given wavelength channel, the DN of a given pixel can be expressed using Equation 1:

$$DN_\lambda = DN_{1\lambda} \cdot f_1 + \ldots + DN_{k\lambda} \cdot f_k + \varepsilon_\lambda \quad \text{and} \quad \Sigma f_k = 1 \qquad 1)$$

where DN_λ is the reflectance observed in wavelength channel λ, $DN_{k\lambda}$ is the reflectance observed for end-member k in wavelength channel λ, f_k is the fraction of the pixel that is in end-member cover type k, and ε_λ is the error for that particular channel. If the number of wavelength channels is at least as large as the number of end-members (k), then end-member spectral values, observed in the field or using image interpretation, can be used to obtain estimates of the fraction of each pixel that is in each end-member. Here the singular value decomposition approach described by Gong et al. (1994) was used to solve for the f_k values. Two constraints were maintained in the solution of f_k values: 1) the fractions across all end-members sum to one, as in Equation 1, and 2) each end-member fraction is in the range [0,1].

The error term for each channel is calculated and summarized across all channels using the root-mean-squared (RMS) error. The RMS error is a measure of the goodness of fit of the model and can be mapped by pixel to examine its geographic distribution.

Although unmixing works best with hyperspectral data, i.e., where the number of independent spectral dimensions is large, Adams et al. (1995) showed that it can work reasonably well with Landsat TM data when the number of end-members is kept small. The focus of this work on the alpine treeline ecotone naturally limits the end-members to the landscape components found in that environment. Therefore, four end-members were identified for this application: trees, tundra, barren surfaces, and shadow. Spectral values for each end-member within the six reflected TM wavelength channels (Figure 3) were identified through image interpretation to identify pixels that were relatively free of contamination from other end-members (Adams et al. 1989). Although spectra were measured in the field using a portable spectro-radiometer, we used image end-members to limit problems associated with image calibration. Although these end-members served the analysis well, the tundra category, which included both wet and dry tundra, was relatively heterogeneous and would be broken into two categories if additional spectral channels were available.

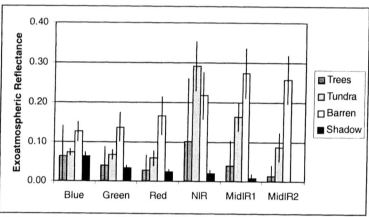

Figure 3. Spectral signatures of four end-members on six TM channels. Error bars indicate 1 standard deviation about the mean value.

The results of the unmixing process were evaluated by comparing the derived tree end-member fraction, estimated at each quadrat location through bilinear interpolation, with the tree end-member fractions calculated: 1) in the field, and 2) from the DOQs. Comparisons were summarized graphically and through the use of the RMS error.

The unmixing process was an interative one in which 1) training pixels were selected on-screen, 2) the unmixing was performed, 3) the RMS errors resulting from the unmixing run were examined, 4) the RMS errors based on comparison with field and photo tree fractions were examined, and 5) the channels and/or end-member training areas were adjusted to achieve the result which minimized both errors. The best result was obtained with relatively small number of the purest training pixels (603, 131, 1236, and 3247 for trees, tundra, bare, and shadows, respectively) and the following input channels: TM3, TM4, TM5, and NDVI.

Adjusting Vegetation Index Values

Having obtained estimates of the fractional composition of each pixel in each end-member, the vegetation index values were adjusted such that they represented the tree-only contribution to the pixel value. This step represents, in essence, a second mixture model, this time for the vegetation index (VI) signal. The VI values were from NDVI and SR. The approach assumes that the VI value for a pixel results from the linear combination of VI values from each of the end-member components, as in Equation 2:

$$VI_{pixel} = f_{tree} \cdot VI_{tree} + f_{tundra} \cdot VI_{tundra} + f_{bare} \cdot VI_{bare} + f_{shadow} \cdot VI_{shadow} + \varepsilon \qquad 2)$$

where VI_{pixel} is the index value observed for the pixel, ε is measurement error, and the inputs are the fs and VIs for each end-member. By assuming an average VI for each of the non-tree components and rearranging the equation, an estimate of the tree-LAI is obtained using Equation 3:

$$VI_{tree} = VI_{pixel} - f_{tundra} \cdot VI_{tundra} - f_{bare} \cdot VI_{bare} - f_{shadow} \cdot VI_{shadow} - \varepsilon \qquad 3)$$

An obvious limitation on the accuracy of this estimate involves the degree to which any of the non-tree components have a greater variability in the VI than does the tree component, i.e., the degree to which the mean is not a good estimate of the non-tree VI value everywhere. Figure 4 illustrates the differences between means and variations in vegetation index values among end-members and the large variability in vegetation index values for tundra, owing to the inclusion of both wet and dry tundra in the class.

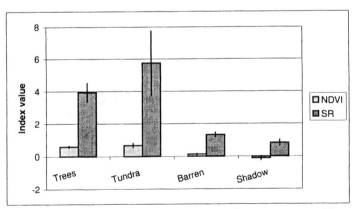

Figure 4. Average vegetation index values by end-member component, with one S.D. bars indicating variation.

VI-LAI Relationships

Original and adjusted vegetation index values (VI and VI_{tree}), obtained in the previous step, were regressed against the leaf area index of trees, which was measured in the field. Curves were fitted to the relationship using an exponential form for NDVI and a linear form for SR. The fitted curves were compared with those generated by White et al. (1997) in their analysis, which was based in Glacier National Park but grouped by homogenous vegetation types. White et al. (1997) analysis produced Equations 4 and 5 for estimating total LAI from NDVI and SR:

$$LAI = 0.2273 \, e^{4.9721(NDVI)} \qquad 4)$$

$$LAI = 1.2565(SR) + 0.069 \qquad 5)$$

The fits of the curves generated here using unadjusted VI values are compared with those generated using VI_{tree} and both adjusted and unadjusted curves are compared among the two indexes (NDVI and SR).

RESULTS AND DISCUSSION

Unmixing for End-Member Fractions

Although there was generally good agreement between the orthophoto- and field-based estimates of percent trees, the fraction in trees tended to be overestimated by unmixing (Figure 5). Most of the estimates of f_{tree} above 0.3 are over-estimates. The RMS errors from the unmixing ranged from 0 to 9.41 across the entire Park, in units of reflectance, with the highest values occurring in lakes and snow patches. Because water and snow were not included as end-members the error in fraction estimates were quite high in areas dominated by these cover types. At snow-free treeline sites the RMS errors were generally well below 1.0.

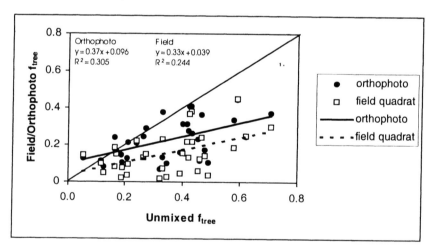

Figure 5. The fraction of trees estimated through unmixing, compared with observed fractions from digital orthophotographs and field observation.

The approach to estimating tree LAI developed here is dependent on acquiring accurate estimates of the fractional composition of each pixel, so that adjustments to NDVI and LAI values can be made. The best estimates of tree, tundra, bare rock, and shadow fractions (Figure 6 – on CD-ROM) were used here to illustrate the LAI estimation process.

Three approaches are considered reasonable for dealing with the observed bias in tree fraction estimates. First, better estimates should be attainable with a hyper-spectral sensor. This would allow for a greater number of purer end-members. Second, the observed bias (Figure 5) can be used to downwardly adjust the percent trees estimate. This approach is only reasonable if other end-members are adjusted upward. The similarities in spectral signatures (Figure 2 – on the CD-ROM) and the observed mixed pixel combinations (Figure 6 – on CD-ROM) suggest that forests were most likely to have been confused with shadow. Therefore any decrement in forest fraction might be attributable to shadow. Finally, some uncertainty in the end-member fraction estimates is inevitable. Propagation of the uncertainty through the estimation of LAI might be conducted through Monte Carlo simulation to acquire, at least, an estimate of the uncertainty in LAI estimates resulting from uncertainty in the end-member fraction estimates. Although none of these approaches is described in more detail here, due to space limitations, using this technique in practice may require that one or several of them be implemented to improve the accuracy of resulting estimates.

VI-LAI Relationships

Four regression models were generated to fit curves to the relationships between the vegetation indexes and LAI_{tree} in the quadrats (Figure 7). A total of six quadrats were eliminated from the analysis. Each of the six were anomalous situations and outliers in the regression. The most obvious outliers were due to quadrats that were heavily shaded and the tree fractions were under-estimated in favor of shadow. The others appeared to be affected by other site conditions that weakened the estimates of either the field LAI or the satellite vegetation indexes (e.g., steep slopes, patch configurations).

The relationship between the unadjusted NDVI ($NDVI_{pixel}$) and LAI_{tree} was relatively weak ($R^2 = 0.028$; Figure 7A). When compared to the equation presented by White et al. (1997), it was clear that $NDVI_{pixel}$ over-estimated LAI_{tree} (RMSE = 3.07). The equation presented by White et al. (1997) was fitted to data collected from several ecosystem types and included LAI from all sources (i.e., canopy and understory). The fit of the NDVI-LAI regression, though still weak, increased ($R^2 = 0.191$) when the influences of non-tree vegetation were removed from the NDVI signal to calculate $NDVI_{tree}$ (Figure 7B). Also, the ability of the White equation to estimate LAI_{tree} was much better using $NDVI_{tree}$ (RMSE = 0.60).

The results for SR suggest a similar pattern, however the fits of the relationships with LAI were not as good as with NDVI. Using the unadjusted SR (SR_{pixel}), the linear relationship with LAI_{tree} had an R^2 of 0.008. Again, the curve fit by White et al. (1997) substantially overestimated LAI_{tree} (RMSE = 3.64). Upon adjusting for non-tree vegetation (SR_{tree}) the relationship improved to a still-weak R^2 of 0.04. The RMSE of the fit of the data to White et al. (1997) curve for SR_{tree} was 2.40. Using White's equation and SR_{tree} resulted in an over-estimation of LAI_{tree}.

NDVI provided a much better estimate of LAI_{tree} than did SR. This is consistent with White et al. (1997) results, in which the fit of the relationship between NDVI and LAI was $R^2 = 0.90$ but the fit for SR

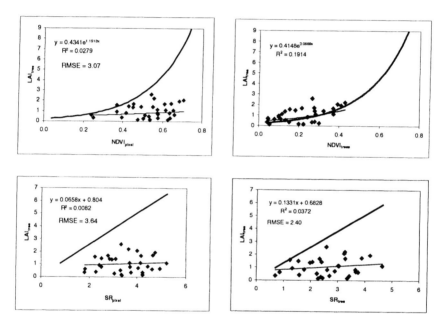

Figure 7. Relationships between vegetation indexes and quadrat LAI measured using the LAI-2000. The equation and R2 refer to the curve fitted to the data (thinner line) and RMSE indicates the fit of Equations 4 and 5 to the data (heavier line), using the original and adjusted VI values to estimate LAI.

and LAI was $R^2 = 0.74$. Although the saturation of the LAI signal in NDVI at a LAI of about 4.0 would normally limit the applicability of NDVI for estimating LAI, LAI_{tree} at treeline tended to be much lower. In fact, all quadrat LAI_{tree} estimates made in this study were below 3.0, suggesting that saturation is not a problem.

Two differences between these studies explain the much lower fit for the data presented here when compared to White et al. (1997) data. First, estimating the vegetation index contribution of trees only was more challenging than estimating the value for all vegetation in a pixel. This is especially true where there is a great deal of uncertainty or bias in the estimates of the end-member fractions. Second, White et al. (1997) field data were more extensive and, more importantly, were aggregated by vegetation cover type. This aggregation across a range of sites averaged out much of the site-to-site variability present in the data used here.

This analysis suggests that, by adjusting the NDVI value to remove the effects of non-tree vegetation, the LAI of trees can be approximated. Because the data used by White et al. (1997) to develop LAI-NDVI relationships were more complete, their equation was applied using the adjusted NDVI values ($NDVI_{tree}$) to estimate LAI_{tree} of pixels within the Park (Figure 8). Although the field data and analysis presented here focused on treeline sites, Figure 8 includes sites across multiple ecosystem types, as did White's analysis.

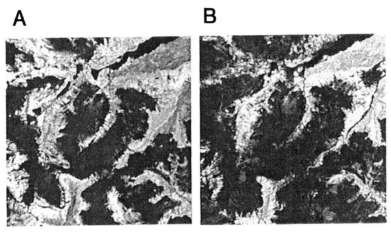

Figure 8. Subset image of Glacier National Park study area showing (A) total LAI estimated using NDVI and Equation 4 and (B) LAI_{tree} estimated using NDVI adjusted to reflect the tree-only component and Equation 4.

CONCLUSIONS

The method presented here permits the quantification of forest canopy characteristics where the forest has an open canopy and mixes spatially with other cover types. The approach combines vegetation index (VI) based estimates of leaf area index (LAI) with spectral unmixing to isolate a particular canopy component. Specifically, by adjusting VI values to represent only the tree component of pixels, maps of the LAI specific to conifer trees were generated. The fit of VI-LAI_{tree} relationships was improved using the unmixing-based adjustments of VI. Although the results for the SR index were weak, the NDVI-LAI relationship developed and published by White et al. (1997) provided reasonably good estimates of LAI_{tree} (±0.60 LAI) when applied to adjusted NDVI values ($NDVI_{tree}$). The resulting maps are useful where estimates of tree distribution and condition are required regardless of the landscape mosaic within which they are found. This approach to combining unmixing and parameter estimation should have applicability for estimating other biophysical properties, including albedo and photosynthetically active radiation (PAR).

Although the results suggest that the technique may be promising for estimating component vegetation characteristics, the approach requires further refinement. The potential for propagation of error in the analysis is particularly troublesome and consistent with the relatively low levels of fit found in this analysis. A sensitivity analysis to determine the most important sources of error has not yet been undertaken. However, potential errors appear at several steps in the process. Errors contributing to uncertainty in the result include: (1) LAI measurements taken with the LAI-2000 on sunny days are less reliable than those taken on cloudy days; (2) error in the unmixed fractions, especially a bias towards over-estimating the tree fraction; (3) variability in the end-member mean VI estimates used to calculate VI_{tree}, especially the inclusion of both wet and dry tundra in the tundra end-member; and (4) the non-synchronous satellite and field data collection efforts. The relative importance of each of these sources of error is unknown at this point.

ACKNOWLEDGEMENTS

This work was funded through collaborative research grants from the National Science Foundation (#97-14347) to Michigan State University and from the USGS Biological Resources Division (#99CRAG0035) to the University of Michigan. Professor Stephen J. Walsh, University of North Carolina at Chapel Hill, provided both assistance in the field and use of the LAI-2000 sensor. The work would not have been possible without the able field and lab assistance of Mark A. Bowersox, Evan S. Hammer, and Sean E. Savage.

REFERENCES

Adams JB, Sabol DE, Kapos V, Filho RA, Roberts DA, Smith MO, Gillespie AR. Classification of multispectral images based on fractions of endmembers: application to land-cover change in the Brazilian Amazon. Remote Sensing of Environment 1995; 52:137-54.

Adams JB, Smith MO, Gillespie AR. Simple models for complex natural surfaces: a strategy for the hyperspectral era of remote sensing. Proceedings of IGARSS 1989, Vancouver, B.C., Canada, 1989; 16-21.

Butler DR, Malanson GP, Cairns DM. Stability of alpine treeline in Glacier National Park, Montana, U.S.A. Phytocoenologia 1994; 22(4):485-500.

Deblonde G, Penner, M, Royer, A. Measuring leaf area index with the Li-Cor LAI-2000 in pine stands. Ecology 1994; 75(5): 1507-1511.

Gong P, Miller JR, Spanner M. Forest canopy closure from classification and spectral unmixing of scene components-Multispectral evaluation of an open canopy. IEEE Transactions in Geoscience and Remote Sensing 1994; 32(5):1067-1080.

Gower, ST, Norman, JM. Rapid estimation of leaf area index in conifer and broad-leaf plantations. Ecology 1991; 72(5):1896-1900.

Leprieur, CE, Durand, JM, Peyron, JL. Influence of topography on forest reflectance using Landsat Thematic Mapper and digital terrain data. Photogrammetric Engineering and Remote Sensing 1988; 54(4):491-496.

Myneni RB, Nemani RR, Running SW. Estimation of global leaf area index and absorbed par using radiative transfer models. IEEE Transactions in Geoscience and Remote Sensing 1997; 35(6):1380-1393.

Nemani R, Pierce L, Running S, Band L. Forest ecosystem processes at the watershed scale: sensitivity to remotely-sensed leaf area index estimates. International Journal of Remote Sensing 1993; 14(13):2519-2534.

Peterson DL, Spanner MA, Running SW, Teuber KB. Relationship of thematic mapper simulator data to leaf-area index of temperate coniferous forests. Remote Sensing of Environment 1987; 22(3): 323-341.

Running SW, Gower ST. FOREST-BGC, a general model of forest ecosystem processes for regional applications. II. Dynamic carbon allocation and nitrogen budgets. Tree Physiology 1991; 9:147-160.

Spanner MA, Pierce LL, Peterson DL, Running SW. Remote sensing of temperate coniferous forest leaf area index: The influence of canopy closure, understory vegetation and background reflectance. International Journal of Remote Sensing 1990; 11(1):95-111.

Waring RH, Running SW. *Forest Ecosystems: Analysis at Multiple Scales*, 2nd Edition. San Diego: Academic Press, 1998.

Welles JM. Some indirect methods of estimating measurement of canopy structure. Remote Sensing Reviews 1990; 5(1):31-43.

White JD, Running SW, Nemani R, Keene RE, Ryan KC. Measurement and remote sensing of LAI in Rocky Mountain montane ecosystems. Canadian Journal of Forest Research 1997; 27:1714-27.

Chapter 3

THE UTILIZATION OF AIRBORNE DIGITAL MULTISPECTRAL IMAGE DYNAMICS AND KINEMATIC GLOBAL POSITIONING SYSTEMS FOR ASSESSING AND MONITORING SALT MARSH HABITATS IN SOUTHERN CALIFORNIA

Douglas Stow, Alice E. Brewster and Brian K. Bradshaw
NASA Affiliated Research Center (ARC), Department of Geography, SanDiego State University (SDSU), California, USA
stow@mail.sdsu.edu

Keywords:	airborne imaging, kinematic global positioning system (KGPS), salt marsh, habitat mapping, light-footed clapper rail, Belding's Savannah sparrow, California.
Abstract	The utility of two data acquisition technologies, airborne digital multispectral imaging, and kinematic global positioning systems (KGPS), for providing spatially detailed and precise habitat data for salt marsh reserves in southern California is assessed. Two case studies demonstrate that by combining these technologies, habitat of endangered bird species that occupy these salt marshes can be mapped and analyzed in an efficient manner.
	The first case study focuses on mapping marsh vegetation types for artificial salt marshes constructed to create habitat for the light-footed clapper rail. High levels of accuracy by image classification of airborne multispectral data alone can be difficult to achieve due to spectral-radiometric similarity of certain plant species. The integration of digital elevation data, generated from KGPS field surveys, with image classification significantly improved the image classification accuracy. A post-classification sorting approach of integrating DEM data by discrete rules, resulted in an improvement of overall accuracy from 54.6% to 71.1% and the Kappa coefficient from 0.45 to 0.65.
	The salt marsh habitat requirements of the Belding's Savannah sparrow are examined in the second case study, also by utilizing data captured by ADAR imaging and KGPS surveying systems. Elevation, vegetation height, and percent middle-high marsh were all found to be significantly greater for territories utilized by the sparrow compared to randomly selected sample areas. The percent Middle-high marsh variable was

derived from image classification of ADAR image data. The overall accuracy of the ADAR-derived map of five habitat types was 86 ± 4%, with a Kappa coefficient of 0.75. For both case studies, a rigorous and precise assessment of map accuracy was achieved because of the positional accuracy of KGPS surveys and very-high resolution georeferenced image data.

INTRODUCTION

Coastal wetlands provide habitat for fauna, such as waterfowl and shorebirds that use salt marshes as nesting and feeding grounds. In southern California, salt marshes have greatly diminished in areal extent and are under constant pressure for development (Zedler 1986). Birds and other fauna that inhabit or use these wetlands are threatened by the reduction or degradation of salt marsh habitat. The effects of development and associated anthropogenic disturbances include the destruction of plants used for food, nesting, and seclusion and the increase of noise from surrounding areas.

Two endangered bird species being affected by the loss and degradation of the southern California wetlands are the Belding's Savannah sparrow (*Passerculus sandwichensis beldingi*) and the light-footed clapper rail (*Rallus longirostris levipes*). Their habitat range is limited to coastal wetlands of southern and Baja California. Unlike coastal wetlands on the east and gulf coast of the U.S., southern and Baja California wetlands are inherently small in extent and are composed of heterogeneous patches of salt marsh vegetation with typical dimensions ranging from 2 to 10 m and open areas known as salt pannes. Preserves have been established to protect and maintain some of the remaining habitat and artificial salt marshes have been constructed or restored on marginal wetlands or uplands adjacent to wetlands. Important for adaptively managing preserves and artificial wetlands is sound biological knowledge of the habitat requirements of protected bird species. However, such knowledge is limited for clapper rails and Savannah sparrows. Mapping and monitoring the distribution of salt marsh vegetation types and understanding bird habitat relationships for the relatively small and heterogeneous salt marshes of southern California may be best accomplished using a combination of airborne multispectral imagery and field observations supported by kinematic global positioning systems (KGPS) (Phinn et al. 1996; Stow et al. 1996) Very-high spatial resolution multispectral images are acquired with digital camera or digital video systems in three or four wavebands within the 0.4 to 1.0 micron region of the spectrum and with ground sampling distances ranging from 0.2 to 3.0 m. KGPS surveying units are capable of making horizontal and vertical positional measurements with precision and accuracy that approaches a few centimeters. KGPS enables the subtle, yet ecologically important topographic relief of salt marshes to be mapped with high detail and precision.

The objective of this chapter is to explore the utility of integrating very-high resolution data from airborne multispectral digital imaging systems and

KGPS to improve the general understanding of and assess the ability to monitor southern California salt marsh habitat utilized by rare and endangered birds. Two case studies are presented that demonstrate the integration of remote sensing, GPS, and Geographic Information Systems (GIS) technologies for quantifying and analyzing environmental properties associated with bird habitat in natural and restored salt marshes within the Sweetwater Marsh National Wildlife Refuge (SMNWR) near San Diego, California (Figure 1).

The first case study illustrates the benefit of integrating airborne multispectral imagery with digital elevation data derived from KGPS surveys to generate reliable maps of habitat types associated with the light-footed clapper rail for a constructed salt marsh. In the second case study, knowledge is gained about nesting territory preferences of Belding's Savannah sparrow by using KGPS to survey territory locations and capture positions and attributes of topography and vegetation that may be important predictors of habitat. Cover proportions estimated from maps of general marsh vegetation types derived from airborne multispectral data are also incorporated into the habitat analysis.

BACKGROUND

Very-High Resolution Multispectral Imagery

Given the limited areal extent and fine-scale patchiness of southern California salt marshes, the spatial resolution of current satellite and most airborne scanner systems are not suitable for mapping and monitoring salt marsh habitat. An alternative source of imagery is provided by very-high resolution digital multispectral camera systems such as the Airborne Data Acquisition and Registration (ADAR) 5500 system (Stow et al. 1996). The ADAR 5500 normally images in three visible (blue, green, red) and one near-infrared waveband. It can provide imagery with ground sampling distances ranging from 0.25 to 2.50 m per pixel, depending on the altitude and airspeed of the aircraft platform. Each frame is approximately 1000 by 1500 pixels. Each waveband image for a given frame is registered by a semi-automatic, image registration procedure. Individual frames can then be georeferenced and stitched together to yield image maps of areas larger than the scene captured by a single frame (Stow et al. 1996).

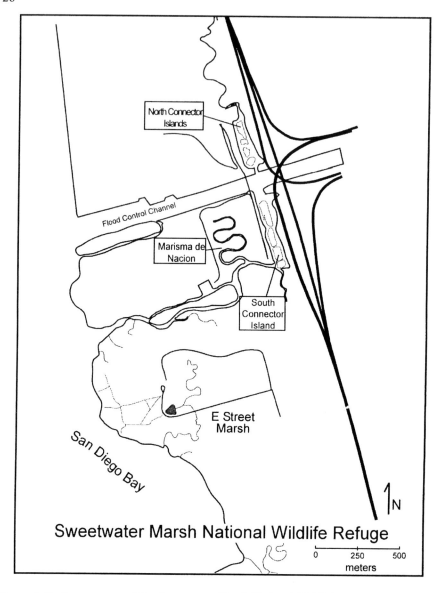

Figure 1. Study area map of the Sweetwater Marsh National Wildlife Refuge (SMNWR). The refuge is located near San Diego, National City and Chula Vista, California and is adjacent to San Diego Bay. The Connector Marsh and E Street Marsh study areas for the two case studies are delimited.

Digital multispectral image data used in the two case studies were acquired with a spatial resolution of 0.75 m by the ADAR 5500 system on August 4, 1994. Positive Systems Inc. captured the image data and performed band-to-band registration and radiometric processing to reduce brightness fall-off across each image frame, normally associated with vignetting. An orthorectified mosaic of these ADAR image frames was produced through a proprietary process by TRIFID, Inc., resulting in an image map with high positional accuracy (RMSE = 0.80 m; n = 30). This high positional accuracy enabled precise co-location of field data generated from the highly precise KGPS.

Kinematic Global Positioning Systems (KGPS)

A KGPS is a type of continuous differential carrier-phase GPS requiring only short periods of observations. Differential positioning is the determination of relative coordinates of two or more receivers that are simultaneously tracking the same satellites. Before KGPS data can be collected, a well-distributed control network consisting of horizontal and vertical control points must be established. We established a control network with initialization points and permanent base stations with known baselines within the Sweetwater Marsh study area. When performing a KGPS survey the surveyor must initialize the survey from a known baseline, while the other receiver is statically located over the base station. If a quality control network has been established for the study area, the positional accuracy for resultant horizontal and vertical measurements should fall within 5 cm of the true positions. Besides the tremendous precision of three-dimensional survey measurements, KGPS units enable a field scientist to navigate in real time to field plot or transect locations that were previously established or randomly selected and with very high location precision. KGPS surveys for the case studies were conducted with a Trimble 4000SE Land Surveyor

Sweetwater Marsh Study Site

The study areas utilized for the two case studies are located within the Sweetwater Marsh National Wildlife Refuge (Sweetwater Marsh), adjacent to San Diego Bay at the boundary between the cities of National City and Chula Vista, California (Figure 1). One study area included several constructed and restored marshes in the northeastern portion of the Refuge. The other study area was the E Street Marsh, a mostly natural marsh. The most prevalent vegetation species in the Refuge are *Batis martima, Frankenia grandifolia, Liminium californicum, Monanthochloe littoralis, Salicornia bigelovii, Salicornia subterminalis, Salicornia virginica, Spartina foliosa,* and *Suaeda californica.*

Restoration and construction of salt marsh habitat was conducted at Sweetwater Marsh to compensate for wetlands destroyed and disturbed during the construction of a freeway overpass and a flood control channel. Wetlands were excavated from fill material from existing marginal wetlands and dredge spoil deposits to create a set of islands and channels. Marsh vegetation was then planted on these newly created islands. Eight salt marsh islands, comprising the North Connector and South Connector islands, were built in 1984 and planted in 1985. These restored marshes connect Paradise Creek Marsh, a natural marsh in the north, to the main part of Sweetwater Marsh to the south and southwest.

E Street Marsh was once connected to the rest of Sweetwater Marsh and is now fragmented by roadways and isolated by a landfill. The marsh covers approximately 9.32 hectares and is dominated by *S. virginica* (a type of pickleweed).

CASE STUDY 1: MONITORING RESTORED CLAPPER RAIL HABITAT

Introduction

The objective of this case study is to demonstrate the effectiveness of integrating very-high resolution digital elevation data acquired by KGPS with ADAR image data to map vegetation types within a restored coastal salt marsh. Some key results are presented pertaining to habitat for the light-footed clapper rail from a more detailed study by Bradshaw (1997). Salt marsh habitat maps derived from computer-assisted image classifications of ADAR image data alone and from a combination of ADAR and topographic data are compared to determine whether classification accuracy is improved by incorporating ancillary data.

Characteristics of the Light-Footed Clapper Rail and Its Habitat

The endangered light-footed clapper rail tends to build its nests in tall dense stands of Pacific cordgrass (*Spartina foliosa*), which only grows in fully tidal salt marshes (Jorgensen 1975). In addition the clapper rail habitat requires an area around the nest for foraging, referred to as the home range. A criterion for suitable clapper rail habitat at the Sweetwater Marsh restoration sites are that seven home ranges of 0.8 to 1.6 hectares in size must contain at least 15% Lower Marsh, some Middle Marsh, and at least 15% Upper Marsh (Zedler 1993). Monitoring and evaluating restored salt marsh habitat for the endangered light-footed clapper rail requires accurate maps of marsh vegetation types to determine whether the requirements established for the restoration efforts have been met.

General Approach

Distribution of coastal salt marsh vegetation is primarily driven by tidal inundation which changes with elevation. This is evident by the groupings of plant species into lower, middle and upper marsh associations related to generalized elevation zones. These differences in elevation were investigated to determine if they could help separate vegetation having similar spectral-radiometric signatures in digital multispectral image data (Phinn et al. 1996).

Field Measurements of Topography and Vegetation Types

Elevation and vegetation type data were sampled along ground transects with the KGPS unit. A modified systematic sampling strategy was implemented, such that the spacing was adjusted, or additional samples were taken to capture significant changes in elevation. Transects were spaced either 3 or 6 m apart and samples were taken every 3 or 6 m along transects. The majority of the surveying was based on 6 m spacing. The southernmost section of the North Connector site was sampled using a 3 m spacing to generate independent data points for assessing the accuracy of digital elevation modeling techniques (Bradshaw 1997). Overall, a total of 681 horizontal coordinates and elevation measurements were surveyed with 426 points at the North Connector site and 255 points at the South Connector site. In addition to positional and topographic measurements, the dominant plant species present within approximately a square meter patch around each sample point was identified and recorded with the KGPS data logger. All of the data from the KGPS survey were used to generate a digital elevation model (DEM) of the study site. The survey data were split into three separate groups for use in image classification, ecological modeling, and accuracy assessment of the image classifications.

Elevation and Ecological Modeling

The elevation data from the KGPS survey were used to generate DEMs and evaluate the relationship between vegetation species and elevation. The relationship between vegetation and elevation was then used to develop rules for modifying the image-based classifications of vegetation types based on elevations corresponding to each image pixel. ARC/INFO was used to generate raster DEMs by kriging the KGPS data points (Figure 2 – on the CD-ROM). Kriging was selected as the optimal interpolation scheme after comparing the results from this approach with a DEM product generated with a triangulated irregular networks (TIN) interpolation scheme (Bradshaw 1997). The DEM was generated with a 0.75 m grid spacing to correspond to the spatial resolution of the ADAR image mosaic.

Differences between marsh type elevations were investigated using the corresponding elevation and vegetation type data sampled during the KGPS survey. Various statistical comparisons of marsh type elevations were performed using the non-parametric Kruskal-Wallis test and the Mann-

Whitney test (Barber 1988; Gilbertson et al. 1985). Four marsh vegetation habitat types were analyzed: Lower Marsh – Cordgrass, Lower Marsh – Cordgrass/Succulent Mix, Middle Marsh, and Upper Marsh (Figure 3). Based on the sample data, the four marsh habitat types were found to occur at different elevations for a given location of the marsh. The elevations of the four marsh types were all significantly different at the North Connector. At the South Connector most of the marsh types occurred at different elevations. However, the elevations of Lower – Cordgrass compared to Lower – Cordgrass/Succulent Mix as well as the elevations of Lower – Cordgrass/Succulent Mix compared to Middle Marsh were not significantly different.

Image Classification

Image data used for classification were subset from the August 1994 ADAR image mosaic (nominal ground sampling distance of 0.75 m). Image subsets of the North Connector and South Connector were extracted from the mosaic based on the DEMs generated for these two sites.

An unsupervised classification approach of the study sites was performed using the ISODATA clustering algorithm in the ERDAS Imagine software package. An initial set of 20 training clusters were generated and labeled based on general knowledge of the study sites gained during the KGPS surveys. Ten separable cluster classes and 10 confused or mixed classes were identified in the labeling process. A second iteration was then performed generating another 20 clusters based only on the pixels corresponding to "confused" cluster classes in the first iteration. This yielded another 12 separable classes and eight classes that were confused. The signature data from the first 10 and second 12 acceptable clusters were then combined and evaluated by analyzing their histograms and statistics. The histograms and statistics of the signature set were considered acceptable and a pixel-by-pixel classification of the two ADAR subsets was performed with a maximum likelihood algorithm that was trained with this signature set. The resulting classes were identified as Channel/Water, Lower-Cordgrass, Lower-Cordgrass/Succulent Mix, Middle Marsh, Upper Marsh, and Bare Dry Soil. A 3 x 3 majority filter was then applied to the resulting image classification products for generalization purposes and to reduce artificial high frequency variability.

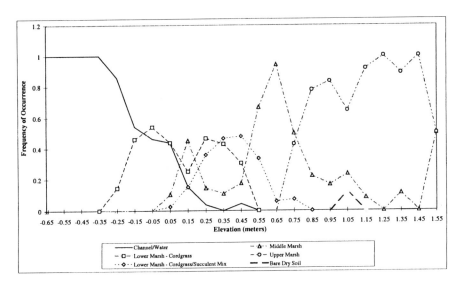

Figure 3: Marsh type frequency of occurrence by elevation. Compiled from frequency values in Table 6.

Incorporating Ancillary Data

A post-classification sorting method of incorporating the DEM as ancillary data was applied to the output of the ADAR-based unsupervised classification (Hutchinson 1982). A set of sorting rules was implemented to correct confused or mixed classes. Defining the sorting rules required identifying the types of confusion that existed in the image-only classifications and utilizing ecological modeling results to select elevation thresholds for separating these confused classes. Thresholds were established from the frequency of occurrence of each marsh type within 10 cm elevation intervals. An elevation range of 10 cm was chosen as the maximum range in error from the GPS data ($\pm$ 5 cm). The frequency of occurrence is simply the percentage of samples within an elevation interval where the presence of a marsh type was recorded. These frequencies are the same as the conditional probability for each marsh type and sum to one for each elevation range. The frequencies of occurrence of marsh types by elevation are plotted in Figure 3 and are the basis of the post-classification sorting rules provided in Table 1. The sorting rules are related to the mid-points between the transition of dominance from one marsh type to the next, or the upper/lower limits of a marsh type where the frequency of occurrence of the marsh type approaches zero.

Table 1. Post-Classification Sorting Rules.

IF MARSH TYPE IS	AND ELEVATION (meters NAVD29) IS	THEN RECLASSIFY AS
Lower Marsh – Cordgrass	> 0.3 m and < 0.79 m	Lower Marsh - Cordgrass/Succulent Mix
Lower Marsh – Cordgrass	> 0.8 m	Upper Marsh
Lower Marsh - Cordgrass/Succulent Mix	< 0.3 m	Lower Marsh - Cordgrass
Lower Marsh - Cordgrass/Succulent Mix	> 0.8 m	Upper Marsh
Middle Marsh	< 0.0 m	Channel/Water
Middle Marsh	> 0.8 m	Upper Marsh
Upper Marsh	< 0.8 m and > 0.51 m	Middle Marsh
Upper Marsh	< 0.50 m	Lower Marsh - Cordgrass

Comparison of Image Classification Results

The resultant maps from the unsupervised image classification and post-classification sorting approaches are shown in Figure 4 (on the CD-ROM). Accuracy assessment for these products was performed using a set of 163 sample points set aside from the GPS survey for this purpose. Accuracy levels were determined by generating error matrices and calculating user's accuracy, producer's accuracy, overall accuracy and the Kappa coefficient (Hudson and Ramm 1987).

A comparison of error matrices in Tables 2 and 3 indicates that significant improvements in classification accuracy resulted upon modifying the image-based classification with the aid of the DEM and simple discrete rules. The overall accuracy increased over 16% and the Kappa coefficient increased over 20%. A statistical z-test was run to determine if there was a significant difference between the Kappa coefficient of the two error matrices (Congalton and Mead 1983, Hudson and Ramm 1987). Using a 95% confidence interval the null hypothesis that the two matrices are similar is rejected for z values greater than 1.96. Incorporating the ancillary data (DEMs and ecological model data) resulted in a significantly more accurate map product.

Major confusion for the unsupervised classification occurred for Lower Marsh - Cordgrass which was often mistakenly classified as Lower Marsh - Cordgrass/Succulent Mix, Lower Marsh - Cordgrass/Succulent Mix which was incorrectly classified as Lower Marsh - Cordgrass, and Upper Marsh which was falsely classified as Middle Marsh. These confusions were consistently corrected by the sorting rules based on the digital elevation data. One type of confusion that was not improved occurred for areas of Middle Marsh that were incorrectly classified as Upper Marsh or Lower Marsh - Cordgrass/Succulent Mix.

Table 2. Error Matrix for Unsupervised Classification.

Image Classification	Reference Data						Total	User's Accur. (%)
	Water	Lower-Cordgrass	Lower-Cordgrass/ Succulent Mix	Mid-Marsh	Upper Marsh	Bare Dry Soil		
Water	31	1	2	0	0	0	34	91.2
Lower - Cordgrass	1	12	17	2	2	0	34	35.3
Lower - Cordgrass/ Succulent Mix	0	13	7	3	2	0	25	28.00
Mid- Marsh	3	1	2	16	11	2	35	45.7
Upper Marsh	1	2	1	8	14	0	26	53.9
Bare Dry Soil	0	0	0	0	0	9	9	100.0
Total	36	29	29	29	29	11	163	
Producer's Accuracy (%)	86.1	41.4	24.1	55.2	48.3	81.8		
Overall Accuracy (%)	54.6							
Kappa Coeff. (%)	44.6							

Table 3. Error Matrix for Post-Classification Sorting of Unsupervised Classification.

Image Classification	Reference Data						Total	User's Accur. (%)
	Water	Lower - Cordgrass	Lower - Cordgrass/ Succulent Mix	Mid-Marsh	Upper Marsh	Bare Dry Soil		
Water	32	1	2	0	0	0	35	91.4
Lower - Cordgrass	2	17	8	2	0	0	29	58.6
Lower - Cordgrass/ Succulent Mix	0	10	17	4	1	0	32	53.1
Middle Marsh	2	1	2	15	2	0	22	68.2
Upper Marsh	0	0	0	8	26	2	36	72.2
Bare Dry Soil	0	0	0	0	0	9	9	100.0
Total	36	29	29	29	29	11	163	
Producer's Accuracy (%)	88.9	58.6	58.6	51.7	89.6	81.8		
Overall Accuracy (%)	71.2							
Kappa Coeff. (%)	64.8							

A significant advantage of post-classification sorting was the ability to correct for confusion from overlapping spectral-radiometric signatures where the differences in elevation of two classes were minimal (such as between Lower Marsh - Cordgrass and Lower Marsh - Cordgrass/Succulent Mix). For example, post-classification sorting resulted in over a 30% improvement in producer's accuracy of Lower Marsh - Cordgrass/Succulent Mix.

Recommendations for Clapper Rail Habitat Mapping

While significant improvements were achieved by incorporating ancillary data, this technique yielded an overall accuracy of 73% and a Kappa coefficient of 67%. The adequacy of this level of accuracy for restoration monitoring depends on mandated requirements for precision, spatial unit of assessment (e.g., home ranges versus entire marsh complex), and other factors. At a minimum, the maps should be sufficiently reliable for determining if compliance to monitoring objectives are sufficiently close to justify more costly and ecologically disturbing field assessments.

For an operational implementation, our results suggest that an elevation survey of a restoration site using a nominal sample interval of 6 m should be conducted after marsh construction and prior to planting. A DEM of the restoration site should be generated using kriging as the interpolation scheme. A random survey of vegetation should be conducted to establish an ecological model and develop sorting rules. An unsupervised image classification should be performed with the output modified by the post-classification sorting rules (Bradshaw 1997).

CASE STUDY 2: ASSESSMENT OF BELDING'S SAVANNAH SPARROW HABITAT

Introduction

The objective of this case study is to examine the nesting habitat preferences of the Belding's Savannah sparrow through the use of KGPS, remote sensing, and GIS. Examples and results are drawn from a more comprehensive study conducted by Brewster (1996).

In 1974 the Belding's Savannah sparrow was placed on the California state endangered list when it was estimated that fewer than 1500 individuals remained (Bradley 1973, Powell 1993). It is also a Category 2 candidate for federal listing (James and Stadtlander 1991). Knowledge of the traits, habitat requirements and distributions of the sparrow is limited (Massey 1977, White 1986, Johnson 1987, Powell 1993). Censuses for the sparrow were conducted for three different years at the E Street Marsh. In 1977,

Massey (1977) counted 18 breeding pairs. By 1986, Zembal et al. (1988) found that the population decreased to only eight breeding pairs. In 1991, there were 15 breeding pairs. James and Stadtlander (1991) attribute this 88% gain to the restriction of foot and vehicle traffic, which allowed the pickleweed vegetation (*Salicornia spp.*) to flourish.

Characteristics of the Belding's Savannah Sparrow and its Habitat

The Belding's Savannah sparrow is a small bird that resides year-round in the coastal salt marshes of southern California. The sparrow resides in the upper littoral zone of the marsh, where it usually nests above the highest spring tide. The vegetation cover of the upper littoral zone is dominated by pickleweed (*S. virginica*) and according to Massey (1977), the sparrow tends to build its nest within these plants. It has also been known to nest in the upper salt marsh, which is usually dominated by *S. subterminalis* and *M. littoralis* (Massey 1977).

Belding's Savannah sparrows are highly territorial birds that seem to re-establish the same nesting territory annually. A nesting territory is a fixed area, which may change slightly over time, is avoided by rivals, and is defended by the possessor (Brown and Orians 1970). The nests seem to be located at sites with minimal tidal inundation and stream flooding, and with adequate vegetation cover and protection from the sun and predators. Thus, plant form and structure play key roles in habitat selection, variations of which can often be remotely sensed. Territory location also seems to be based on a balance between the ease of defending it with the need to attract a mate. The size of the territory seems to vary with the density and height of the vegetation. Powell (1993) found that the territories ranged from 80 to 420 m^2 at Los Penasquitos Lagoon and from 308 to 936 m^2 at the Tijuana Estuary. It has been suggested that the Belding's Savannah sparrows need only a small territory because they use it mostly for nesting and forage elsewhere (Bradley 1973, Massey 1977).

General Approach

The spatial distribution of three potential habitat variables: elevation, vegetation height, and percent middle-high marsh, were examined for territory sites and compared to random sites where sparrows were not observed. Elevation may be an important habitat variable because the sparrows generally nest above the highest spring tide to insure the nest's safety. In the past, elevation could not be measured in an efficient, precise and accurate manner in the marsh due to subtle topographic relief. With the high precision accuracy of KGPS this is now possible. Vegetation height may be important because the sparrows require vegetation that is sufficiently tall to keep their nest above the tide and provide adequate protection from predators. The percent cover of middle-high marsh type was examined because certain cover types such as low marsh and bare/sparsely covered areas do not offer protection from the tides and/or do not provide adequate cover for sparrow nests. Estimates of the percent cover of middle-high marsh type were derived from the classification of ADAR image data.

Field Measurements of Topography and Vegetation Properties

Perch location, topography, and vegetation property data were collected using KGPS surveying. The control networks of known horizontal and vertical points necessary for kinematic surveying were established in 1994. Using a systematic, unaligned spatial sampling scheme, we collected data every 2 to 5 m along transects spaced between 3 and 5 m. At each measurement location horizontal coordinates, terrain height, vegetation height and vegetation type were recorded. Vegetation types present within a 1 m square area around the GPS measurement sites were interpreted and recorded. The vegetation type data provided a basis for an accuracy assessment of the classified ADAR imagery.

Perch locations of the Belding's Savannah sparrow were surveyed on April 25, 1994 at E Street Marsh. The sparrows were located by listening for their song, then visually locating them with binoculars. If they were perched and singing, which are signs of territorial defense (Zembal et al. 1988), the location was noted. We then proceeded to the location from which songs were thought to have emanated and surveyed the coordinates with the KPGS unit. We assumed that each perching sparrow indicated the presence of a territory.

Because territory size was not measured in this study, the size of the area examined in the habitat analysis was based on previous findings (Powell 1993, Massey 1977) and on the distribution of surveyed perch sites. Powell (1993) and Massey (1977) found that territory size in three southern California marshes varied from 80 to 936 m^2, with areas of denser vegetation having smaller size territories. To examine the perch site within the nesting habitat at different spatial extents, we examined environmental variables sampled within circular areas of three radii: 2 m (12.6 m^2 area), 5 m (78.5 m^2 area), and 10 m (314.0 m^2 area) (Figure 5). The 2, 5 and 10 m radius circles were not meant to represent territories and were intended to characterize a perch site, which is part of a territory, and capture environmental characteristics associated with the nesting habitat. Twenty-five additional sample sites were randomly located and enabled statistical comparison of properties of the observed sites to those where sparrows were not observed. This does not imply that the random sites were not part of an actual territory, but it does indicate that no sparrows were observed at those locations. Locations of observed and random sites are depicted on an ADAR image map of E Street Marsh in Figure 6 (on the CD-ROM).

Image Classification of Habitat Types

A similar unsupervised image classification approach and clustering algorithm (ISODATA) used for the first case study was applied to a subset of the 1994 ADAR image mosaic covering the E Street Marsh study area to generate a map of habitat vegetation types. Multiple iterations were run with varying clustering parameters in an attempt to optimize the classification accuracy. Fifty cluster classes were generated in each iteration and a pixel-by-pixel classification based on a maximum likelihood decision rule was implemented. The resulting cluster classes were assigned the following habitat labels: Water, Bare, Low marsh, Middle marsh, and High marsh utilizing field knowledge and data collected with the KGPS.

Stratification of the study area into two subareas corresponding to an inner zone of predominantly low and middle marsh types and an outer, high marsh zone proved to be useful in separating confused spectral signatures of vegetation types. The iterative clustering and masking process utilized in the first case study was less successful for reducing these confusions. Upon clustering, classifying, and labeling the two strata separately, the subsets of the outer and inner portions of the marsh study area were combined into a single map portraying the distribution of the five marsh and surface cover types. The resultant marsh habitat map is shown in Figure 7 (on CD-ROM).

Assessment of Image Classification Results

To quantify the accuracy of the image classification product 270 field samples were randomly selected from the KPGS survey. Vegetation attributes identified in the field were compared with the image-derived category for corresponding pixels. An error matrix, overall, user's, and producer's accuracy statistics (Story and Congalton 1986), and the Kappa coefficient were derived to quantify the accuracy of the classification results and are shown in Table 4 (Rosenfield and Fitzpatrick-Lins 1986, Hudson and Ramm 1987).

The overall accuracy of the marsh habitat type map from image classification was $86 \pm 4\%$. The Kappa coefficient of 0.75 is at the threshold between 'good' and 'very accuracy (Fitzgerald and Lees 1994). Particularly encouraging was the high classification accuracy of Middle marsh class (user's = 89.2%, producer's = 88.1%), since Middle marsh cover data used for statistical analyses of habitat characteristics were extracted from the image classification product. The High marsh and Water categories were also classified with a high degree of accuracy. The Low marsh and Bare categories were found to be less accurate and were confused with the Middle and High marsh classes, respectively. Areas of misclassification most commonly consisted of mixed pixels.

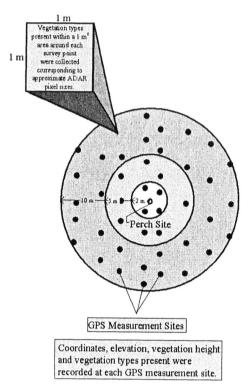

Figure 5: Schematic demonstrating testing territory delineations around observed/random sites.

Environmental Statistics for Sparrow Territories

Statistics were analyzed for environmental variables sampled within potential territories (i.e. the circular areas encompassing locations where sparrows were sighted) and random areas. KGPS and image-classified data were converted to ARC/INFO GIS layers for convenience in extracting samples. Circular zones of varying radii (1, 5 and 10 m) were generated around each perch observation location and random points. Mean elevation and plant height was computed from the KGPS data for sample points falling within each zone. Percent Middle and High marsh cover was derived as the percentage of pixels within buffer zones classified as Middle and High marsh.

Table 4. Error matrix and accuracy statistics for ADAR-based classification of habitat types.

Image Classified	Reference Data					Total	User's Accuracy (%)
	Low	Middle	High	Bare	Water		
Low	59	17	0	0	0	76	77.6
Middle	16	141	1	0	0	158	89.2
High	0	2	17	0	0	19	89.5
Bare	0	0	3	8	0	11	72.7
Water	0	0	0	0	6	6	100
Total	75	160	21	8	6	270	
Producer's Accuracy (%)	78.7	88.1	81.0	100	100		
Overall Accuracy	86.4						
Kappa Coefficient	75.0						

Table 5 provides descriptive summary statistics for environmental variables measured throughout the E Street Marsh study area. The observed territory density was 2.68 territories per hectare. Much of the vegetation at the study site was not tall and dense enough to hide sparrow nests.

To compare habitat characteristics of observed and random sites we used univariate statistics to determine if there were significant differences between each continuous variable (e.g., vegetation height, elevation, percent of mid-high marsh) (Li and Martin 1991, Frederick and Gutierrez 1992, Shiraki 1994). For each environmental variable, skewness and kurtosis measures were computed and the histograms were viewed to determine if they were normally distributed. The distributions for all variables except elevation were weakly normal and skewed. Thus, both Mann-Whitney (non-parametric) and F-statistic (parametric) statistical tests were applied.

Vegetation Height

Since Belding's Savannah sparrows are thought to need vegetation of sufficient height to provide adequate protection for their nests, we tested the hypothesis that the vegetation height for observed sites would be significantly ($\alpha = 0.05$) taller than that of random sites. We also hypothesized that the standard deviation for observed sites would be smaller than that of random sites, because there should be less variability at the desired territory locations. Table 6 summarizes the results of the Mann-Whitney and F tests.

Table 5. Environmental Statistics for E Street Marsh.

Environmental Variable	Value
extent of marsh (hectares)	9.32
known territories	25
territory density (per hectare)	2.68
mean vegetation height (cm)	25.26
median vegetation height (cm)	25.4
minimum vegetation height (cm)	2
maximum vegetation height (cm)	83.82
mean elevation (m)	0.970
minimum elevation (m)	0.235
maximum elevation (m)	2.796
% low marsh	28.08
% middle marsh	53.70
% high marsh	10.73
% water	4.15
% bare/sparse	5.41
no. of KGPS points collected	4265

Vegetation height at observed sites was significantly taller than that at the randomly selected sites for all three radii sizes. The standard deviation was also smaller for all circular areas. Much of the E Street Marsh site was covered by stunted vegetation in low lying areas. These low lying areas are often covered by water, provide no cover and therefore, are unsuitable as nesting sites. The maximum difference between means for observed and random sites was 9 cm. While this may seem like a small difference, the total range in vegetation heights was about 80 cm over this study area. For the 2 and 5 m radius sizes the vegetation height ranges were smaller for the observed sites, but the ranges for the observed and random sites were identical for the 10 m radius size. As area around the perch site increased for observed sites the mean vegetation height decreased, while the standard deviation and range increased. This may indicate that patchiness increased as the areal extent increased. It could also indicate that the perch site contained taller vegetation.

Elevation

The Belding's Savannah sparrows are thought to build nests above the highest spring tide, therefore, we hypothesized that the elevation of the observed sites would be greater than that of the random sites. The elevation of the observed sites was significantly ($\alpha = 0.05$) higher than that of the random sites (Table 7). The standard deviation for the observed sites was smaller for all territory sizes except the 5 m radius size, which suggests that there was less topographic variability in the observed sites.

Table 6. Results of one-tailed Mann-Whitney and F tests for vegetation height.

	2 m radius		5 m radius		10 m radius	
	observed	random	observed	random	observed	random
n	26	44	119	134	484	451
mean (cm)	35.46	27.2	33.79	24.76	30.18	23.17
S.D.	6.44	12.7	9.16	12.37	12.03	12.49
range	22.86	50.8	45.72	58.42	63.5	63.5
M-W test	1142*		18616.5*		175702*	
M-W ρ	0.004		0.00		0.00	
F	3.62*		6.65*		8.73*	
ρ	0.0003		0.00		0.00	
direction	obs. >random		obs. >random		obs. >random	

Significant at $\alpha = 0.05$
M-W test = Mann-Whitney test statistic
M-W ρ and ρ = level at which test becomes significant
obs. = observed sites random = random sites

Although the elevations for observed sites were significantly greater, the magnitude of differences was small. The maximum difference between means for observed and random sites was 9 cm. The maximum difference in means between observed sites was 2 cm. As with vegetation height, the range in elevations across the study area is also small (2.5 m).

Percent Middle and High Marsh

We tested the hypothesis that the observed sites have a higher percentage of middle and high marsh types than the randomly selected sites. The observed sites for all radii sizes did contain a higher percentage of middle and high marsh, the standard deviations were lower and the range was larger for random sites (Table 8). Only the 5 and 10 m radii zones contained significantly ($\alpha = 0.05$) more middle and high marsh than the random sites.

Recommendations for Future Assessment of BSS Habitat

The use of remote sensing for assessing Belding's Savannah sparrow habitat is warranted, because it provides a non-invasive and non-destructive means of acquiring wall-to-wall information on some aspects of habitat. The use of spectral vegetation indices (SVIs) to quantify vegetation height and/or cover should be investigated. Also worth exploring is the use of landscape ecological pattern indices applied to image-derived maps of SVIs or vegetation types to examine known territory sites and search for spatial patterns (Cressie 1991).

Table 7. Results of one-tailed Mann-Whitney and F tests for elevation.

	2 m radius		5 m radius		10 m radius	
	observed	random	observed	random	observed	random
n	51	44	144	134	474	484
mean (m)	0.983	0.916	1.003	0.925	0.999	0.945
S.D.	0.154	0.194	0.215	0.206	0.199	0.213
range	0.894	0.975	1.310	1.079	1.310	1.594
M-W test	2772.5*		22707*		251995*	
M-W ρ	0.0078		0.000		0.000	
F	1.83*		3.11*		4.04*	
ρ	0.035		0.001		0.00	
direction	obs.>random		obs.>random		obs.>random	

*Significant at $\alpha = 0.05$
M-W test = Mann-Whitney test statistic
M-W ρ and ρ = level at which test becomes significant
obs. = observed site; random = random sites

Table 8. Results of one-tailed Mann-Whitney and F tests for percent middle-high marsh.

	2 m radius		5 m radius		10 m radius	
	observed	random	observed	random	observed	random
n	25	25	25	25	25	25
mean	0.694	0.619	0.719	0.574	0.703	0.605
S.D.	0.217	0.309	0.179	0.256	0.134	0.201
range	0.714	1.000	0.660	0.948	0.509	0.760
M-W test	612		531.5*		531.0*	
M-W ρ	0.311		0.020		0.019	
F	0.942		0.267*		2.52*	
ρ	0.178		0.007		0.009	
direction	obs.>random		obs.>random		obs.>random	

* Significant at $\alpha = 0.05$; ** Test not performed
M-W test = Mann-Whitney test statistic
M-W ρ and ρ= level at which test becomes significant
obs. = observed site; random = random sites

More reliable and precise information on habitat requirements may be derived if the true size and position of the territories are known. KGPS could be used to map actual territory locations as part of a more rigorous observation of bird activity and movements within territories.

SUMMARY AND CONCLUSIONS

Managers of habitat reserves containing rare and endangered animal species require spatial information on the type and condition of the habitat that is utilized by these animals. For many of these reserves, the areal extent is limited by the fragmentation caused by human land use activities and because the establishment of the reserve may have required land purchases which are expensive. Also, the grain or spatial resolution at which habitat information is required is often very small, due to the fine-scale patchiness of vegetation distributions and because many of the disturbance effects of interest to managers occur at fine scales. Thus, data collection schemes that satisfy management information needs must enable spatially detailed and continuous measurements to be captured over limited spatial extents.

In the first case study, we focussed on vegetation mapping of restored salt marshes constructed to create light-footed clapper rail habitat. We showed that high levels of accuracy by image classification of ADAR multispectral data alone can be difficult to achieve due to spectral-radiometric similarity of certain plant species. Through the analysis of vegetation associations with the very subtle topographic relief (on order of 1 to 2 m range) of the salt marsh, we found that vegetation types generally occur within elevation zones. These differences in elevation can help separate vegetation with similar spectral characteristics. The integration of digital elevation data, generated from precise KGPS field surveys with image classification significantly improved the accuracy of classification results. The post-classification sorting approach of integrating DEM data by discrete rules, resulted in an improvement of the overall accuracy from 54.6% to 71.1% and the Kappa coefficient from 0.45 to 0.65.

The salt marsh habitat requirements of the Belding's Savannah sparrow were examined in the second case study, also by utilizing the detailed and precise data captured by ADAR imaging and KGPS surveying systems. Our analyses showed that elevation, vegetation height, and percent middle-high marsh were all significantly greater for territories utilized by the sparrow compared to randomly selected sample areas. The percent Middle-high marsh variable was derived from image classification of ADAR image data. The overall accuracy of the ADAR-derived map of five habitat types was 86 $\pm$ 4%, with a Kappa coefficient of 0.75. Our results correspond with previous findings by Massey (1977) and Powell (1993) and support the hypothesis that the Belding's Savannah sparrow tend to build nests in taller vegetation above tidal levels and where they are protected from predators.

The tools assessed in this study, ADAR image data, KGPS surveys, and GIS, have proven valuable. Overall accuracy for both case studies were high given the heterogeneous landscape and the very precise and rigorous accuracy assessment procedures. These results suggest that ADAR System 5500 or other high resolution multispectral data from similar airborne digital camera systems have merit for generating baseline inventories of wetland vegetation and for monitoring changes that might affect endangered bird species that utilize salt marshes as habitat (Stow et al. 1996).

KGPS surveys provided precise and accurate positions and elevations and were convenient for recording site-specific environmental attributes. Also, KGPS surveys allowed an efficient and precise means of checking the accuracy of image classification derived from very high-resolution images. However, these surveys are invasive, destructive and costly. In future studies of this nature, the damage to the marsh caused by human intrusion could be minimized by exclusively collecting KGPS data at known territory sites and pre-selected random sites, or in the case of restored and constructed sites, before marsh vegetation is planted. In conjunction with airborne imaging and KGPS surveys, GIS and image processing software proved to be valuable in the display, manipulation, extraction, and analysis of the large amounts of data that were collected for these studies.

ACKNOWLEDGMENTS

Funding for components of this research was provided by the National Aeronautics and Space Administration (NASA) Earth Observations and Commercialization Program (EOCAP), Southern California Edison, Inc., and the State of California Department of Transportation. Professor Joy Zedler contributed significantly to establishing the biological and resource management goals of this research and contributed to data interpretations of clapper rail habitat. Dr. Abby Powell sighted Belding's Savannah sparrows perches and supported interpretation of environmental habitat data relative to sparrow sightings. Dr. Stuart Phinn and Bruce Nyden, supported field data collection and image processing efforts. KGPS equipment was made available by Coastal Resources Associates, Inc. of Vista, California. Elizabeth Witztum assisted with production of graphical material.

REFERENCES

Barber GM. *Elementary Statistics for Geographers*, New York: The Guildford Press, 1998.

Bradley RA. A population census of the Belding's Savannah sparrow, *Passerculus sandwichensis beldingi*. Western Bird Bander 1973; 48:40-43.

Bradshaw BK. Integrating high resolution digital imagery and digital terrain data for mapping restored salt marsh habitat, Master's Thesis, San Diego State University, 1997.

Brewster AE. Utilizing geographic technologies to analyze the nesting habitat preferences of the Belding's Savannah Sparrow, Master's Thesis, San Diego Sate University, 1996.

Brown JL, Orians GH. Spacing patterns in mobile animals. Annual Review of Ecology and Systematics 1970; 1:239-262.

Congalton RG, Mead RA. A quantitative method to test for consistency and correctness in photointerpretation, Photogrammetric Engineering & Remote Sensing 1983; 49:69-74.

Cressie NAC. *Statistics for Spatial Data*, New York: John Wiley and Sons, 1991.

Fitzgerald RW, Lees BG. Assessing the classification accuracy of multisource remote sensing data. Remote Sensing of Environment 1994; 47:362-368.

Frederick GP, Gutierrez RJ. Habitat use and population characteristics of the White-tailed Ptarmigan in the Sierra Nevada, California. The Condor 1992; 94:889-902.

Gilbertson DD, Kent M, Pyat FB. *Practical Ecology for Geography and Biology: Survey, mapping and data analysis*, London: Hutchinson and Co., 1985.

Hudson WD, Ramm CW. Correct formulation of the Kappa coefficient of agreement, Photogrammetric Engineering & Remote Sensing 1987; 53:421-2.

Hutchinson CF. Techniques for combining Landsat and ancillary data for digital classification improvement, Photogrammetric Engineering & Remote Sensing 1982; 48:123-30.

James R, Standtlander D. *A survey of the Belding's Savannah sparrow (Passerculus sandwichensis beldingi) in California, 1991*. Sacramento: California Department of Fish and Game, Nongame Wildlife Program, 1991.

Johnson J. Correlations between vegetative characteristics of six salt marshes and Belding's sparrow densities. M.S. Thesis. California State University, Los Angeles, 1987.

Jorgensen P. Habitat preferences of the light-footed clapper rail in Tijuana Estuary Marsh, California. M. S. Thesis. San Diego State University, 1975.

Li P, Martin, TE. Nest-site selection and nesting success of cavity-nesting birds in high elevation forest drainages. The Auk 1991; 108:405-418.

Massey BW. A census of the breeding population of the Belding's Savannah sparrow in California. Sacramento: California Department of Fish and Game, Nongame Wildlife Investigation, Federal Aid for Endangered Species, Final Report, E-1-1, Study IV, Job 1.2, 1977.

Phinn S, Stow D, Zedler J. Monitoring wetland restoration using airborne multispectral video data in southern California, Restoration Ecology 1996; 4: 412-422.

Powell AN. Nesting habitat of Belding's Savannah sparrows in coastal salt marshes. The Society of Wetland Scientists 1993; 13:219-223.

Rosenfield GH, Fitzpatrick-Lins K. A coefficient of agreement as a measure of thematic classification accuracy, Photogrammetric Engineering & Remote Sensing 1986; 52:223-7.

Shiraki S. Characteristics of White-tailed Sea Eagle nest sites in Hokkaido, Japan. The Condor 1994; 96:1003-1008.

Story M, Congalton RG. Accuracy assessment: a user's perspective. Photogrammetric Engineering & Remote Sensing. 1986; 52(3): 397-399.

Stow DA, Hope A, Nguyen AT, Phinn S, Benkelman CA. Monitoring detailed land surface changes using an airborne multispectral digital camera system, IEEE Transactions on Geoscience and Remote Sensing, 1996; 34:1191-1202.

White AN. Effects of habitat type and human disturbance on an endangered wetland bird: Belding's Savannah sparrow. Masters Thesis. San Diego Sate University, 1986.

Zedler JB. "Restoring diversity in salt marshes: Can we do it?." . In *Biodiversity* EO Wilson d. Washington DC: National Academy Press, 1986.

Zedler JB Canopy architecture of natural and planted cordgrass marshes: Selecting habitat evaluation criteria, Ecological Applications 1993; 3: 123-128.

Zembal R, Kramer KJ, Bransfield RJ, Gilbert N. A survey of Belding's Savannah sparrows in California. American Birds 1998; 42: 1233-1236.

Chapter 4

SPATIAL VARIABILITY IN SATELLITE-DERIVED SEASONAL VEGETATION DYNAMICS

Simon D Jones[1], Andrew C Millington[1], and Barry K Wyatt[2]
1 Department of Geography, University of Leicester, Leicester, LE1 7RH, England
sdj2@le.ac.uk acm4@le.ac.uk
2 Centre for Ecology and Hydrology, Monks Wood, Abbots Ripton, Huntingdon, England.
bkw@ceh.ac.uk

Keywords: NOAA-AVHHR, vegetation indices, seasonal vegetation dynamics, phenology, local image variance.

Abstract Spatial patterns of seasonal vegetation dynamics were derived from coarse-spatial resolution (NOAA14-AVHRR) image data for three ecosystems with contrasting levels of soil moisture stress in the Bolivian lowlands over a 14-month period. These data were compared to ground measurements of parameters that represent seasonal vegetation dynamics (vegetation phenology), local climatic conditions, and the pathways of red and near-infrared (NIR) radiation through the vegetation canopy. Internal spatial variability in the image sub-scenes representing "homogenous" areas of the vegetation in the three ecosystems was measured using a transient 3x3 pixel kernel by four standard measures of variance (mean local image Euclidian distance, variance, skewness, and kurtosis). The seasonal behaviour of these measures of patterns of spatial variation in the imagery is the opposite of the behaviour of frequently applied vegetation indices (VIs). When canopy greenness is low, spatial variance is high; and when the canopy greenness is high, spatial variance is low. These temporal patterns of spatial behaviour of VIs are readily explained by a combination of topographic and edaphic and vegetation structure in semi-arid ecosystems, but the explanation is more complex in wetter ecosystems.

INTRODUCTION

From the mid 1980s onwards, remote sensors have conducted research into monitoring the greening up and dieback of vegetation over very large areas using coarse-spatial resolution remotely sensed imagery. Studies were initially conducted in Africa (Tucker et al. 1985), but rapidly other continents were studied.

This line of research has informed a number of important ecological areas, namely:

➢ Visualisation of vegetation greening up and dieback at continental scales (e.g., Achard et al. 1988).
➢ It has enabled spatial patterns of greening-up and dieback to be researched in relationship to spatial variations in phenological triggers by analysing patterns in contemporaneous meteorological data (e.g., rainfall and temperature) (Justice et al. 1985, Lloyd 1990).
➢ Runs of satellite and meteorological data over a number of years have enabled inter-annual variability in vegetation responses to climate fluctuations to be researched (Davenport and Nicholson 1993).
➢ The synoptic view afforded by satellite data has enabled comparisons of phenological phenomena over large areas.
➢ The "remote sensing phenologies" have been used to map land cover classes using multi-temporal, rather than multi-spectral, classification (Defries and Townshend 1994).

There is no doubt such research constitutes a clear example of an application of GISc (Geographic Information Science) that has advanced ecological understanding of seasonal and inter-annual vegetation dynamics.

Whilst the results of such analyses are visually appealing when viewed at the continental and regional scales, many vegetation scientists are able to identify relatively small areas where their "local ground knowledge" of seasonal (and sometimes inter-annual) vegetation dynamics does not fit with the products derived from remotely-sensed data. This is to be expected because of mismatches in spatial scale between ground observations and the synoptic scale of image products. The majority of studies on tropical forest phenology have been conducted at the species level (e.g., Umana-Dodero 1988, Bullock and Solismagellenes 1990, Ghate and Kumbhojkar 1991) or community scale (e.g., Martinez-Yrizar and Sarukhan 1990, Wright and Cornejo 1990, Bierregaard et al. 1992). Relatively few researchers (e.g., Bray and Gorham 1964, Da Cruz Alencar et al. 1979, Sarmiento and Monsaterio 1983, Fleming and Partridge 1984, Lonsdale 1988) have studied ecosystem-scale phenology. Whilst in some areas GISc provides new insights into spatial patterns of seasonal vegetation dynamics (which may cause us to change our minds about vegetation responses to seasonal changes in bioclimatic parameters), we must be cautious in ascribing all such "mismatches" as GISc generating new ecological knowledge. Spatial patterns of seasonal vegetation dynamics derived from the image data may be incorrect representations of what is happening on the ground. Erroneous spatial patterns arise because of the high contribution to pixel reflectance from soil in arid and semi-arid areas (Huete et al. 1992), atmospheric effects (Pinty and Verstraete 1992a), sun-sensor geometry (Qi et al. 1994), sensor characteristics (Elvidge and Chen 1995) and spatial sampling issues that arise when image products such as GAC and GVI data are produced. Internal spatial variability in the vegetation (or land cover) types being monitored is also a potential cause of the "mismatch" between ground-based and satellite-based observations.

Internal spatial variability within an "homogenous" parcel of a particular vegetation type can arise because of many factors. For natural vegetation communities they are: (i) differences in the architecture of the vegetation (e.g., the size and frequency of forest gaps); (ii) the effects of topographic variation on microclimate (e.g., slope aspect and receipt of solar radiation), and moisture availability (e.g., edaphic conditions); and (iii) the influences of vegetation architecture and topography on species composition. Less attention has been paid to the internal spatial variability of vegetation communities in studies of seasonal vegetation dynamics than the five factors listed above.

We attempt to address this imbalance in this chapter by reporting research into the relationships between ground-based measurements of vegetation (biological) phenology, ground-based microclimate measurements, the receipt of visible and near-infrared (NIR) radiation in the vegetation canopy, and estimates of vegetation greenness derived from satellite imagery. Observations were made at three sites along a vegetation/land cover gradient in central Bolivia that is strongly determined by climate. Measurements were made over a full seasonal cycle from mid 1995 to mid 1996 (Table 1). Our aims in this chapter are to:

➢ analyse the spatial variability in the vegetation indices that are commonly used by remote sensors to map the seasonal dynamics of vegetation;
➢ explain ecologically how this spatial variability arises;
➢ consider the robustness of remote sensing approaches to mapping and monitoring seasonal vegetation dynamics in the light of our findings.

These aims are important given the studies of the seasonal dynamics of tropical ecosystems that have been conducted using coarse spatial resolution imagery (e.g., Tucker et al. 1985, Townshend and Justice 1986, Justice et al. 1986, Townshend et al. 1987, Achard et al. 1988, Gregoire 1990, Achard and Blasco 1990, Arino et al. 1991, Cross 1991, Gond et al. 1992, Singh and Singh 1992, Davenport and Nicholson 1993, De Almedia et al. 1994, Lambin and Strahler 1994a 1994b, Boyd and Duane 1997, Jones et al. 1997).

METHODOLOGY

Three sites were studied along a climatic gradient that extends from east to west in central lowland Bolivia. The sites were selected so that human disturbance was minimised as far as was possible because we were interested in studying spatial variability in natural ecosystems. Details of the sites are provided in Table 1. They are presented in order of decreasing seasonal moisture stress, starting with *savanna abroizada* – a grass and tree savanna typical of the southern Amazon basin – to seasonally-inundated tropical forest – classical lowland rainforest which experiences periodic flooding.

At each site we measured parameters that were either aspects of seasonal vegetation dynamics, the local climatic conditions, or the pathways of red and NIR radiation through the vegetation canopy. All measurements were taken between June 1995 and October 1996. Contemporaneously, image data were acquired for the sites from NOAA14-AVHRR and ESA-ATSR 2 sensors, although only the NOAA14–AVHRR images were used to analyse the spatial variability of seasonal vegetation dynamics.

The following meteorological data were obtained from automatic weather stations we installed at Campo de Buffalo (the closest open ground to the Vallé de Sajta site, and 6 km to the northwest) and at Las Trancas (adjacent to the Las Trancas site, and approximately 13 km to the east of the Lomerio site): net solar radiation, total solar radiation, air temperature, wet bulb depression, wind speed and direction, and precipitation. Two further parameters were calculated from these data - relative humidity and potential evaporation.

In terms of biological phenology, we measured the following every two weeks.

(i) *Litterfall* using 0.5 mm mesh basket traps emptied every two weeks (Proctor 1987, Parker et al. 1989) and 2 x 2 m areas of ground cleared of surface litter at every. We were unable to measure litterfall using either of these methods at the *savanna aborizada* site because of disturbance by cattle.

(ii) Visual assessments of the phenological status of leaves on selected trees at each site were made using a four-point scale comprising absent, emerging, expanded, and senescing. Flowering and fruiting were simply recorded as absent or present. We made observations in the canopy, the sub-canopy, and in the ground layer.

(iii) We measured canopy closure as the percent open sky for four fixed points at each site using hemispherical photography taken 170 cm above the soil surface.

(iv) We measured leaf area index (LAI) using a ceptometer. A ceptometer is a linear array of photo-diodes sensitive to PAR. It measures the distribution of sunflecks (i.e., direct solar radiation which has passed through the canopy unattenuated) as a proportion of the PAR absorbed by the canopy. These data are then inverted to retrieve LAI. The length of sunfleck ceptometer required in tropical forests is too long to be manufactured. Therefore we used a 1 m ceptometer and made measurements every 10 m along regular 250 m transects through the three sites to obtain mean LAI values for each site.

Table 1. Environmental characteristics of study sites.

Study site	Las Trancas	Lomerio	Vallé de Sajta
Latitude (°S)	16° 35' 35"	16° 31' 30"	17° 04' 15"
Longitude (°W)	61° 51' 48"	61° 50' 45"	64° 45' 49"
Elevation (m.a.s.l.)	400	400	212
Vegetation Type	Savanna abroizada	Seasonal semi-deciduous tropical forest	Seasonally-inundated tropical forest
Meteorological Base Station	Las Trancas	Las Trancas	Campo de Buffalo (UMSS research station)
Dominant Geology	Pre-Cambrian (Brazilian Shield) Granitoid	Pre-Cambrian (Brazilian Shield) Gneiss, schist	Recent alluvium
Soil Type	Dystric Ferralsols & Nitosols	Ferralsols, Lixisols & Nitosols	Fluvisols and Cambisols
Climate Zone (and number of dry months)	Sub-humid (3-5)	Sub-humid (3-5)	Humid tropical (1-2)
Mean Annual Precipitation (mm)	1000-1500	1000-1500	4500-5000
Mean Annual Temperature (°C)	23	23	25

Red and NIR radiation at up to seven locations in the vegetation canopies were measured daily over the entire period of the experiment, at times which corresponded to local overpass times for the ESA-2 (10.15hrs) and NOAA-14 (14.15hrs) satellites. Texas Instruments silicon photodiodes were used to measure red and NIR radiation which approximates to the 580-680 and 735-1100 nm AVHRR bands, and the 545-565, 649-669 and 855-875 nm ATSR-2 bands. To simulate the AVHRR 580-680 nm band, the output from a TSL260 photodiode was subtracted from the output from a TSL250 photodiode which had a LEE106 primary transmitter filter fitted. To simulate the AVHRR 735-1100 nm band, a TSL260 photodiode without a filter was used. The pairs of diodes were calibrated, fitted to a foam-backed board, covered with a neutral density filter, and had their fields-of-view restricted to 40°. They were placed at various locations in the canopies of the forest and savanna ecosystems at the sites monitored and connected to Grant Instruments Squirrel data logger and power systems. Seven locations were monitored in the two forest ecosystems and three at the savanna site.

AVHRR image data were acquired from NASA. They were cloud screened manually using the 11.5-12.5µm band and NDVI images. Of the 271 images downloaded 67 were suitable for comparison with ground data, and these were sorted according to sensor view angle into 0-10°, 10.1-20° and 20.1-30° from nadir classes. One of the images was geometrically corrected using map and GPS coordinates. All other images were registered to this image and RMSE values ranged from 0.122 to 0.244 pixels. All images were atmospherically corrected using 5S code (Tanré et al. 1990).

THEORETICAL CONSIDERATIONS

A major element of the research, which is not reported in detail in this chapter, was to verify that the spectral indices which are frequently applied to coarse spatial resolution imagery to monitor vegetation dynamics can be related to a comprehensive set of ground observations of vegetation phenology and the environmental factors controlling phenology. These results are detailed in Jones (1999). The results are summarised in Table 2, because they are required to appreciate the spatial variability in the image data we report in this chapter. For the *savanna abroizada* and semi-deciduous seasonal tropical forest ecosystems, both of which exhibit pronounced vegetation phenologies in response to strong seasonal differences in moisture availability, the correlation coefficients between the bioclimatic triggers for phenological change and the phenological measurements are generally very high, though there are often time lags between the climate triggers and the phenological responses. The equivalent correlations are significantly lower for the seasonally-inundated tropical forest (Table 2).

The correlations in Table 2 represent the correlations between spectral indices and geo-referenced ground observations of single pixels. This is important, because the experimental basis for using spectral vegetation indices to monitor vegetation dynamics over large areas using coarse spatial resolution is based on laboratory-scale leaf reflectance and field plot studies are limited. Verification at the landscape scale is rare, and has not been attempted for most ecosystems. In this chapter, we wish to extend the spatial scope of this analysis and consider the seasonal behaviour of spectral indices over (supposedly) homogenous area of vegetation – we term this spatial-temporal monitoring.

Table 2. Summary of significant correlation coefficients between the meteorological measurements) and the phenological measurements for the three ecosystems studied. Values of the correlation coefficients are not given here (but can be obtained from Jones 1999), rather an indication of which correlations are significant and not significant is given. Correlations were calculated for contemporaneous and lagged relationships between the parameters, the most significant correlations for each bivariate pair are indicated below.

Ecosystem	Bioclimatic Parameters	Litterfall	Leaf flushing	Presence of PAL8	Leaf senescence
Savanna aborizada	Temperature1	-15	0, -1	0, -1	0, -1
	Precipitation2	+1	0, -1	0, -1	0, -1
	R. Humidity3	+1	0, -1	0, -1	0, -1
	Pot. Evap4	+1	0, -1	0, -1	0, -1
Seasonal semi-deciduous tropical forest	Temperature1	nd7	*	0	0
	Precipitation2	nd	*	0	0
	R. Humidity3	nd	*	0	0
	Pot. Evap4	nd	*	0	0
Seasonally-inundated tropical forest	Temperature1	*6	*	*	*
	Precipitation2	*	*	*	*
	R. Humidity3	*	*	*	*
	Pot. Evap4	*	*	*	*

Ecosystem	Bioclimatic Parameters	Canopy Closure	LAI	Ground Reflectance	Lower canopy Reflectance
Savanna aborizada	Temperature[1]	0, -1	1	*	*
	Precipitation[2]	0, -1	0	0	*
	R. Humidity[3]	-1	0	0	*
	Pot. Evap[4]	*	0	0	*
Seasonal semi-deciduous tropical forest	Temperature[1]	+1	nd	-1	-1
	Precipitation[2]	+1	nd	0	0
	R. Humidity[3]	0	nd	0	0
	Pot. Evap[4]	0	nd	0	0
Seasonally-inundated tropical forest	Temperature[1]	*	0	*	*
	Precipitation[2]	-1	*	+1	*
	R. Humidity[3]	*	*	*	*
	Pot. Evap[4]	*	*	*	+1

Key to Table 2.
1. Temperature = Mean monthly temperature
2. Precipitation = Mean monthly precipitation
3. R. Humidity = Mean monthly relative humidity
4. Pot. Evap. = Mean monthly potential evaporation
5. Significant correlation coefficient (at significance levels > 0.095) are provided in the following format: 0 significant correlation between parameters for same month; −1 significant correlation between phenological parameter and meteorological parameter for previous month; +1 significant correlation between phonological parameter and meteorological parameter for subsequent month.
6. no significant correlation (at significance level > 0.09)
7. nd = no data collected for the phonological parameter
8. PAL = photosynthetically active leaves.

Spatial-Temporal Monitoring

Our approach to spatial-temporal monitoring of satellite-derived vegetation dynamics is based on the hypothesis that phenological changes in the vegetation contributing to the reflectance of groups of pixels induce changes in the spatial structure of an image. Seasonal changes are also characteristic of other surfaces that contribute to pixel reflectance, for example, soil surfaces. The seasonal changes in the biotic and abiotic elements contributing to reflectance of a group of pixels can be combined, and if the relationships between landscape elements change seasonally and at different spatial scales, heterogeneity will exist within the landscape or ecosystem.

Landscape ecology provides a suitable concept to investigate this - domains of scale (Turner et al. 1989). A domain of scale is a range of spatial scales across which processes that influence the spatial pattern will be relatively stable, that is, a range of spatial scales associated with a set of environmental processes. Since the scale for monitoring over time in this study is constant at approximately 1 km (the size of an AVHRR pixel), the question becomes – what varies seasonally at a spatial scale of 1 km? If the spatial resolution (1 km) is considerably less than the average landscape element size in the scene, most of the pixel radiance values will be highly correlated with neighbouring pixels, resulting in low local variance. However, if the mean landscape element size approximates to that of the scene resolution (pixel size) the likelihood of pixel values being similar decreases and local variance rises. Woodcock and Strahler (1987) demonstrate that this peak in local variance occurs at 0.5 to 0.75 times the size of the fundamental scene elements. If the size of the scene elements falls further, many landscape elements will be found within a single pixel and the local variance will decline.

Turning to three sites, the fundamental scene elements (at 1 km scale) that spatial-temporal monitoring may detect are:

1. Differences in the rate and amplitude of seasonal vegetation change between, and within, the ecosystems. These differences can be subdivided by cause, namely:
   a. differences in microclimates within the vegetation formations; or
   b. different phenological responses of vegetation which, to a certain extent is a function of physiological differences between the species.
2. Differences in phenological variations associated with gap size, frequency, and distribution. Structural variations within forests modify microclimates and light environments. These, in turn, may alter leaf phenology at some spatial scales.
3. Seasonal differences in canopy leaf area and density (LAI and canopy closure) result in seasonal variations in the soil-background contribution to pixel reflectance.

Variance Analysis

Spatial-temporal monitoring of seasonal vegetation dynamics for the three ecosystems researched was achieved by moving a 3 x 3 pixel kernel over 15 x 15 pixel areas of "homogenous" vegetation in soil-adjusted vegetation index (SAVI) images (Huete 1988). SAVI was selected for this monitoring rather than NDVI or GEMI (the global environmental monitoring index, Pinty and Verstraete 1992b) images because it yielded the largest dynamic range of values in analyses of single pixels (although the seasonal trends in SAVI, NDVI and GEMI were very similar) (Jones 1999). Four measures of mean local image variance were calculated (a) mean Euclidean distance, (b) variance, (c) skewness, and (d) kurtosis. In each 15 x 15 pixel area, all pixels form the centre of the moving kernel once with the exception of the edge pixels. Therefore the four variance measures were computed nine times for each "homogenous" area and the mean values taken as an indication of local variability within the image (Woodcock and Strahler 1987, Lambin and Strahler 1994a,b). When interpreting changes in spectral variance within the images it is the relative rather than absolute changes in reflectance values in the 15 x 15 km areas that are important.

Two preconditions were necessary to analyse temporal changes in spatial variance so that the variance could be attributed to vegetation dynamics rather than *ex-situ* factors. They were that no cloud (or cloud shadow) is present in the image and that no significant human disturbance of the vegetation had occurred. A further point, not relevant to this study however, is that this technique cannot be used on composite images (e.g., GAC or GVI) as view geometries could be different for each pixel in the image. These preconditions precluded the analysis of areas of "homogenous" vegetation greater than 15 x 15 km because of natural changes in ecosystems across the lowland Bolivia and because of anthropogenic disturbances. Moreover, as the size of the area studied increased, the ability to meet cloud precondition was compromised by increased cloud occurrence. This was especially problematic at Vallé de Sajta. To obtain enough image data for spatial-temporal monitoring, the viewing angle criteria that were applied to single-pixel phenological analysis (20° either side of nadir) were relaxed to 30° (Table 3).

Table 3. Numbers of AVHRR images available for spatial-temporal analysis of image variance at different angles from nadir.

Site	Number of AVHRR images available at the following range of viewing angles from nadir		
	+10 to –10	+20 to – 20	+30 to -30
Lomerio	5	12	20
Las Trancas	4	11	20
Vallé de Sajta	13	13	26

RESULTS

Seasonal Variations in Image Variance

At each of the three areas, the four measures of image variance had similar trends (Jones 1999). The data in Figure 1 shows the temporal distribution of the four measures of variance for the seasonal semi-deciduous tropical forest ecosystem around the Lomerio site. All four measures of image variance are high in the dry, austral winter and low during the wet, austral summer. A similar seasonal pattern is characteristic of the *savanna aborizada* around Las Trancas. Surprisingly, the seasonal trends in image variance are also similar in the much wetter seasonally-inundated tropical forest ecosystem at Vallé de Sajta (Figure 2).

Image Variance and SAVI

Seasonal trends in image variance are in direct opposition to the trends in the vegetation index (SAVI) for the same images and over the same 15 x 15 km areas of "homogenous" vegetation (Figure 3). For example, in the seasonal semi-deciduous tropical forest around Lomerio, when the image variance statistics are high in the dry season SAVI is low, and, during the wet season the situation is the opposite (Table 4).

Table 4. Comparison between the ranges of image variance and SAVI values in the wet and dry seasons in seasonal semi-deciduous tropical forest and Savanna aborizada ecosystems.

Parameter		Ecosystem and Season			
		Savanna aborizada		Seasonal semi-decid. tropical forest	
		Wet	Dry	Wet	Dry
SAVI		0.40-.59	0.03-0.54	0.53-6.08	0.10-0.47
Image variance statistics	E. distance	1.56-5.68	0.53	5.53-9.30	2.01
	Variance	1.11-7.34	1.50	1.56-7.77	0.93-1.04
	Skewness	0.67-5.13	0.91	0.62-5.11	3.06
	Kurtosis	7.60-1.70	2.05	1.47-6.79	1.50

RELATING IMAGE VARIANCE TO SEASONAL VEGETATION DYANMCIS

The fact that the behaviour of SAVI and all of the measures of variance in the two ecosystems with strong seasonal water deficits is out of phase by a whole season, suggests that water availability has a strong influence on both the greening-up and dieback of vegetation, and on within-ecosystem spatial heterogeneity. Whilst the role of water availability in seasonal vegetation dynamics in tropical savanna and deciduous forest is clear, its influences on spatial variability of these phenomena are much less well known.

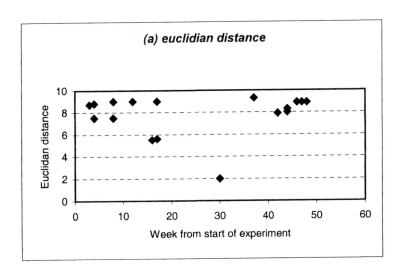

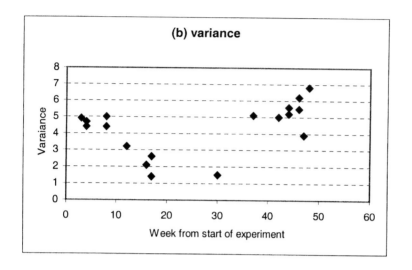

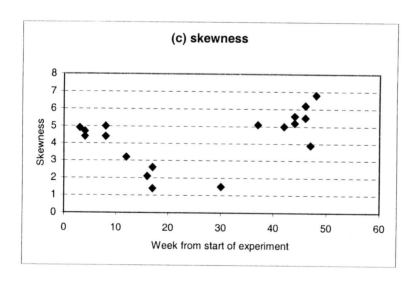

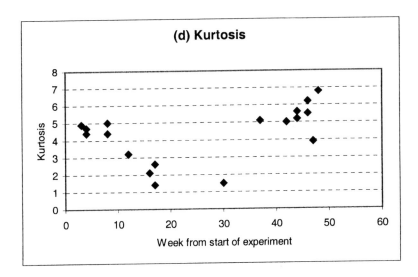

Figure 1. Four measures of spatial variance in AVHRR-SVAI imagery calculated for a 15x15 pixel block of Seasonal Semi-deciduous Tropical Forest at the Lomerio site. All four graphs use the same time axis in which Week 1 is the first week of June 1995. All viewing angles from nadir to 30° from nadir are shown. The behaviour of spatial variance at this site contrasts with that of SAVI (Figure 3), being out-of-phase by a whole season.

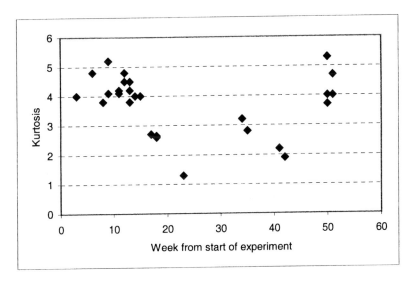

Figure 2. Spatial variance represented by the measurements of kurtosis in AVHRR-SVAI imagery calculated for a 15x15 pixel block of Seasonally-inundated Tropical Forest at the Vallé de Sajta site. As for Figure 1, Week 1 is the first week of June 1995.

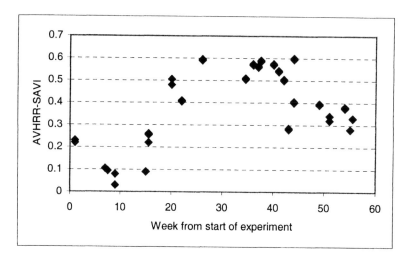

Figure 3. SAVI calculated for a single AVHRR pixel in the centre of the 15x15 block pixels which represent Seasonal Semi-deciduous Tropical Forest at the Lomerio site. All viewing angles from nadir to 20° from nadir are shown. NDVI and GEMI show strongly similar patterns of behaviour with respect to climate seasonality. The behaviour of SAVI contrasts with that of variance at this site (Figures 1a-1d) in that it is out-of-phase by an entire season.

Differences in water availability - associated with spatial variations in topography, hydrology and micrometeorology - exist in the seasonal semi-deciduous tropical forest and *savanna aborizada* ecosystems. In the dry season, when soil moisture is at a premium, these differences are exaggerated and it is this that leads to significant spatial variations in photosynthetic activity. Seasonal and spatial variations in soil moisture are only one of the potential bioclimatic triggers for vegetation phenology. To broaden the analysis we correlated the bioclimate measurements with SAVI and a combination of the four image variance measures (mean local image variance or MLIV) (Table 5).

Savanna aborizada

The *savanna aborizada* landscape comprises a small semi-evergreen tree component in a matrix of grasses. The grasses dominate the seasonal dynamics of vegetation in this ecosystem, and it is their dominance over trees that explains the relative contributions of different landscape elements to the spectral behaviour of images of this ecosystem.

Table 5. Correlation coefficients of bioclimate parameters and SAVI and a composite of local image variance measures (mean local image variance – MLIV) for the three ecosystems studied. For the Savanna aborizada and Seasonal Semi-deciduous Forest data the correlation coefficients are Pearson's product moment, and for Seasonally-inundated Tropical Forest are Spearman's ρ. For each bivariate pair correlations where attempted at four time lags (+1, 0, -1, -2 months), in each case only the most significant correlation is shown.

	Savanna aborizada		Seasonal Semi-decid Tropical Forest		Seasonally-Inundated Tropical Forest	
	SAVI	MLIV	SAVI	MLIV	SAVI	MLIV
Temp	0.76**	-0.712*	0.661*	0.703*	-0.401	-0.705
	-1	-2	+1	0	-2	0
Precip	0.785**	-0.822	0.75**	0.782*	-.0457	-0.795*
	-2	-1	+1	0	-2	-1
Rel Hum	0.921**	-0.822*	0.879*	0.492	0.205	-0.732*
	0	-1	0	+1	-1	-1
Pt Evap	-0.905**	0.778*	-0.909	-0.418	-.0299	-0.663
	0	-1	0	+1	-1	-1
P-E_p	0.923**	-0.869**	0.883**	0.625	-.0265	-0.772*
	0	-1	0	+1	-2	-1

Key to Table 5.
1. Temp = monthly mean temperature
2. Precip (P) = monthly mean precipitation
3. Rel Hum = monthly mean relative humidity
4. Pt Evap (E_p) = monthly mean potential evaporation
5. * = significant at 95% significance level
6. ** = significant at 95% significance level

In the wet season both the grass layer and the tree canopy green up and vegetation cover is high. SAVI is high at this time (Table 4) and the high rates of photosynthetic activity across the landscape mean little spatial variation in spectral properties – image variance is low.

As soil moistures reserves are depleted during the dry season, SAVI declines. The few actively photosynthesising green elements are some of the trees and vegetation in topographic depressions, which have better soil moisture availability than the interfluves. The range of landscape patches – from those with green vegetation to those where all the vegetation has died back - leads to high image variance. In the late dry season image variance remains high, despite almost all patches having experienced dieback, because burning by cattle grazers leaves spectrally-distinct burn scars in the landscape. Shadows also make substantial contributions to the recorded reflectance measurements under dry season conditions in savannas (Pech et al. 1986). Spatial heterogeneity at this time is therefore a function of (i) soil moisture availability, (ii) the size and distribution of forest islands, (iii) soil and rock exposure; and (iv) the impacts of grazing management.

The correlations presented in Table 5 show a wide time frame of lags between bioclimate triggers and vegetation responses. Some responses are rapid, for example, fire phase dieback greening up of grasses, whilst others are slower, for example, the greening up of the tree stratum (Killeen and Hinz 1992a,b). As a consequence of this wide (>2 month) time frame of vegetation responses, ecosystem and landscape heterogeneity, as indicated by MLIV, is highest during the greening up phase in September and October when the contrasts between soil and vegetation are greatest.

Seasonal Semi-Deciduous Tropical Forest

In the seasonal semi-deciduous tropical forest seasonal water deficits lead to an open forest canopy when the upper canopy trees shed their leaves. However, the situation is different in the understory and ground layer. Some plants are synchronous with the upper canopy - shedding their leaves in the dry season - whilst others remain green. Some of these plants are succulents and are well adapted to seasonal moisture deficits, but other plants remain green by virtue of their spatial locations close to adequate soil moisture reserves. Leaf shedding by upper canopy trees and some understory trees and shrubs leads to a marked decline in SAVI during the dry season. This decline is gentle, with the SAVI decreasing from about 0.6 at the end of the wet season (March to April) to between 0.1 and 0.3 during the main dry season months (July to August). Greening up is more rapid. At Lomerio dry season SAVI values of around 0.15 had risen to a little over 0.5 in six weeks. It has already been noted that the variance measures show the opposite behaviour. High spatial variance in imagery of seasonal semi-deciduous tropical forest in the dry season is mainly a function of the relatively "transparent" upper canopy leading to (i) increased shadowing, (ii) multiple scattering, (iii) increased contributions from the spatially variable lower canopy, ground vegetation and substrate elements, and (iv) differences in the number of phenological stages present in the vegetation because the spatial heterogeneity in the edaphic conditions is maximised. In addition, as canopy openness increases the amount of electromagnetic radiation received at the ground increases. In response soil temperatures increase and the diurnal water and temperature cycles are amplified. This causes vegetation changes, for example, the ratio of dicotyledons to monocotyledons changes, as does the ratio of pinnate to palmate plants. This in turn can lead to changes in photosynthetic pigments and cell structures which affect reflectance (Chazdon and Fletcher 1984a,b, Canham 1988).

Many different amplitudes and phases of phenological development of vegetation in this forest ecosystem are present during the dry season, and they provide a small grain matrix of vegetation patches with widely different contributions to the overall reflectance of the pixels that represent this forest ecosystem. Therefore, all measures of local variance are high.

In the wet season spatial contrasts in soil moisture availability are lower. The forest greens up, forming a dense upper canopy which intercepts much incident radiation. Single scattering reflections are more common than in the dry season, and the ground and understory contributions to pixel reflectance are reduced. Spatial variations in the understory and ground vegetation are mainly masked by the upper canopy contribution. Local variance in images of seasonal semi-deciduous tropical forests at this time will be low when the canopy is fully developed but, like the *savanna aborizada* ecosystem, MLIV is high during the greening up phase. In common with other semi-deciduous tropical forests, in this ecosystem the vegetation responses to bioclimate triggers are often lagged by a month (Malaisee 1974, Hubble and Foster 1986). However, some degree of anticipatory phenology (i.e., vegetation responding before a phenological trigger) is evident, and this has been found in other semi-deciduous forest as a response to change relative humidity levels in advance of rains (Alvim and Alvim 1978, Borchert 1980, Borchert 1994a). This range of responses (lagged, anticipatory and, sometimes, opposed) during the greening-up period creates differential responses in SAVI across the ecosystem. It is these spatial variations in phenostages that probably contribute most to MLIV. The mosaic of phenostages varies according to microclimatic, topographic and edaphic conditions.

Seasonally-Inundated Tropical Forest

The correlations between SAVI, MLIV, and bioclimate triggers are low. This suggests leaf flushing and abscission combine to create an integrated phenology and little variation in seasonal vegetation dynamics– a situation that has been found elsewhere in Amazonia (Batistia et al. 1997). However, seasonal patterns of local image variance can be clearly seem in imagery of the seasonally-inundated tropical forest (Figure 2). They show the same seasonal trends as the other two ecosystems studied – high variance during the winter season and low variance during the summer.

In seasonal forest and savanna ecosystems such seasonal patterns can be explained mainly in terms of the response of different vegetation elements to spatial and seasonal variations in soil water availability. This is not the case for inundated forest, where there is less direct influence of climate than in the other two ecosystems. Leaf phenology at Vallé de Sajta is a function of various of exchange mechanisms, and though there is no period of pronounced leaflessness, there is some degree of synchronisation in deciduousness (Medway 1972, Gordon et al. 1974, Putz 1979, Hopkins and Graham 1989).

There are a number of possible ecological explanations for this unexpected seasonal behaviour of local image variance in these forests. The variation most probably results from differences in reflectance between inundated and *terra firme* forest communities within the ecosystem, both of which exist in a very fine grain mosaic created by terrain variations. The area studied at the foot of the Andes is a dissected outwash plains and the variations in discharge from mountain catchments has led to shifting channels, differential erosion and the incision of river terraces. The seasonally inundated forests are found in the lower topographic positions and exhibit an evergreen behaviour in the drier winter period than the *terra firme* forests on the high terraces. This is because the soils retain more water, which lessens the impact of the winter dry period on leaf-abscission. In the wet summer months flooding of these forests, again because of their topographic position compared of to *terra firme* forests, leads to inundation and a severe wet season leaf fall.

Both leaf fall periods lead to increases local image variance when SAVI is relatively low, because the variations in greenness and understorey reflectance is at a maximum when the canopy of the inundated forests are most open.

CONCLUSIONS

In the three ecosystems studied, strong seasonal trends in vegetation greening up and dieback lead to strong seasonal trends in three vegetation indices – GEMI, NDVI and SAVI. However, the use of these indices to monitor seasonal (and inter-annual) variations in vegetation dynamics needs to acknowledge the differential responses within any particular ecosystem at any one time. Seasonal changes in this heterogeneity can be monitored using mean local image variance statistics.

Spatial heterogeneity in spectral responses at the landscape scale are a function of biotic and abiotic factors. Of particular importance are the controls on leaf flushing and abscission. The most important abiotic factors are the edaphic and topographic controls on soil water availability; whilst the most important biotic factors appear to be the differential responses of different layers in the vegetation structure. An appreciation of the dynamic nature and ecophysiology of the three-dimensional structure of the vegetation is important to understanding the relative contributions of the ground, understory, and upper canopy to the integrated reflectance signal and mean local image variance.

What do these findings indicate for the robustness of remote sensing approaches to mapping and monitoring seasonal vegetation dynamics? In ecosystems with strong seasonality and a relatively simple of bioclimatic triggers for vegetation phenology, the use of vegetation indices is quite robust. Seasonal trends can easily be recognised and explained ecologically. However, there is within-ecosystem spatial heterogeneity in these ecosystems despite their seemingly simple vegetation responses. This needs to be recognised because it could lead to spurious conclusions about ecosystem degradation or recovery in inter-annual analyses. The situation for humid tropical forests remains unclear (Millington et al. 1995). There is evidence of seasonal patterns in vegetation indices, but there is also a dearth of strong correlations between leaf phenology and bioclimate triggers. The fine scale and subtle environmental variations in rainforests (cf. Furley et al, Chapter 10 in this book) mean that ecological explanation of these seasonal patterns in image data remain unclear. Vegetation indices derived from coarse spatial resolution data in humid tropical forest should only be used with caution.

ACKNOWLEDGEMENTS

This research was funded by the UK Natural Environmental Research Council and the UK Institute for Terrestrial Ecology. The support of the Universidad Mayor San Simon (Cochabamba), in particular Marithza del Castillo, Raúl Arquipino, Tatiana Sanabria, and Jannette Maldonado, and the University of Leicester is greatly appreciated.

REFERENCES

Achard F, Blasco F. Analysis of Vegetation Seasonal Evolution and Mapping of Forest Cover in West Africa with the use of NOAA AVHRR HRPT data. Photogrammetric Engineering and Remote Sensing 1990; 56: 1359-1365.

Achard F, Blasco F, Lavenu F, Podaire A. Study of Vegetation Dynamics at a Forest-Savanna Contact in West Africa by Remote Sensing. Proceedings of the 22nd International Symposium of Remote Sensing of Environment, Abidjan. 1988; II: 517-528

Alvim, P deT, Alvim R. "Relation of Climate to Growth Periodicity in Tropical Trees." In *Tropical Tress in Living Systems*, PB Tomlinson, MH Zimmerman, Cambridge: Cambridge University Press, 1978.

Arino O, Viovy N, Belward AS. Vegetation Dynamics of West Africa Classified Using AVHRR NDVI Time Series. Proceedings of the 5th AVHRR Data Users' Conference, Tromso;441-447.

Bierregaard RO, Lovejoy TE, Kapos V, Dos Santos AA, Hutchings RW. The Biological Dynamics of Tropical Rain Forest Fragments 1992; 42: 859-865.

Borchert R. Phenology and Ecophysiology of Tropical Trees: *Erythina peoppigiona* O.F.Cook. Ecology 1980; 61:1065-1074.

Borchert R. Soils and Stem Water Storage determine Phenology and Distribution of Dry Tropical Trees. Ecology 1994; 75: 1437-1449.

Boyd DS, Duane WJ. Seasonal Effects on the Relationships between NOAA AVHRR Data and the Biophysical Properties of Tropical Forests. Proceedings of the 23rd Remote Sensing Society Conference, Reading UK. Nottingham: Remote Sensing Society, 1997.

Bray JR, Gorman E. Litter Production in Forests of the World. Advances in Ecological Research 1964; 2: 101-57.

Bullock SH, Solismagallenes JA. Phenology of Canopy Trees of a Tropical Deciduous Forest in Mexico. Biotropica 1990; 22: 22-35.

Canham, CD. An Index for Understorey Light Levels in and around Canopy Gaps. Ecology 1988; 69: 1634-1638.

Chazdon RL, Fletcher N. Photosynthetic Light Environments in a Lowland Tropical Forest in Costa Rica. Ecology 1984a; 72: 5553-5564.

Chazdon RL, Fletcher N. "Light Environments in Tropical Forests" In *Physiological Ecology of Plants in the Wet Tropics,* E Medina, H Mooney, C Vasquez-Yanes, The Hague: Dr W Junk, 1984b.

Cross AM. Tropical Forest Monitoring Using AVHRR Data: Towards and Automated System for Change Detection. Final Report to UNEP-GRID. Swindon UK: Natural Environmental Research Council, 1991.

Da Cruz Alencar J, De Almeida RA, Fernandes NP. Fenologia de Especies Florestáis em Floresta Tropical Umida de Terra Firme na Amazonia Central. Acta Amazonica 1979; 9: 163-198.

Defries RS, Townshend JRG. "Global Land Cover: Comparison of Ground based Data Sets to Classifications with AVHRR Data. " In *Environmental Remote Sensing from Regional to Local Scales,* PJ Curran, GM Foody, eds. Chichester UK: John Wiley and Sons, 1994.

Davenport ML, Nicholson SE. On the Relation between Rainfall and the NDVI for Diverse Vegetation Types in East Africa. International Journal of Remote Sensing 1993; 14: 2369-2389.

De Almeida ES, Batista GT, Dos Santos JR. Temporal Analysis of NDVI and Precipitation Data of Selected Vegetation Covers of Amazonia. ISPRS International Archives of Photogrammetry and Remote Sensing Commission VII 1994; 30: 507-510.

Elvidge CD, Chen Z. Comparison of Broad-Band and Narrow-Band Red and Near-Infrared Vegetation Indices. Remote Sensing of Environment 1995; 54: 38-48.

Fleming MD, Partridge BL. On the Analysis of Phenological Overlap. Oecologica 1984; 62: 344-50.

Ghate VS, Kumbhojkar MS. Phenology of Deciduous Ornamental Trees from Western Maharashtra. Indian Journal of Forestry 1991; 14: 181-189.

Gond V, Hubschman J, Meste C. Analyse par Teledetection Spatiale du Rythme Bioclimatique et du Compartiment Phenologique de la Vegetation dans la Nordeste du Brasil. Serchesse 1992; 3: 97-102.

Gordon W, Herbert GB, Opler PA. Comparative Phenological Studies of Trees in Tropical Wet and Dry Forests in the Lowlands of Costa Rica. Ecology 1974; 62: 881-919.

Gregoire JM. Effect of the Dry Season on the Vegetation Canopy of some River Basins of West Africa as Deduced from NOAA AVHRR data. Hydrological Sciences Journal 1990; 35:323-338.

Hopkins, MS, Graham AW. Community Phenological Patterns of a Lowland Tropical Rainforest in Northwest Australia. Australian Journal of Ecology 1989; 14: 399-413.

Hubbell SP, Foster RB. "Canopy Gaps and the Dynamics of a Neotropical Forest." In *Plant Ecology,* Crawley ed., Oxford, Blackwells, 1986.

Huete AR. A Soils Adjusted Vegetation Index (SAVI). Remote Sensing of Environment 1988; 25: 213-232.

Huete AR, Hua G, Chehbouni A, van Leaven WJD. Normalisation of Multidirectional Red and Near Infrared Reflectance with the SAVI. Remote Sensing of Environment 1992; 41: 143-154.

Jones SD. *Understanding Tropical Forest Phenology using Remotely Sensed and Ground Data Sources.* Unpublished PhD thesis. Leicester: University of Leicester, 1999.

Jones SD, Millington AC, Wyatt BW. An Evaluation of ERS-2 ATSR2 and NOAA 14 AVHRR Multitemporal Data for Phenological Studies Using Comprehensive Ground Verification Procedures in Humid and Sub-Humid Environments. Proceedings of the 23rd Remote Sensing Society Conference, Reading UK. Nottingham: Remote Sensing Society, 1997.

Justice CO, Holben BN, Gwinn MD. Monitoring East African Vegetation Using AVHRR Data. International Journal of Remote Sensing 1986; 7: 1453-1474.

Justice CO, Townshend JRG, Holben BN, Tucker CJ. Analysis of the Phenology of Global Vegetation Using Meteorological Satellite Data. International Journal of Remote Sensing 1985; 6: 1271-1318.

Killeen, TJ, Hinz PN. Grasses of the Precambrian Shield Region in Eastern Lowland Bolivia: I Habitat References. Journal of Tropical Ecology 1992a; 8: 409-433.

Killeen, TJ, Hinz PN. Grasses of the Precambrian Shield Region in Eastern Lowland Bolivia: I Life forms and C^3-C^4 Photosynthetic Types. Journal of Tropical Ecology 1992b; 8: 409-433.

Lambin EF, Strahler AH. Indicators of Land Cover Change for Change Vector Analysis in Multitemporal Space at Coarse Spatial Scales. International Journal of Remote Sensing 1994a; 15:2099-119.

Lambin EF, Strahler AH. Change Vector Analysis in Multi-Temporal Space: a Tool to Detect and Categorise Landcover Change Processes Using high Temporal Resolution Satellite Data. Remote Sensing of Environment 1994b; 48: 231-244.

Lloyd D. A Phenological Classification of Terrestrial Vegetation Using Shortwave Vegetation Index Data. International Journal of Remote Sensing 1990; 11: 2269-2279.

Lonsdale WM. Predicting the Amount of Litterfall in the Forests of the World. Annals of Botany 1988; 61: 319-324.

Malaisse F. "Phenology of the Zambezi Woodland Area with Emphasis on the Miombo Ecosystem." In *Phenology and Seasonality Modelling*, H Leith ed. New York: Springer-Verlag, 1974.

Martinez-Yrizar A, Sarukhan J. Litterfall Patterns in a Tropical Forest in Mexico over a Five-year Period. Journal of Tropical Ecology 1990; 6: 433-444.

Medway FLS. Phenology of a Tropical rainforest in Malaya. Botany Journal of Linnean Society 1972; 4: 117-146.

Parker GG, O'Neill JP, Higman D. Vertical Profile and Canopy Organisation in a Mixed Deciduous Forest. Vegetatio 1989; 85: 1-11.

Pech RP, Graetz RD, Davis AW. Reflectance Modeling and the Derivation of Vegetation Indices for an Australian Semi-Arid Shrubland. International Journal of Remote Sensing 1986; 7: 389-403.

Pinty B, Verstraete M. On the Design and Validation of Surface Bidirectional Reflectance and Albedo. Remote Sensing of Environment 1992a; 41: 155-167.

Pinty B, Verstraete M. GEMI: A Non-linear Index to Monitor Global Vegetation from Satellites. Vegetatio 1992b; 101:15-20.

Proctor J. Nutrient Cycling in Primary and Old Secondary Rainforests. Applied Geography 1987; 7: 135-152.

Putz FE. Aseasonality in Malaysian Tree Phenology. 1979 Malay Forester; 42:1-24.

Qi J, Chehbouni A, Huete AR, Kerr YH, Sorooshian A. A Modified Soil Adjusted Vegetation Index. Remote Sensing Envt. 1994; 48:119-26

Sarmiento G, Monasterio M. "Life Forms and Phenology." In *Tropical Savannas*, F Bourliere ed. Amsterdam:Elsevier, 1983.

Singh JS, Singh VK. Phenology of Seasonally Dry Tropical Forest. Current Science 1992; 63: 648-689.

Tanré D, Derdoo C, Duhaut P, Herman M, Morcrette JJ, Perbos J, Deschamps PY. Description of a Computer Code to Simulate the Satellite Signal in the Solar Spectrum: the 5S code. International Journal of Remote Sensing 1990; 11: 659-668.

Townshend JRG, Justice CO, Kalb V. Characterisation and Classification of South American Land-Cover Types Using Satellite Data. International Journal of Remote Sensing 1987; 8: 1189-1207.

Townshend JRG, Justice CO. Analysis of the Dynamics of African vegetation Using the Normalised Difference Vegetation Index. International Journal of Remote Sensing 1986; 7: 1435-1445.

Tucker CJ, Townshend KRG, Goff TE. African Land Cover Classification Using Satellite Data. Science 1985; 227: 369-375.

Turner MN, O'Neill RV, Gardner RH, Milne BT. Effects of Changing Scale on the Analysis of Landscape Pattern. Landscape Ecology 1989; 3&4: 153-162.

Umana-Dodero G. Fenologia de *Conostegia oerstedianna* y *C. xalapensis* (Melastomataceae) en el Bosque del Nino, Reserva Forestal Grecia, Costa Rica. Brenesia 1988; 30: 27-37.

Woodcock CE, Strahler AH. The factor of Scale in Remote Sensing. Remoet Sensing Envt. 1987; 21: 311-332.

Chapter 5

DOCUMENTING LAND COVER HISTORY OF A HUMID TROPICAL ENVIRONMENT IN NORTHEASTERN COSTA RICA USING TIME-SERIES REMOTELY SENSED DATA

Jane M. Read[1], Julie S. Denslow[2] and Sandra M. Guzman[2]
[1]*Department of Geography, Syracuse University, Syracuse, NY-13244, USA*
jaread@maxwell.syr.edu
[2]*Department of Biological Sciences, Louisiana State University, Baton Rouge, LA-70803-4743, USA*

Keywords: land use/cover change, Costa Rica, tim- series, landscape metrics.

Abstract A time-series of aerial photography and Landsat TM data were compiled for an area of the Caribbean lowlands of northeastern Costa Rica from 1960-1996. Geo-referenced, ground-based information was collected in 1996 and 1997. Changes in land cover were mapped and landscape fragmentation was examined using landscape pattern metrics. By 1996 the area was characterized by a complex mosaic of forests, pastures, and crop lands of different ages and disturbance histories. The most important forces driving land use changes were colonization, infrastructure development, and changes in export markets, but the spatial patterns of land use change were determined by the physical landscape. In terms of long-term forest health and conservation, there are three considerations: (1) whilst forests remain an important component of the unprotected landscape in the region, and deforestation rates have declined, the forests are highly fragmented and the area exposed to edge effects is high; (2) the secondary forests in the area are young and transient in nature; and (3) never cleared forests are in good condition, but they are susceptible to disturbance in the future.

INTRODUCTION

Land cover and land use change (LUCC) is one of the most important global change issues (Vitousek 1992, Walker and Steffen 1996). Of major concern are the widespread and rapid changes in the distribution and characteristics of tropical forests as a result of human activities. Tropical deforestation, forest degradation, and secondary forest regeneration represent major foci of research because tropical forests are important for global biodiversity, biogeochemical cycles, and climate patterns. Although much attention has been focused on quantifying forest extent and rates of deforestation, it is clear that there are many potential trajectories of LUCC in the tropics (Lambin 1997). Land-use histories may include forest clearing, cryptic deforestation (e.g. high-grading and other forms of degradation), abandonment of agricultural land uses and regeneration of secondary forests, as well as conversion from one agricultural land use to another. Forest regeneration patterns depend on time since abandonment, the length of time of prior land use, and the degree, nature and duration of the disturbance (Uhl 1982, Uhl et al. 1982, Denslow 1996). An understanding of the history of land use provides an important context for the study of tropical forest successional processes and their impact on local, regional and global earth systems.

The purpose of this research was to reconstruct the land cover and land use history of a lowland tropical forest landscape and to examine the spatial and temporal patterns of anthropogenic disturbance and forest regrowth. This was achieved through identification and mapping of historical and current land cover using aerial photographs and Landsat Thematic Mapper (TM) images to generate 1960-1996 time-series data for a site in northeastern Costa Rica with extensive existing field data and historical information.

LAND USE AND LAND COVER CHANGE IN THE TROPICS

The increasing human population in the tropics, combined with changing socio-economic conditions, changing policies, and the local impacts of globalization, promote rapid changes in land use and land cover. Deforestation and forest degradation remain the key elements of LUCC in the tropics (Myers 1993, Houghton 1994, McGuffie et al. 1995, Palubinskas et al. 1995). Data compiled by the Food and Agriculture Organization (FAO) of the United Nations and interpreted by Whitmore (1997), demonstrate that between 1981 and 1990 the annual loss of natural tropical forests worldwide was 15.4 million ha, an annual rate of 0.81% of the 1980 forest extent. Almost half of this deforestation occurred in the Americas. During the same period an additional 5.6 million ha of tropical forests worldwide were altered through logging activities (Whitmore 1997).

Tropical forests are cleared or modified primarily by logging, cattle ranching, shifting cultivation, and permanent agricultural activities (Singh 1986, Malingreau and Tucker 1988, Dale et al. 1993, Myers 1993). These proximate causes of deforestation and alteration of forests vary geographically, although shifting cultivation is widespread and has been identified as the primary proximate cause of tropical deforestation worldwide (Myers 1994). Rapid conversion of forests for cattle pastures has been associated primarily with colonization frontiers of Central America and Amazonia (Myers 1994), whereas logging activities have affected mostly the South American and Asian forests (2.45 and 2.15 million ha annually from 1981 to 1990 respectively (Whitmore 1997). Often, logging activities promote access to frontier regions contributing to subsequent deforestation as farmers follow logging roads (Whitmore 1997).

The fates of deforested areas vary with physical, biotic, and socio-economic conditions. Case studies of LUCC throughout the tropics demonstrate that many different trajectories of change occur as environmental and socio-economic conditions change (e.g. Collier et al. 1994, Skole et al. 1994, Virgo and Subba 1994, Hiraoka 1995, Dimyati et al. 1996, Sader et al. 1997, Kull 1998, Sierra and Stallings 1998). Many tropical landscapes are characterized by a mosaic of forest and scrub in different stages of regeneration following cessation of agricultural and pastoral land uses.

Prior land use affects the rate and characteristics of subsequent forest regeneration processes. Repeated burning, clearing, or grazing may compact the soil, deplete soil organic matter and nutrients, change the composition of the seed bank, and increase the abundance of weed seeds and propagules (Nye and Greenland 1960, Sanchez 1976, Uhl et al. 1982, Hecht 1982, Reiners et al. 1994, Denslow 1996). Regeneration following clearcutting is likely to contain many primary forest species that have germinated from shortlived seeds or sprouted from stumps. Regeneration following years of agricultural or pasture use are often dominated by fast growing weedy species, including grasses and other forbs, which suppress the establishment of tree seedlings. Proximity of seed sources, soil conditions, early dominance by grasses and forbs, and the prevalence of repeated disturbances such as fire all affect rates of forest recovery (Denslow 1996).

STUDY SITE

The study site is in the Caribbean lowlands of northeastern Costa Rica (Sarapiquí Canton, Heredia Province), where the foothills of the Cordillera Volcanica Central meet the Caribbean coastal plain (Figure 1). The study area was approximately 20 km by 30 km (60,000 ha), and included extensive lowland forests as well as agricultural lands. Elevations range from less than 50 m asl in the north to nearly 500 m asl in the south. Rivers drain northward into the Río Sarapiquí. Holdridge life zones in the study area vary from tropical wet forest in the north to premontane wet forest in the south (Hartshorn 1983). Soils are volcanic in origin; relatively infertile weathered soils (ultisols and oxisols) that developed on lava flows and lahars occur in upland sites, while younger, more fertile alluvial soils (inceptisols) characterize the river terraces, with the youngest occurring on the present-day floodplains (Sollins et al. 1994).

Two major roads connect the town of Puerto Viejo at the northern edge of the study area with San José. Several small communities exist along these routes. La Selva Biological Station of the Organization for Tropical Studies OTS) and the northern section of Braulio Carillo National Park lie within the study area (Figure 1). La Selva and Braulio Carrillo National Park together (encompass an area of 47,000 ha, 12% of which lie within the study area, and make up part of the Cordillera Volcánica Central Biosphere Reserve.

Rapid human population increases with accompanying changes in land cover and land use have taken place in the Sarapiquí region since the agricultural frontier opened in the 1950's. The region is now a major contributor to the nation's agricultural export economy. Land uses include protected and unprotected natural forests, subsistence farming, small-scale commercial agriculture, plantation agriculture, and cattle ranching.

DEVELOPMENT OF REMOTE SENSING LAND COVER TIME-SERIES

This research employed geographic information systems (GIS) with remotely sensed data to generate a time-series of land cover information for the period 1960-1996. The resulting time-series consisted of medium-scale land cover maps (minimum mapping unit = 3 ha) suitable for analysis of forest, pasture, and agricultural crops, for 1960, 1983, 1986, 1992 and 1996 (Table1).

Land cover maps were created based on panchromatic aerial photographs and Landsat Thematic Mapper (Landsat TM) data (see detailed methods sections below). Ground-based information for driving the classifications was collected during 1996 and the summer of 1997. Spatially-referenced land cover characteristics, including information on specific crops and conditions of pastures and forests, were recorded using differentially-corrected global positioning system coordinates (yielding accuracies of ±2-5 m under average conditions) with a Trimble Pathfinder unit and Community Base Station. Historical topographic maps and literature sources were used for interpreting and classifying the land cover maps.

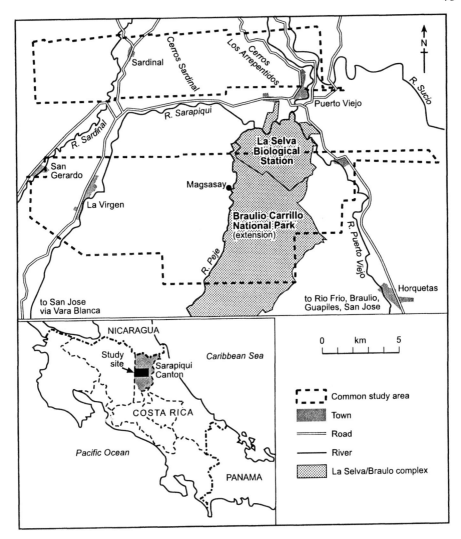

Figure 1. Location of study site, Sarapiquí, Costa Rica.

Table 1. Data sources.

Date Acquired	Processing	Resulting data layer
Landsat TM		
Dec 1997	Classification	1996 land cover
Nov 1996	Classification and replacement of no-data pixels with 1997 classified pixels	
Feb 1986	Classification	1986 land cover
B/W aerial photographs 1:60,000		
Jan/Feb 1992	Interpretation and digitization	1992 land cover
Jan 1983	Interpretation and digitization	1983 land cover
March 1960	Interpretation and digitization	1960 land cover

The classification scheme adopted for the final land cover classifications included forest, pasture, agricultural crops (including plantation agriculture), and scrub (Table 2).

Table 2. Classification scheme.

Class	Description
Pasture	Pastures used primarily for cattle. May include pastures with some degree of tree cover (mostly scattered individual trees, or groups of trees), although not closed canopy.
Crops	Include annual and permanent crops; plantation agriculture.
Forest	Closed canopy forest. May include undisturbed old growth, selectively-logged, secondary.
Scrub	Young scrubby vegetation resulting from fallowed pastures and agricultural fields.
Other	Urban (buildings, roads); barren (i.e. river beaches)
No data	No data (mixed classes of clouds, cloud shadows, lakes, rivers)

Aerial Photograph Processing

Sets of panchroamtic aerial photographs at 1:60,000 scale from 1960, 1983, and 1992 were used to derive land cover maps of the site. Approximately 20,000 ha were represented on all three sets of aerial photographs (Figure 1), and were used as the basis for the land-cover analyses presented here. Land cover patches, roads, and rivers were visually interpreted from the aerial photographs, and drawn onto mylar at a scale of approximately 1:25,000 using a zoom transfer scope. The resulting line maps were subsequently digitized.

Landsat TM Image Processing

Land cover maps of the site for 1986 and 1996 were generated from Landsat TM data (28.5 by 28.5 m pixels) for an area approximately 1400 by 1200 pixels (1360 km^2). Due to the lack of available recent cloud-free Landsat TM data, the 1996 classification was derived from two images from 1996 and 1997.

The 1986, 1996, and 1997 data were georeferenced and rectified to yield a between-date rectification accuracy of less than half a pixel. To reduce the differential effects of haze and other external factors, the 1986 and 1996 data were normalized using a linear regression technique based on features that were consistent in both images (old growth forest, deep lakes, rivers, and banana plantations). The lack of strong seasonality operating in the site, combined with the similarity in time of year of data acquisition (November and February), minimized year-to-year differences in reflectances typical of the selected features. Pixels affected by clouds and their shadows for the 1986, 1996 and 1997 images were masked out and excluded from further analyses.

The three cloud-free georeferenced images were classified using an unsupervised classification technique (ISODATA routine in IMAGINE (ERDAS 1995)) using the six visible, near- and mid-infrared bands, in addition to the thermal infrared band, which despite coarser spatial and radiometric resolutions was considered to provide useful additional information for the classification. The 1996 map was generated by substituting 'no data' pixels resulting from clouds and their shadows in the 1996 image with classified pixels from the 1997 image. The final 1986 and 1996 land cover classifications were then converted from raster to vector format to conform to the format of the aerial photograph maps.

Final Processing of the Data Layers

The aerial photograph-derived classifications (1960, 1983 and 1992) were georeferenced and rectified to the Landsat TM classifications. A common minimum mapping unit was set for all data layers by eliminating land cover patches of less than 3 ha from the final vector data layers. Thus, isolated buildings, houses with gardens, small farms of mixed agriculture, fields and patches of scrub and forest less than 3 ha were not identifiable on the final land-cover maps. The time-series was checked for logical consistency, and inconsistencies in the Landsat TM classifications were edited using the aerial photograph data as truth.

We were not able to determine the accuracy of the historical aerial photograph-derived land cover maps quantitatively, because we lacked alternative detailed land cover information to serve as reference data. Where possible the aerial photograph interpretations were visually checked using the available satellite imagery, and field and literature information. The overall quality of the land cover classifications from the aerial photographs appeared good.

The Landsat TM classifications were assessed using the more detailed aerial photograph interpretations and ground information as reference data. This assumed that the aerial photograph interpretations were correct. The overall accuracy for the 1986 and 1996 classifications were 85% and 89% (Kappa statistic 0.79 and 0.83) respectively. An examination of the contingency matrices demonstrated that the forest class was generally well classified (1986 and 1996 User's accuracy 88% and 93%, and Producer's accuracy 92% and 95%, respectively).

ANALYSES

We focus on LUCC that occurred in the area from 1960 to the early 1990s based on the land-cover maps derived from aerial photographs. A basic assumption of these analyses is that no intermediate changes in land cover occurred between the beginning and ending dates of each period, thus it was important to consider the length of each period when interpreting the results. Landscape fragmentation during this period was examined using landscape pattern metrics calculated with FRAGSTATS spatial pattern analysis program (McGarigal and Marks 1995). Forest patterns in 1996, determined from the 1996 classification of Landsat TM images, were analyzed in the context of their disturbance histories.

Landscape Patterns

The time-series land cover maps show that the study area historically has been, and continues to be, primarily a forested environment (Figure 2). Prior to 1950 the region was relatively isolated, and occupied by shifting agriculturists and a few permanent subsistence farmers. Colonists began to move into the region during the 1950s, but it was not until the early 1960s, when intensive colonization of the area began, that forests were cleared principally for establishment of cattle pastures. In 1960 old growth forest was extensive; cleared areas generally were located close to major rivers and roads (Figure 2).

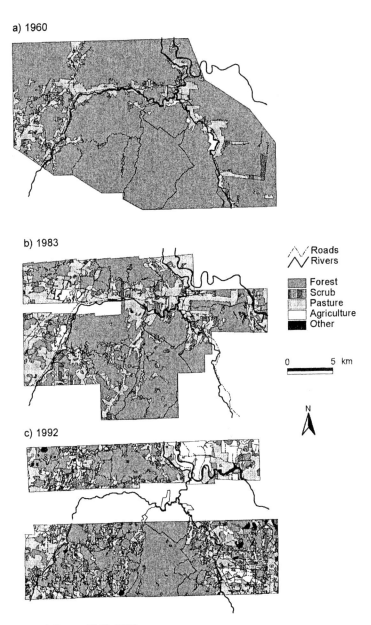

Figure 2. Land Cover: 1960-1992.

Our results show that in 1960 there was strong association between roads and rivers; most roads, and thus clearings as well, followed river routes (Figure 2). The first road connecting the region with San Jose, and along which most colonists arrived, was completed in the 1950s and closely followed the Río Sarapiquí as far north as the town of Puerto Viejo de Sarapiquí. By 1960 Horquetas and the town of Puerto Viejo were connected via a road along the Río Puerto Viejo, with a ferry crossing over the Río

Sarapiquí at Puerto Viejo (Butterfield 1994a). Similarly, several roads followed the Río Sardinal (Figure 2). The relatively rich alluvial soils and gentle topography of the river valleys were particularly suitable for agriculture, and the rivers provided a transportation route to the Caribbean coast. In 1960 only the road to Magsasay, a penal colony established in the late 1950's (Butterfield 1994a), did not closely follow a major river (Figure 2). Magsasay is located on the opposite side of the Río Sarapiquí from the main routes, and even today is not easily accessible (Figure 1). Little agricultural activity had taken place along this route by 1960.

A large percentage of the landscape (36%) changed during the 22 yr period from 1960 to 1983. Much of this change represented clearing of forest for the establishment of cattle pastures (Figure 3). The annual rate of

Figure 3. Land cover change dynamics.

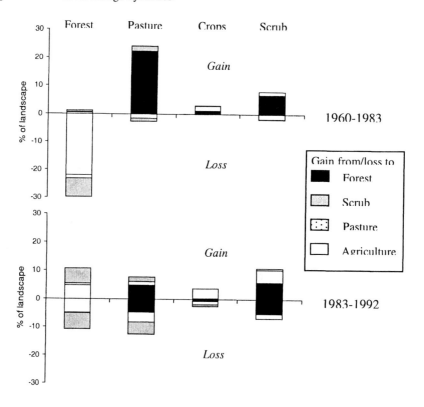

decrease in forest area during this 22-year period was 1.43% of the 1960 forest area (Table 3), with almost three-quarters (74%) of the deforested area converted to pasture. Rapid deforestation during this period and creation of cattle pastures coincided with a period of intensive colonization as a result of population increase nationally, and demand for land throughout Costa Rica (Hall 1985). At the same time there was a move toward establishment of large cattle ranches; during 1960-1983, the mean size of forest clearings for pasture was 50 ha, compared to smaller 10-ha clearings created later (1983-1992). This large scale conversion to cattle pasture was driven by the boom in the beef export market, originally promoted by United States' interests (Augelli 1987) and encouraged through Costa Rican government incentives (see also, Harrison 1991, Pierce 1992, Butterfield 1994a). Costa Rican land tenure laws served to encourage clearing of forest lands by farmers wanting to gain or retain title to land. At the same time these laws encouraged land speculation by squatters, which facilitated the creation of large ranches by wealthy land owners (Schelhas 1996).

Table 3. Mean annual percentage rates of land-cover change from 1960-1992 as a percentage of 1960 land-cover estimates

	1960-1983	1983-1992
Forest	-1.43	-0.14
Pasture	+10.9	-1.98
Crops	+25.48	+4.46
Scrub	+6.33	+5.44
Total landscape change (% area)	1.6	3.7

Twenty-two percent of the total area deforested between 1960 and 1983 was scrub land in 1983. These areas had been cleared (probably for pasture) and subsequently fallowed. Most occurred in areas marginal for farming due to inaccessibility, poor upland soils, or steep slopes. Fallowed areas mostly were located away from roads; many patches were abandoned in the relatively inaccessible area between the Río Sarapiquí and the La Selva-Braulio corridor where upland soils are also relatively poor. West of Puerto Viejo and north of the Puerto Viejo-Vara Blanca road two relatively large patches of scrub lands occurred on steep slopes. Fallowed areas were also found at the southern boundary of La Selva and within the narrow corridor linking Braulio Carrillo National Park (BCNP) with La Selva Biological Station. This strip of land was declared a protection zone in 1982, prohibiting any further forest clearing (McDade and Hartshorn 1994). The majority of these scrub lands regenerated to forest and remained as forest through 1996.

Areas of crop lands and plantations (citrus groves and African oil palm) were established and maintained during 1960-1983, but were not important sources of deforestation (Figure 3), because they were established predominantly on old pasture land. Crop lands established during this period were located mostly near the major roads and rivers on relatively fertile alluvial soils along the western part of the study site.

The rate of decrease in forest area slowed dramatically, as did the rate of pasture establishment, between 1983 and 1992 (Table 3). This coincided with a dramatic decline in beef exports in response to falling prices

(Lehmann 1992). The Protection Zone linking Braulio Carrillo National Park with La Selva Biological Station was declared in 1982, and formally added as an extension to BCNP in 1986 (McDade and Hartshorn 1994). The inaccessibility and poor upland soils of this strip of land meant that in 1983 only 3% of deforested lands lay within the newly protected corridor where further forest clearing and intensification of land use was prohibited. Thus, by 1986 20% of the study site lay within the protected areas of the La Selva-Braulio complex, and there was no longer unclaimed land available for further frontier colonization.

Although the total amount of forest in the landscape remained approximately constant from 1983 to 1992 (Figure 4), clearing of old growth, selectively-logged, and secondary forests for new pastures continued. During this period forest conversion was balanced by forest regeneration from abandoned pastures (and scrub) (Figure 3). Many of the areas that returned to forest cover were located between the Río Sarapiquí and the Río Puerto Viejo where both access and soils are poor. However, others were located close to roads and on alluvial soils. New pastures created from forest clearings likewise were established throughout the site, on both alluvial and upland soils, and in areas of varying accessibility. Despite low financial returns from pastures, cattle ranching remains an attractive option for both small and large farmers as a low risk activity requiring low labor inputs (Schelhas 1996).

Between 1983 and 1992 less than 5% of the study area comprised agricultural land. However, in line with the institution of government incentives to diversify the export economy, there was a renewed interest in export crops in the late 1980s (Lehmann 1992), resulting in the establishment of large-scale plantations. Our data show banana plantations were established on the alluvial plains of the Río Sucio, and have continued to expand with accompanying road and bridge improvements. At approximately the same time, pineapple plantations were established in the vicinity of Sardinal and San Gerardo. Most of the new crop lands established since 1986 within the common study area were previously in pasture. However the Landsat TM images show that between 1986 and 1996 at least 500 ha of forests were converted to banana plantations in the vicinity of Río Sucio, but lay outside breakup of large cattle ranches and creation of many small holdings where subsistence farmers produced a diversity of crops (Butterfield 1994a), the area covered by these analyses. In addition to large-scale plantation enterprises, land redistribution since the mid-1980s led to th changes in land use and land cover not only affected the areas of differen cover types (Figure 4), but also led to an overall increase in habitat fragmentation and landscape diversity (Table 4). Most forest fragmentation occurred from 1960-1983, and thereafter remained approximately constant between 1983 and 1992 (Figure 5). In contrast, fragmentation of pastures occurred after 1983 with the breakup of large ranches to smaller farms.

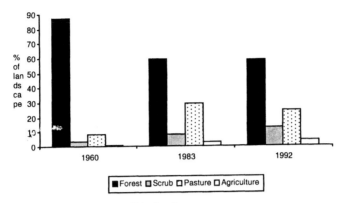

Figure 4. Land cover as a percentage of the landscape.

Table 4. Standard landscape indices: general trends 1960-1992.

# patches	Increase
Mean patch size	Decrease
Edge density (edge length/area)	Increase
Total area at least 100 m from an edge (change in cover)	Decrease
Shannon's diversity index	Increase

Since 1992 the trend toward agricultural diversification and land redistribution has continued. The 1996 Landsat TM data showed that small-scale farming and plantation enterprises, including bananas, pineapples, ornamental plants, and hearts-of-palm continued to expand, mostly at the expense of pastures. Recently small scale (mostly two to five hectares) reforestation projects have been initiated as a result of tax incentives set out in the 1996 Costa Rican forestry law, which encourages reforestation projects in the private sector (Diario Oficial 1996). These incentives serve to offset the high initial costs, the long wait for financial returns, and the medium rates of return associated with reforestation activities (Schelhas 1996). Tourism has become an important source of revenue, with visitors attracted to various small private reserves and BCNP.

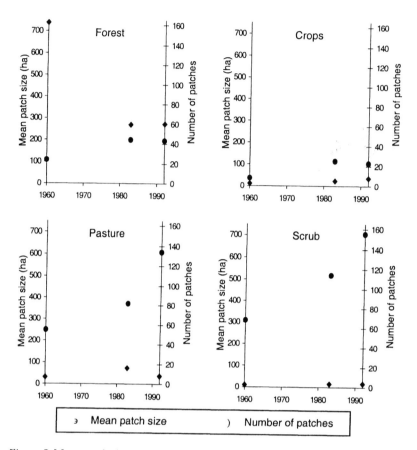

Figure 5. Mean patch size and number of patches by cover type: 1960-1992.

The aerial photographs showed a strong relationship between land cover and the road network from 1960 to 1992. The initial pattern of roads and deforested land that had developed by 1960 remained evident in 1996, with later clearings having expanded the early fields. In 1992 the road network was more than five times as dense as the 1960 network. All agricultural cropland lay within one kilometer of the nearest road, presumably driven by the need for transportation to carry cash crops to market. Forests were located primarily away from roads, in protected areas, or in locations not favored for agriculture. Other researchers also have found a strong relationship between the location of roads and forest clearings at both local and national scales (e.g., Sader and Joyce 1988, Veldkamp et al. 1992, Sader et al. 1994, Sader 1995).

Forest Patterns

The distribution and age composition of forests in 1996 in the study area reflect the different land-cover and land-use patterns operating since the 1950s. In 1996 forest remained an important feature, representing 61% of the landscape (Figure 6), reduced from 88% in 1960. Very little forest remained close to the major roads and rivers. Over a third of these forests had been protected within the La Selva-Braulio complex, and not subsequently affected by changing land-use patterns after 1982.

Most of the forests lying within the La Selva-Braulio complex in 1996 were mature old growth forests with no evidence of prior clearing. Poor soils, difficult access, and availability of land elsewhere in the area all contributed to the preservation of these forests prior to their legal protection in 1982. The effectiveness of this protection is demonstrated by post-1982 forest regeneration on fields that had been cleared prior to the declaration of the protection corridor; all these sites were deforested after 1960, fallowed prior to 1992, and have remained in forest.

Outside the protected areas of the La Selva-Braulio complex human activities continued. Half of this unprotected landscape was forested in 1996. Forests that had existed throughout the period of study and presumably not cleared prior to 1960 represented almost a third of the unprotected landscape, and 58% of its forests. The largest patches of uncleared forest occurred in the relatively inaccessible upland area between the Río Sarapiquí and La Selva-Braulio corridor where soils are relatively poor, and in the northern *cerros* (hills) of Sardinal and Arrepentidos where steep slopes are less suitable for agriculture than are the flat river valleys (Figure 1).

While patches of old growth forest undoubtedly remain in particularly inaccessible portions of the landscape, it is likely that many of these patches of previously uncleared forest had been logged selectively for large trees (>60 cm dbh) one or more times over the study period. Such practices can leave a substantial portion of the canopy intact, making selectively logged forests difficult to distinguish from unlogged forest in aerial photographs or satellite images. Nevertheless the impact on the forest is often severe because soils are compacted and gullied by heavy machinery, drainage is altered, and remaining trees and saplings are damaged in the felling and extraction processes. In some cases construction of logging roads and opening of the forest canopy are the first steps in conversion of forest to pasture (Butterfield 1994a).

The remaining 42% of forests existing in 1996 in the unprotected area were in secondary vegetation (forest and scrub) growing on previously cleared land. Most of these forests (62%) were young in 1996, with 36% recently fallowed and still in the scrub stage in 1992, and 26% only fallowed

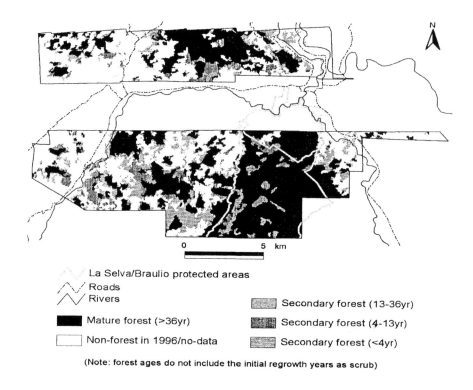

Figure 6. Distribution of forests in the study area in 1996.

after 1992. Over a third of secondary forests in 1996 were 5-13 yr old, with 20% having been fallowed just prior to 1983, and 15% fallowed between 1983 and 1992. Forests older than 13 years accounted for less than 4% of secondary forests, and existed in small fragments scattered throughout the unprotected portion of the study area. Increasing habitat fragmentation after 1960 (Figure 5) means that by 1996 almost half of all forests outside La Selva and Braulio Carrillo National Park were located within 100 m of a forest edge. This has serious implications for these unprotected forests and the biodiversity they represent, in terms of vulnerability to further encroachment and degradation.

Our ability to detect instances of repeated clearing and regrowth was limited by the combination of rapid regrowth of fallow vegetation and lack of aerial photo coverage between 1960 and 1983. Our limited data suggest that only 1% of secondary forests had histories of repeated disturbances involving two cycles of clearing and fallow. All others had regenerated following only a single disturbance. The majority of secondary forest patches revealed change trajectories of forest – pasture – scrub – forest. Few areas converted to crop land were subsequently abandoned, but remained in use for crop production.

CONCLUSION

In 1996 the Sarapiquí landscape was characterized by a complex mosaic of forests, pastures, and crop lands of different ages and disturbance histories (see also Schelhas 1996). Most secondary forests had regenerated following abandonment of pasture lands, and tended to be young. While colonization processes, infrastructure development, and changes in export markets represented important forces driving land-use changes, the physical landscape (rivers, soils, topography) was important in determining the spatial patterns of land use and development and structure of the road network. The long-term presence of La Selva Biological Station and the extension of Braulio Carrillo National Park in the 1980s undoubtedly have preserved forest cover on a substantial portion of the landscape. It seems likely that, without protection status, the forests of Braulio Carrillo National Park would have met fates like those on similar soils and with similar problems of access, where now only isolated patches of older forest remain.

Our analyses suggest several implications for the long-term forest health and conservation in Sarapiquí. Forest remains an important component of the unprotected landscape: net rates of deforestation have decreased from earlier highs, and forest cover has returned to large parts of the landscape following abandonment of pasture. While there would seem to be reason for optimism regarding the extent of the forest cover in Sarapiquí, existing forests have been affected strongly by logging and agricultural activity. Forests are also highly fragmented, with substantial areas of forest exposed to edge effects, lowering their suitability for the maintenance of plant and animal species dependent on forest interior conditions (Laurance et al. 1997).

Secondary forests are predominantly young forests. Studies of secondary forests in the region (Finegan 1996, Guariguata et al. 1997) suggest considerable economic and conservation potential in some stands based on the composition of both canopy trees and sapling size classes. However, we lack experience in management of young naturally regenerated tropical forest stands for optimum economic and conservation values, and their futures are uncertain. The Costa Rican timber industry historically has depended on large-diameter logs of select species felled in the process of land conversion to agriculture or pasture (Butterfield 1994b). Few incentives have existed to promote the management of secondary forests or their maintenance in the landscape. Where stands of secondary forests were accessible, they were often heavily exploited for cattle forage or for such forest products as firewood and hearts-of-palm, and they were likely candidates for pasture conversion when economic conditions warranted. The scarcity of old secondary forests in the study area is an indication of their transient nature. The future impacts of the 1996 forestry law (Diario Oficial 1996) on reforestation activities and secondary forests in the site remain to be seen.

Finally, it is likely that few never-cleared forests are in good condition. Forest exploitation rates in Sarapiquí are high and logging practices are wasteful (Butterfield 1994b). Few sawmills own or manage timber reserves so incentives for good forest management practices have been few. Moreover relative rates of deforestation in Costa Rica have been some of the highest in Central America (Leonard 1987). As the few remaining timber reserves are exhausted, pressures on preserves and remnant stands on private land will likely increase. The high degree of fragmentation and percentage of forests lying within 100 m of the forest edge make these forests more susceptible to future disturbances. In spite of the dominance of forest in the landscape of Sarapiquí, the fate of forests on unprotected land is uncertain and should continue to be monitored for future changes.

The findings of this research suggest three factors which may provide potential predictors of future LUCC sites in the region: 1) road networks, examined within their historical context; 2) local physical environmental factors which indicate land suitability characteristics for specific land uses; and 3) proximity to intensive land uses and locations of recent land-use changes. Studies investigating the spatial nature of LUCC for countries in Central America, other than Costa Rica, are limited. More landscape scale studies are needed to understand the processes of change in the region.

ACKNOWLEDGMENTS

Funding for this research was provided by the U.S. National Science Foundation (grants SBR-9627957 and DEB-9208031), Space Imaging EOSAT Co., and the Robert C. West Field Research Award program (Department of Geography and Anthropology, Louisiana State University). We are grateful for logistic support provided by the Organization for Tropical Studies and La Selva Biological Station. We thank Dr. Nina Lam (Louisiana State University) for her advice on different phases of this project, and two anonymous reviewers for their helpful comments.

REFERENCES

Augelli JP. Costa Rica's frontier legacy. The Geographical Review 1987; 77: 1-16.

Butterfield RP. "The regional context: land colonization and conservation in Sarapiqui." In *La Selva: Ecology and Natural History of a Neotropical Rain Forest*, LA McDade, KS Bawa, HA Hespenheide, GS Hartshorn eds., Chicago: University of Chicago Press, 1994.

Butterfield RP. "Forestry in Costa Rica: status, research priorities, and the role of La Selva Biological Station." In *La Selva: Ecology and Natural History of a Neotropical Rain Forest*, LA McDade, KS Bawa, HA Hespenheide, GS Hartshorn eds., Chicago: University of Chicago Press, 1994.

Collier GA, Mountjoy DC, Nigh RB. Peasant agriculture and global change. BioScience 1994; 44: 398-407.

Dale VH, O'Neill RV, Pedlowski M, Southworth F. Causes and effects of land-use change in central Rondônia, Brazil. Photogrammetric Engineering and Remote Sensing 1993; 59: 997-1005.

Denslow JS. "Functional group diversity and responses to disturbance." In *Biodiversity and Ecosystem Processes in Tropical Forests*, GH Orians R Dirzo, JH Cushman eds., Berlin: Springer-Verlag, 1996.

Diario Oficial. Ley No 7575 Ley Forestal. Diario Oficial, 04/16/96, 1-8, 1996.

Dimyati M, Mizuno K, Kobayashi S, Kitamura, T. An analysis of land-use/cover change using the combination of MSS Landsat and land use map - a case study in Yogyakarta, Indonesia. International Journal of Remote Sensing 1996; 17: 931-944.

ERDAS Erdas Field Guide. Atlanta, GA: Erdas, Inc., 1995.

Finegan B. Pattern and process in neotropical secondary rain forests: the first 100 years of succession. Trends in Ecology and Evolution 1996; 11: 119-123.

Guariguata MR, Chazdon RL, Denslow JS, Dupuy J, Anderson M. Structure and floristics of secondary and old-growth forest stands in lowland Costa Rica. Plant Ecology 1997; 132: 107-120.

Hall C. Costa Rica: a Geographical Interpretation in Historical Perspective. Boulder, CO: Westview Press, 1985.

Harrison S. Population growth, land use and deforestation in Costa Rica, 1950-1984. Interciencia 1991; 16: 83-93.

Hartshorn GS. "Plants." In *Costa Rican Natural History*, DH Janzen ed. Chicago: University of Chicago Press, 1983.

Hecht SB. Deforestation in the Amazon Basin: magnitude and dynamics and soil resources effects. Studies in Third World Societies 1982; 13: 61-101.

Hiraoka M. Land use changes in the Amazon estuary. Global Environmental Change 1995; 5: 323-336.

Houghton RA. The worldwide extent of land-use change. BioScience 1994; 44: 305-313.

Kull CA. Leimavo revisited: agrarian land-use change in the highlands of Madagascar. Professional Geographer 1998; 50: 163-176.

Lambin EF. Modelling and monitoring land-cover change processes in tropical regions. Progress in Physical Geography 1997; 21: 375-393.

Laurance WF, Laurance SG, Ferreira LV, Rankin-de Merona JM, Gascon C, and Lovejoy, TE. Biomass collapse in Amazonian Forest Fragments. Science 1997; 278: 1117-1118.

Lehmann MP. "Deforestation and changing land-use patterns in Costa Rica." In *Changing Tropical Forests: Historical perspectives on today's challenges in Central and South America*, HK Steen, RP Tucker eds. Durham NC: Forest History Society, 1992.

Leonard HJ. Natural Resources and Economic Development in Central America. International Institute for Envrionment and Development eds. Rutgers, NJ: Transaction Books, 1987

Malingreau J-P, Tucker CJ. Large-scale deforestation in the southeastern Amazon basin of Brazil. Ambio 1988; 17: 49-55.

McDade LA, Hartshorn GS. La Selva Biological Station. In La Selva: Ecology and Natural History of a Neotropical Rain Forest, McDade LA, Bawa KS, Hespenheide HA, Hartshorn GS, Chicago: University of Chicago Press, 1994.

McGarigal K, Marks BJ. FRAGSTATS: spatial pattern analysis program for quantifying landscape structure. Portland, OR: U.S. Department of Agriculture, Forest Service, Pacific Northwest Research Station, 1995

McGuffie KA, Henderson-Sellers A, Zhang H, Durbidge TB, Pitman AH. Global climate sensitivity to tropical deforestation. Global and Planetary Change 1995; 10: 97-128.

Myers N. Tropical forests: the main deforestation fronts. Environmental Conservation 1993; 20: 9-16.

Myers N. "Tropical deforestation: rates and patterns." In The Causes of Tropical Deforestation, K Brown, DW Pearce. London: UCL Press, 1993.

Nye PH, Greenland DJ. The Soil Under Shifting Cultivation. Technical Communication No. 51. Harpenden UK: Commonwealth Bureau of Soils, 1960.

Palubinskas G, Lucas RM, Foody G, Curran PJ. An evaluation of fuzzy and texture-based classification approaches for mapping regenerating tropical forest classes from Landsat TM data. International Journal of Remote Sensing 1995; 16: 747-759.

Pierce SM. La Selva Biological Station History: Colonization/Land Use/Deforestation of Sarapiqui, Costa Rica. Masters thesis, Colorado State University, 1992.

Reiners WA, Bouwman AF, Parsons WFJ, Keller M. Tropical rain forest conversion to pasture: changes in vegetation and soil properties. Ecological Applications 1994; 4: 363-377.

Sader SA. Spatial characteristics of forest clearing and vegetation regrowth as detected by Landsat Thematic Mapper imagery. Photogrammetric Engineering and Remote Sensing 1995; 61: 1145-1151.

Sader SA, Joyce AT. Deforestation rates and trends in Costa Rica, 1940-1983. Biotropica 1998; 20: 11-19.

Sader SA, Reining C, Sever T, Soza C. Human migration and agricultural expansion: a threat to the Maya tropical forests. Journal of Forestry 1997; 95: 27-32.

Sader, SA, Sever T, Smoot JC, Richards M. Forest change estimates for the Northern Peten region of Guatemala - 1986-1990. Human Ecology 1994; 22: 317-332.

Sanchez PA. Properties and Management of Soils in the Tropics. New York: Wiley, 1976.

Schelhas J. "Land-Use Choice and Forest Patches in Costa Rica." In *Forest Patches in Tropical Landscapes*, J Schelhas, R Greenberg eds. Washington, D.C.: Island Press, 1996

Sierra R, Stallings J. The dynamics and social organization of tropical deforestation in northwest Ecuador, 1983-1995. Human Ecology 1998; 26: 135-161.

Singh A. "Change detection in the tropical forest environment of Northeastern India using Landsat." In *Remote Sensing and Tropical Land Management*, MJ Eden, JT Parry eds. New York: John Wiley & Sons Ltd, 1986

Skole DL, Chomentowski WH, Salas WA, Nobre AD. Physical and human dimensions of deforestation in Amazonia. BioScience 1994; 44: 314-322.

Sollins PF, Sancho MR, Mata C, Sanford Jr RL. "Soils and soil process research." In *La Selva: Ecology and Natural History of a Neotropical Rain Forest*, LA McDade, KS Bawa, HA Hespenheide, GS Hartshorn eds., Chicago: University of Chicago Press, 1994.

Uhl C. Recovery following disturbances of different intensities in the Amazon rain forest of Venezuela. Interciencia 1982; 7: 19-24.

Uhl C, Jordan C, Clark K, Clark H, Herrera R. Ecosystem recovery in Amazon caatinga forest after cutting, cutting and burning, and bulldozer clearing treatments. Oikos 1982; 38: 313-320.

Veldkamp E, Weitz AM, Staritsky IG, Huising EJ. Deforestation trends in the Atlantic Zone of Costa Rica: a case study. Land Degradation and Rehabilitation 1992; 3: 71-84.

Virgo KJ, Subba KJ. Land-use change between 1978 and 1990 in Dhankuta District, Koshi Hills, Eastern Nepal. Mountain Research and Development 1994; 14: 159-170.

Vitousek PM. Global environmental change: an introduction. Annual Review in Ecology and Systematics 1992; 23: 1-14.

Walker BH, Steffen WL. "GCTE science: objectives, structure and implementation." In *Global Change and Terrestrial Ecosystems*, BH Walker, W Steffen eds. Cambridge: Cambridge University Press, 1996.

Whitmore TC. Tropical forest disturbance, disappearance, and species loss. In *Tropical forest remnants: ecology, management, and conservation of fragmented communities*, WF Laurance, BO Bierregaard Jr eds. Chicago: The University of Chicago Press, 1997.

Chapter 6

PATTERNS OF CHANGE IN LAND USE, LAND COVER, AND PLANT BIOMASS: SEPARATING INTRA- AND INTER-ANNUAL SIGNALS IN MONSOON-DRIVEN NORTHEAST THAILAND

Stephen J. Walsh[1*], Kelley A. Crews-Meyer[2], Thomas W. Crawford[3], William F. Welsh[4], Barbara Entwisle[5] and Ronald R. Rindfuss[6]
[1]*Department of Geography, University of North Carolina, Chapel Hill, North Carolina 27599-3220 swalsh@email.unc.edu*
[2]*Department of Geography, University of Texas, Austin, Texas 78712-1098*
[3]*Environmental Studies Program, Gettysburg College, Gettysburg, Pennsylvania 17325*
[4]*Department of Geography, University of North Carolina, Greensboro, North Carolina 27402*
[5]*Department of Sociology and Carolina Population Center, University of North Carolina Chapel Hill, North Carolina 27516-3997*

Keywords: Land use and land cover change, vegetation indices, pattern metrics, NDVI.

Abstract A satellite time-series was used to assess inter- and intra-annual landscape changes in northeast Thailand, hypothesized to have occurred as a consequence of a number of social, biophysical, and geographical drivers of land use and land cover change (LUCC). Such drivers of change functioned over long- and short-term temporal scales, were locally and regionally mediated, and internally and exogenously rooted. Deforestation began centuries ago to support in-migration into the region and subsistence cultivation of rain-fed rice in the lowlands. Less than 25 years ago deforestation was being concentrated in the uplands to support the cultivation of cassava as a cash crop for their emerging market economy. LUCC continued to occur as evidenced by shifts in landscape composition, spatial organization, and plant productivity set to the rhythms of monsoonal rains at the inter-annual scale and crop phenology at the intra-annual scale. Topography was indirectly examined relative to variations in the composition and spatial structure of land use and land cover (LULC) and plant biomass at both time scales. Landsat Thematic Mapper data were used to assess the spatial and temporal variations in classified LULC types derived through a hybrid classification approach, and plant biomass levels were computed using the Normalized Difference Vegetation Index

(NDVI). Site and situation and principles in landscape ecology were used as the organizing concepts in interpreting pattern and compositional shifts occurring at the inter- and intra-annual scales.

INTRODUCTION

Throughout northeastern Thailand, the pressure placed on the land to yield greater agricultural returns for subsistence and market agriculture has resulted in a rapid expansion of land under cultivation, increases in agricultural inputs (Hirsch and Lohmann 1989), and increased land fragmentation because of the shrinking availability of suitable land for agriculture. Beginning in the 1960s, forested lands were cleared for conversion to cash crop production in the upland sites, agricultural intensification occurred in the alluvial plains and lowlands, and agricultural extensification occurred in the lower, middle, and upper terraces (Rundel and Boonpragop 1995). Relative to the floodplains and lower terraces, expansion of agriculture into the middle and upper terraces was riskier because of their poor environmental suitability and reliance on favorable monsoonal rains for extended and sustained cultivation (Polthanee and Marten 1986). Therefore, LULC tended to oscillate between early-maturing rice and drought-resistant cash crops in the middle and upper terraces depending upon the volume and timing of rainfall. Deforestation of the middle and high terraces has occurred as a reaction to an increasing demand for subsistence agriculture, shrinkage of suitable land, and expanding market opportunities for agricultural products (Fukui 1993). As a consequence, the concepts of site suitability and sustainability are dynamic, reflecting inter-annual variations in LULC patterns as well as variation in plant productivity levels linked to moisture availability.

Upland sites have traditionally been considered areas of pronounced marginality because of their limited resource endowments. Recently, however, greater emphasis has been placed on the uplands resulting in significant LUCC, generally transitioning from forest to agriculture. Beginning in the 1960s, a market demand for cassava developed, fueled through a European need for high calorie animal feed. This, coupled with a redefinition of upland forests as sites amenable to agricultural extensification (because of the limited moisture requirements of cassava) and the expansiveness of uplands not dedicated to other subsistence or commercial endeavors combined to reduce the relative marginality of the uplands, resulting in LUCC. As a result, the landscape matrix in the uplands has been fragmented depending upon the nature of the forest-agricultural transition, terrain, resource endowments, and village topologies (Walsh 1999). In the lowlands, most of the forests were removed, except riparian forests, forest maintained in and around the villages, and individual trees and isolated forest patches that were retained in and around rice paddies and upland fields. Over time, the landscape matrix has cycled from a highly homogenous landscape composed of forests, to a fragmented landscape composed of a forest-agriculture mix, and to a moderately to homogenous landscape dominated by rice paddies (Walsh 1999). This chapter examines inter- and intra-annual change in LULC in this context.

Geographic site and situation are important concepts that influence LULC patterns across the landscape. Site and situation are spatially and temporally mediated. The spatial vagaries of terrain, hydrography, and village locations and their interactions create an environment influenced by a complex set of biophysical, social, and geographic factors. LUCC occurs by people forcing change on the environment, environment influencing human behavior, and some human induced changes resulting in feedbacks on subsequent human behavior (Entwisle et al. 1998). This research were framed around the interplay between population and environment, as well as the change in LULC composition and structure across a recent decade and between periods within a single season defined by the local water year (April 1 to March 31).

To anticipate, the findings are (1) inter-annual (December 1989 and December 1997) satellite change-detections indicate a decrease in vegetation biomass over time due to deforestation and reforestation in upland sites as well as rainfall differences between the image dates (wetter in 1997 than in 1989), whereas intra-annual (December 1989 and March 1990) change detections indicate a decrease in vegetation biomass over time associated with phenological shifts linked to the planting and harvesting of rice generally during the months of June and December respectively; (2) NDVI (Normalized Difference Vegetation Index) was examined to define statistically significant estimates of landscape conditions through an index sensitive to plant biomass and plant canopy structure; (3) vegetation change (type and condition) occurred in different combinations associated with topographic settings (i.e., lowlands, terraces, and uplands) – inter-annual LUCC occurred in response to deforestation, reforestation, and agricultural extensification in the uplands, a somewhat steady state condition in the lowlands, and temporary or periodic expansion into the terraces through favorable economic and climatic opportunities that reflected a diverse LULC matrix of forests, upland crops, and lowland rice; and (4) pattern analysis of LULC and NDVI-derived plant biomass levels suggested a sorting of composition and structure by LULC type, long- to short-term cycles (inter- and intra-annual variation), and site and situation characteristics – the landscape has become less fragmented in the uplands and terraces; in the lowland rice areas the landscape has become less fragmented because the household paddies coalesce into large, rather uninterrupted extents interdigitated by topographic changes that support forest or upland crops because of their general marginality for rice cultivation.

STUDY AREA

Northeast Thailand (or Isaan) contains approximately one-third of the country's area, generates about one-fifth of the GNP, and has a population of more than 18 million people. Nang Rong district, our study site, is located in Buriram province in the southern part of Isaan (Figure 1 – on the CD-ROM).

The district is approximately 1300 square km in size, dominated by wet rice cultivation in the shallow depressions of the alluvial plains and lowlands, and dry dipterocarp "monsoon" forest or drought resistant crops (e.g., cassava, kenaf, and sugarcane) on the upland sites. Agricultural products, the basis of the Nang Rong subsistence and market economies, are highly dependent on rain-fed irrigation, with alternative water infrastructure practically non-existent within the region. Monsoonal rains serve as the primary mechanism for delivering water to the district. These monsoonal rains are subject to a high degree of intra-annual and inter-annual variability in both time and space. Approximately 80 percent of the total annual precipitation within the northeast falls during a five-month period from May to September. The rains may come in early April, or they may not arrive until well into June or July. September may be a dry month, or it may be so wet that extensive flooding occurs. In the period from July to September, the middle of the rice season, there is commonly a period of drought when rainfall may be as little as one-fourth of the seasonal average.

Reliance on a rain-fed system to support the wet rice and upland cash crops has developed in-concert with the occurrence of droughts, floods, and a marginal physical resource base (Kaida and Surarerks 1984). The harsh environment has caused irregular planting schedules, low and inconsistent agricultural yields, and the development of risk abatement strategies (e.g., plot distribution patterns) in an attempt to stabilize production from to year to year (Rigg 1991). The fragmentation and consolidation of LULC types, within and between years can be partly understood in this context.

Farm households own an average of three hectares of land (Ghassemi et al. 1995). Per capita income in Isaan is the lowest in the country, largely because of low and unstable agricultural production resulting from erratic rainfall and generally poor soils (Arbhabhirama et al. 1988, Parnwell 1988, Ghassemi et al. 1995).

Deforestation associated with agricultural expansion has been underway in northeast Thailand for a century or more (Feeny 1988). Prior to World War II, this expansion coincided with increased production of paddy rice.

Since then, it has also included upland crops, particularly cassava and sugar cane, cultivated at the expense of forests. In the recent period, agricultural intensification occurred in the rice producing areas of the alluvial plains and lower terraces in the form of increased labor, mechanization, fertilization, pesticides, and herbicides to generate greater yields (as opposed to double cropping). Agricultural extensification occurred in the middle and high terraces and upland sites of this low relief, undulating, and marginalized environment.

In addition to terrain, village location also influences LULC as a consequence of the ability and propensity of people to impart LUCC through land conversion efforts including deforestation, reforestation, and agricultural extensification and intensification. Areas that tend to focus population on the landscape through a spatial coalescence of villages and their functional territories offer a greater potential for changing the landscape through a consolidation of labor, capital, and technology. However, since we do not have data on the entire period of human settlement in Nang Rong, it is also important to recognize that terrain likely influenced the pattern of village location.

DATA AND METHODS

Three Landsat TM images were chosen for this analysis: 23 December 1989, 29 March 1990, and 29 December 1997 (Figure 2 - on the CD-ROM). The pair of December images were used to assess inter-annual changes, while the December 1989 and March 1990 images were used to assess intra-annual changes. Rainfall in December 1989 and 1997 was similarly and seasonably low, whereas rainfall in March 1990 was unusually plentiful. Overall, 1989 was a dry year, whereas 1997 was a wet year.

The December 1997 image was classified through a hybrid unsupervised/supervised approach that incorporated the visible (primarily influenced by plant pigmentation), near-infrared (primarily influenced by chlorophyll content), and middle-infrared (primarily influenced by plant water content) channels of Landsat TM (Jensen 1996). Initially, an unsupervised classification was run using the ISODATA decision-rule to define "naturally" occurring spectral classes that were subsequently reduced to approximately 30 classes through the interpretation of transformed divergence and divergence statistics. Then, a supervised classification approach was followed using a maximum likelihood classifier to relate study area pixels to the approximately 30 spectral classes (i.e., the training data) defined through the unsupervised classification. The 1989 and 1990 images were combined into a single December 1989 classification to reflect wet and dry conditions for the same water year for class separation through the same hybrid classification approach.

Four LULC types were used for this analysis: rice (predominantly in lowland areas), mixed agriculture (rice and cash crops, often located in middle and low-middle terraces), cash crops (cassava and sugar cane, often in uplands), and forest (primarily village, upland and riparian zones). In addition, for each of three images, the NDVI (Normalized Difference Vegetation Index) was calculated to indicate plant biomass variation throughout the study area for the selected time periods. The NDVI has historically been very effective in assessing plant biomass variation in a host of environments including northeast Thailand (Crews-Meyer 1999), and it has been correlated with a number of vegetation characteristics (e.g., leaf area index, biomass, NPP) (Rouse et al. 1973; Jense, 1996).

NDVI = (NIR − RED) / NIR + RED)   1)
NIR = near infrared brightness values
RED = red visible brightness values

Pattern metrics based on the processed Landsat TM images were used to assess the spatial/temporal nature of landscape structure. Pattern metrics quantify the structure and spatial organization of the landscape across LULC types (class metrics) and space (patch statistics). Use of pattern metrics recognizes that landscape properties are influenced by the surrounding neighborhood and the regional spatial context in which local sites are juxtaposed, and that the nature of spatial patterns and relationships is a function of social and biophysical systems and their interactions. A set of pattern metrics were used to represent the composition and structure of plant biomass levels. Since pattern metrics are usually applied to thematic (non-continuous) data, the continuous NDVI values were classed on an equal-interval decile scheme whereby each class contained one-tenth of the possible data range. Pattern metrics are commonly used to characterize the structure of LULC types defined through a digital classification of remotely-sensed spectral responses. Here, it was important to evaluate NDVI classes associated with landscape and class categorizations so that both landscape state (i.e., LULC) and condition (i.e., plant biomass) variation could be examined.

Terrain impacts LULC types by focusing long-term and short-term deforestation practices on various topographic strata to take advantage of centuries-old cultivation of lowland rice, first for subsistence and now also for market opportunities, and more recently, the concentration of deforestation, reforestation, and agricultural extensification in upland sites that has occurred to support the cultivation of cash crops. Figure 3 (on the CD-ROM) shows LULC plotted against elevation for a profiled sample of

1997 landscape conditions along an east-west trending transect positioned across the district. Figure 3 shows the general trend of LULC by elevation – rice predominately in the lowlands with extents apparent in the low and middle terraces; cassava and sugar cane in the uplands and high to middle terraces; forests in the lowlands (near rivers, streams, and surrounding village compounds), and the most extensive forest tracts at the highest elevations within the district.

ANALYSIS

The following section describes the changes in NDVI and LULC over 8-years (1989-1997) and over 3-months (December 1989 – March 1990) to assess the nature of inter-annual and intra-annual variations imposed by long-term factors (e.g., deforestation, reforestation, and agricultural extensification) and short-term factors (e.g., phenological cycles and precipitation patterns) on landscape dynamics. Satellite change-detections of LULC and NDVI as well as pattern metrics are presented to examine the nature of compositional and structural shifts in the landscape over the study periods. Observations regarding the pattern of LULC and plant productivity over inter- and intra-annual time periods are made relative to the topographic settings derived from the DEM and recoded soil series data as well as population density suggested through village locations and the generation of overlapping village territories.

LULC and Plant Biomass

Table 1 shows the nature of LUCC between 1989 and 1997. The values indicate that 36% of the study area (from the 300,000 randomly sampled pixels) consistently was rice, 8% cash crops, and 10% forest. Areas devoted to rice and cash crops were least volatile, whereas those devoted to forest or mixed agriculture were more volatile. For example, of the 20% of sample pixels in mixed agriculture in 1989, a major proportion (almost 90%) were no longer in mixed agriculture in 1997. We interpret the change in terms of expanded rice cultivation. The year 1997 was a wet year, whereas 1989 was a dry year. The transition to more mixed agriculture in the middle and low-middle terraces would likely occur as marginal lands would become less so as precipitation conditions permitted the cultivation of rice in largely unsuitable settings. Figure 4 (on the CD-ROM) shows the pattern of classified LULC for 1989/90 (a), 1997 (b).

Table 1. Percent LULC change during the period December 1989 and 1997.

	Gained	Lost	Net Change	Stable
Rice	22.2	12.2	+10.0	36.4
Mixed Agriculture	2.3	17.6	-15.3	3.2
Cash Crops	4.7	2.1	+2.6	8.3
Forest	12.8	10.1	+2.7	10.1

Table 2 shows the mean of the rescaled NDVI associated with the cross-classification of the 4 LULC types between 1989 and 1997. Figure 5 (on the CD-ROM) shows the spatial pattern of rescaled NDVI variation for the 1989 (a), 1990 (b), and 1997 (c) TM images.

Table 2. NDVI rescaled to range from 1-200. Mean NDVI change and standard deviation (in parentheses) for each of the 16 possible landscape changes shown; positive values indicate an increase in NDVI, while negative values indicate a decrease in NDVI. NDVI values for stable LULC classes are shown in bold. * Indicates statistical significance.

Mean NDVI, 12/23/89	LULC, 12/29/97			
	Rice	Mixed Ag.	Cash Crops	Forest
Rice	**6.7**	2.5	5.7	15.5*
	(8.1)	(7.9)	(17.0)	(11.3)
Mixed Agriculture	6.8	**2.8**	10.0	17.6*
	(9.0)	**(7.9)**	(18.7)	(13.1)
Cash Crops	0.1	-3.4	**-4.7**	13.7
	(14.3)	(14.6)	**(19.9)**	(16.0)
Forest	4.4	-3.4	1.6	**12.7**
	(11.3)	(14.6)	(19.6)	**(15.1)**

Most of the changes represented in the table indicate an increase in NDVI values, except for the negative changes in NDVI values related to upland sites. This likely reflects precipitation differences between the two years, which appear particularly relevant to forest growth.

Table 3 shows the percent change in the rescaled NDVI values for the 4 LULC types for the December and March dates for the 1989-90 water year allowing assessment of intra-annual change. Positive values indicate an increase in the NDVI and negative values indicate a decrease in NDVI from 1989 to 1990. Figure 6 (on the CD-ROM) shows NDVI change for the two time periods within the same water year. The rescaled NDVI values show a slight increase for lowland rice and middle terrace rice and other agriculture, little change in forest, but a pronounced decrease in NDVI values in upland cash crops.

This is probably because of the harvesting of those crops in the early months of 1990. March 1990 was our unusually wet month, and in the absence of harvest, a major decline in NDVI would be quite unexpected. Bare soil and stubbled rice associated with a post-harvested rice landscape captured in the March 1990 satellite image accounts for this change, as well as the concentration of labor in upland cassava and sugar cane fields following the more time-sensitive rice harvest that is generally completed by December of each year.

From an ecological and land management perspective, forests may also be in some form of transition – deforestation to accommodate new plantings of upland crops, reforestation through government imposed land conservation programs, reforestation through secondary plant succession of sites where yields have substantially declined through use, and the planting of forests representing silviculture such as orchards and eucalyptus trees (grown for construction scaffolding and paper pulp).

Table 3. Inter-annual change from 1989 to 1990 in the rescaled NDVI by LULC type.

	Rice	Mixed Ag.	Cash Crops	Forest
Mean	2.7	2.8	-25.9	-0.1
St. Dev.	9.2	9.9	20.0	15.5

Table 4 shows the mean and standard deviation of the rescaled NDVI for each of the 4 LULC types between December 1989, March 1990, and December 1997. Figure 7 (on the CD-ROM) shows the spatial pattern of plant biomass variation for the intra-annual comparison. A pronounced intra-annual pattern is clearly evident for cash crops, consistent with the interpretation of Table 3, and a pronounced inter-annual pattern is evident for forest.

The standard deviation of the NDVI is generally greater for the March 1990 period as compared to the December 1989 period across LULC types except for cash crops. December is the period that the rice harvest is generally initiated. The exact timing is dependent upon a number of factors including crop physiology, degree of surface wetness and/or inundation, and available labor. By late December, the rice harvest has mostly been completed. By March 1990, the landscape had dried considerably (although March 1990 monthly precipitation totals were abnormally high), with rice stubble predominating the lowland landscape and upland crops having matured leading to harvest. The 1997 water year was quite wet, suggesting a landscape of pronounced greenness and high biomass through harvest. Since monsoonal rains have a relatively large geographic footprint, the effects of above-average rainfall would boost NDVI values across the entire landscape recorded by the end of December 1997 TM image, although there was a slight decrease in NDVI of cash crops.

Table 4. Vegetation biomass variation by LULC type; NDVI rescaled to range from 1-200. Mean and standard deviation of rescaled NDVI (in parentheses) are shown for each LULC type and date.

	Rice	Mixed Ag.	Cash Crops	Forest
December 1989	112.0	112.0	129.7	119.6
	(7.5)	(7.7)	(14.6)	(12.2)
March 1990	114.6	114.8	103.8	119.4
	(9.0)	(9.2)	(14.5)	(14.6)
December 1997	117.7	112.1	127.0	133.9
	(7.9)	(7.0)	(17.5)	(13.3)

PATTERN METRICS AND LANDSCAPE STRUCTURE

FRAGSTATS (McGarigal and Marks 1993) software was used to assess landscape structure at the landscape-level for Nang Rong district and at the class-level for rice, mixed agriculture, cash crops, and forest. Landscape-level represents LULC patterns within a preset district boundary, whereas the class-level is set by the distribution of each of the LULC classes defined through Landsat Thematic Mapper digital classifications. The same five pattern metrics were computed at the landscape- and class-levels: (1) interspersion and juxtaposition index is the observed interspersion over the maximum possible interspersion for the given number of patch types, a measure of adjacency; (2) number of patches is the sum of the total number of patches on the landscape, a measure of fragmentation; (3) mean patch size is the sum of the area of all patches of the corresponding patch type divided by the number of patches of the same patch type, a measure of contiguity; (4) total edge is the sum of the lengths of all patch edge segments on the landscape, a measure of connection and insularity; and (5) shape index is the sum of the landscape boundary and all edge segments divided by the square root of the total landscape area, a measure of configuration (McGarigal and Marks 1993).

Table 5 shows the landscape metrics for the inter-annual comparison of NDVI variation between 1989 and 1997. Over the 8 years between the two TM images, the landscape has become less fragmented. By 1989, the beginning of the observation period, substantial areas of forest have already been converted to agriculture in the upland sites (primarily in the southwest part of the study area), and trees have been largely removed from the rice-producing lowlands. An above average rainfall in 1997, the endpoints of the observation period, followed drought conditions from 1990-1994 that combined with economic factors apparently promoted further agricultural expansion into the topographically transitional middle terraces for commercial rice and cassava cultivation. Collectively, these actions tended to homogenize the landscape -- rice in the lowlands and broad interconnected areas of cassava and sugar cane in the uplands. The landscape showed (a) reduced adjacency of cover types because of this amalgamation of LULC into large extensive monocultures, (b) decrease in the number of patches through deforestation of outliers or forest fragments, (c) consolidation of individual fields in larger tracts and an increase in the mean patch size of forest and agricultural crops, (d) decrease in the total edge through patch reductions resulting in a smoother, less varied landscape by 1997, and (e) a decrease in shape index also suggests a smoother or less fragmented dry season landscape in 1997 as compared to 1989.

Table 5. Landscape metrics from rescaled and classed NDVI values for 1989 and 1997 Landsat TM images.

Selected Landscape Metrics	23 December 1989	29 December 1997
Interspersion/Juxtaposition Index	16.88	13.77
Number of Patches	181,500	152,498
Mean Patch Size	1.70	1.98
Total Edge (m)	64,236,720	50,704,952
Shape Index	1.42	1.37

Table 6 compares class metrics as in Table 5 for rice, mixed agriculture, cash crops, and forests for 1989 and 1997. The table indicates (a) consolidation of rice in the lowlands by the removal of tree clusters and remnant forest patches from previously developed paddies, (b) expansion of rice and cash crop cultivation in the marginalized middle terraces, (c) expansion of cash crops in the uplands through deforestation, and (d) general increase in the spatial complexity of forests through deforestation and agriculture-forest fragmentation, primarily in the uplands, in the lowlands along riparian zones, and in established paddies where trees were previously retained for shade for people and animals.

DISCUSSION

The theoretical framework to explore population-environment interactions in our research builds on work in the social sciences attempting to model population and environment dynamics and work in the biophysical sciences. The research is aimed at understanding landscape composition and spatial patterns across space and time that reflect the effects of human forcing factors, exogenous environmental and social contexts, and environmental endowments. The biophysical endowments reflect what is available in the environment upon which humans can act, as well as current

LULC patterns. Ultimately, the influence of these factors upon LULC depends upon the environmental gradients, geographical accessibility, and population-environment interactions of the region.

Table 6. Pattern metrics computed at the class level for NDVI classes for 1989 and 1997 Landsat TM image dates.

Selected Class (Rice) Metrics	23 December 1989	29 December 1997
Interspersion/Juxtaposition Index	19.70	17.76
Number of Patches	42,053	22,634
Mean Patch Size	3.48	7.98
Total Edge (m)	49,835,672	44,738,672
Shape Index	1.45	1.44
Selected Class (Mixed Ag.) Metrics		
Interspersion/Juxtaposition Index	13.77	11.66
Number of Patches	56,625	42,621
Mean Patch Size	1.20	0.39
Total Edge (m)	31,526,250	10,916,640
Shape Index	1.45	1.24
Selected Class (Cash Crops) Metrics		
Interspersion/Juxtaposition Index	19.75	16.64
Number of Patches	21,195	10,852
Mean Patch Size	1.48	3.56
Total Edge (m)	12,693,030	9,069,930
Shape Index	1.27	1.32
Selected Class (Forest) Metrics		
Interspersion/Juxtaposition Index	19.86	11.56
Number of Patches	60,893	75,656
Mean Patch Size	1.02	0.88
Total Edge (m)	34,379,700	36,645,872
Shape Index	1.44	1.43

The productivity of lands associated with villages and households is an element of landscape discrimination and a possible indicator of household decision-making associated with agricultural extensification and planting options. The available labor force also is likely to have an important impact on LULC. Competing work opportunities and migration patterns might alter the feasibility of certain types of agricultural activities, and make it more difficult to use marginal land. From 1989 (the beginning of the present analysis) to the economic crisis of 1997, Nang Rong had been an area of net out-migration particularly of young adults, which leads to a change in the age structure of the population with a lower proportion of the population able to engage in some of the more physically demanding aspects of agriculture. [With social survey data being collected in spring and summer 2000 and processed in 2001 for subsequent analyses, we will be able to address more carefully the effects of patterns of migration, both permanent and circular, as well as the possibility that the 1997 economic crisis changed the flow of the migration stream]. Since cassava production tends to be more physically demanding than rice, the effect of the changing age structure of the population will vary by crop. Further, out-migration might lead to reforestation through plant succession, either because land is left fallow or because the decline in the available labor force has led to the planting of long-term crops, such as fruit trees. Household and village decisions are influenced by biophysical aspects of the land around them, geographical relationships between villages and markets, and transportation and hydrographic accessibility (Crews-Meyer 1999).

Also important are mechanisms whereby social processes constrain land use choices made within a context of competition for resources and in response to environmental endowments and constraints. Precipitation and landform (terrace position) are key to understanding landscape dynamics. In this research, 1989 was substantially drier than 1997, and March 1990 had a substantial amount of rain (183 mm) relative to the December 1989 image date. Topography is a key factor in sorting LULC and LULC change because of the land suitability restrictions of uplands for rice cultivation, the long-term conversion of lowland forests to rice paddies, and the site suitability of upland forests to cassava and sugar cane production to address commercial crop opportunities and agricultural extensification needs.

LULC in the district changed in some important ways between 1989 and 1997. The largest areal change occurred from mixed agriculture to lowland rice. While both LULC categories are likely dominated by rice, they appear spectrally dissimilar because both rice and cash crops have been partly or mostly harvested by the 29 December 1997 image, the middle terrace rice is populated by a greater degree of upland cover-types (i.e., forest and cassava and sugar cane) because of its topographic transition, and the middle terrace rice reflects a difference in the timing of the rice harvest because of the more marginal (i.e., drier) site conditions for rice cultivation in areas other than lowlands. Also, the December 1997 image captured cover-type information developed over a much wetter period than did the December 1989 image, thereby affecting plant biomass levels and harvest schedules for rice paddies that were unduly wet or submerged. Figure 8 (on the CD-ROM) shows a plot of precipitation variability for a single station in the center of the district for a selected time-series (1989, 1990, 1997, and average conditions) that substantiates the moisture availability interpretations.

The second largest observed change in LULC between 1989 and 1997 was from lowland rice to forest, primarily occurring in riparian zones along the three main rivers. The change may be related to economic opportunities for commercial rice cultivation near high moisture sites, or be the artifact of a pronounced precipitation change between the two years. The third largest change in LULC was from forest to rice, followed by a change from forest to cash crops. The magnitude of conversion to rice was twice that of the conversion to cash crops indicating an extensification into rice.

The rescaled NDVI values computed for the three image dates, that characterized seasonal and near decadal changes in plant biomass levels, indicate that little intra-annual change in plant biomass occurred between December 1989 and March 1990 except for the upland cash crops that decreased significantly as a consequence of pre- and post-harvest conditions. By December 1989 rice had been harvested in the lowlands and by March 1990 rice stubble predominated the landscape and hence the satellite measures of plant biomass reflected a reduced biomass magnitude and a more restricted areal pattern of high biomass levels throughout the district. The intra-annual assessment seeks to understand the phenology shifts in plant conditions related to household decisions to harvest crops in the rice paddies and the upland cassava and sugar cane fields. The timing of the harvest varies according to local rainfall patterns, available labor, and site conditions, which in turn is related to on-site wetness and the position of the fields relative to household locations – further out sites are generally harvested later than closer in sites. The 1997/98 water year was wetter than the 1989/90 water year and hence the 1997 image had higher plant biomass values. While precipitation-induced changes were important, deforestation, albeit at a reduced rate, continues to alter the landscape matrix. Land clearings are focused in the upland sites and terraces, but removal of individual trees and small clumps of trees continues in the lowlands and more areally significant deforestation is occurring in riparian environments. Streams corridors offer the farmer a source of nearby water and relatively high soil moisture potentials, but nearness to streams also introduces risk of economic loss through flooding and/or late planting and harvest schedules due to wetness and/or inundation. Also noteworthy is the government's attempt at reforestation, designation of forest conservation areas, and the planting of fruit trees (e.g., Mango), Para-rubber, and Eucalyptus in a plantation style that are occasionally inter-cropped with cassava or other field crops.

The compositional and spatial organization of the landscape, defined through the pattern metrics, suggests a population-environment interaction that has implications for sustainability. Landscapes compositionally-dominated by agriculture indicate a region in which extensification has occurred and where further enhancements in crop productivity will likely occur through intensification efforts, because "developable" lands have already transitioned from forest to agriculture. Landscapes dominated by forest suggest a site potential for agricultural productivity through extensification, and landscapes of mixed landuse suggest a region that is undergoing LULC change where inputs of labor are being focused for deforestation and agricultural cultivation and inputs of chemicals and technology are likely supporting agricultural sustainability of sites to counteract site degradation through declining soil fertility through use and soil loss through erosion. Landscape organization suggests that pattern and use are interconnected in some complex arrangement in which factors such as (a) field size, (b) nearness to water-bodies and perennial rivers and streams, (c) site location relative to other agricultural sites, roads, and site suitabilities, and (d) village to village, and village to land, and land to land topologies are interwoven into the decision-making process of household members electing to migrate or cultivate in various patterns and through various strategies. For example, Rindfuss et al. (1996) found that village territories that were highly fragmented experienced significant out-migration of young adults suggesting that human behavior and landscape pattern are related through a feedback mechanism that may be bound to certain thresholds of fragmentation.

Villages are generally organized along the north flowing rivers that occur within the district. Because of the reliance on rain-fed rice cultivation, the juxtaposition of nuclear village settlements and perennial rivers and streams is to be expected. As a consequence of their close proximity to water and hence to other villages, some lands are worked by a number of villagers from a number of different villages, thereby increasing the density of people on the landscape at those overlapping village territories and making the competition for land and corresponding resources necessary for the cultivation of crops to be more competitive.

Villages having high populations and large numbers of households, populations dominated by young adults, and a population having a high male-female sex ratio suggest a population that is associated with greater potential to alter the landscape (Walsh et al. 1999). In addition, villages with larger populations tend to influence LULC over a broader geographic extent through greater land holdings, and villages that are linked to other villages through geographic proximity and/or kinship ties tend to expand their areal extents of influence or their geographic reach on the landscape as a consequence of scalar relationships involving multiple villages (Walsh et al.1999; Walsh 1999).

CONCLUSIONS

The application of a satellite time-series to examine long- and short-term changes in both LULC (e.g., composition and spatial structure) and plant biomass variability across a landscape and for mapped cover-types offers the ability to track change on a x, y, and z perspective where terrain and population potentials exert interactive forces on the landscape and are influenced by exogenous forces such as crop prices and climatic variability. While not explored here, time lags in crop prices, precipitation patterns, and subsequent household decisions about LULC change and plant and harvesting schedules are important and have impact on labor and migration patterns – either to participate in agricultural endeavors through return migration or to subsidize the household through non-farm employment and out-migration.

Pattern metrics are a useful set of description algorithms that quantify the composition and spatial organization of the landscape at multiple levels or scales. They reflect the importance of landscape structure and landscape function. In Nang Rong district, the landscape has been in the process of pattern cycling in which forest was replaced by lands under cultivation as the areally dominant cover-type. Reforestation through government programs and secondary plant succession are the counter-balances to agricultural extensification through deforestation, but have yet to significantly impact the landscape.

The examination of terrain and village point locations within a GIS was useful to link LULC and NDVI variation to specific landscape strata, 3-D for topography and 2-D for village territories. Terrain data were used here to imply the relationship of LULC and elevation through a graphic that profiles their co-variation across a random swath across the district, whereas village locations and territory delineation's were used to assess the density of the Nang Rong population on the consolidation of labor and resource use to support agricultural cultivation and the antecedent LULC change through deforestation and agricultural extensification. The research illustrates the value of an integrated GISc (Geographic Information Science) approach that combines data from satellite imagery, social surveys, and analog maps representing biophysical, social, and geographic domains for assessing the inherently complex interactions between humans and the environment.

ACKNOWLEDGMENTS

The work reported in this paper is supported by the National Aeronautics and Space Administration under its Land-Cover and Land-Use Change program (NAG5-6002). This paper is part of a larger set of interrelated projects funded by the National Institute of Child Health and Human Development (R01-HD33570 and R01-HD25482), the National Science Foundation (SBR 93-10366), the EVALUATION Project (USAID Contract #DPE-3060-C-00-1054), and the MacArthur Foundation (95-31576A-POP). The larger set of projects involves various collaborations between investigators at the University of North Carolina, Carolina Population Center, Department of Sociology, and the Department of Geography, and investigators at the Institute for Population and Social Research (IPSR), Mahidol University, Bangkok, Thailand. We recognize the efforts of Joseph P. Messina, Department of Geography, Landscape Characterization & Spatial Analysis Lab for his work on the LULC classifications.

REFERENCES

Arbhabhirama A, Phantumvanit D, Elkington J, Ingkasuwan P. *Thailand Natural Resources Profile*. Singapore: Oxford University Press, 1988

Crawford TW. A comparison of region building methods used to examine human-environmental interactions in Nang Rong district, northeast, Thailand. *Proceedings, Applied Geography Conference* (FA Schoolmaster ed.), 1999; 22: 366-373.

Crews-Meyer KA. Modeling landcover change associated with road corridors in northeast Thailand: integrating normalized difference vegetation indices and accessibility surfaces. *Proceedings, Applied Geography Conference*, (FA Schoolmaster ed.) 1999; 22: 407-416.

Entwisle B, Walsh SJ, Rindfuss RR, Chamratrithirong A. Land use/Land-cover and Population Dynamics, Nang Rong, Thailand. In *People and Pixels*, D Liverman, EF Moran, RR Rindfuss, PC Stern eds. Washington DC: National Academy Press, 1998.

Feeny D, "Agricultural expansion and forest depletion in Thailand, 1900-1975." In *World Deforestation in the Twentieth Century*, JF Richards, RP Tucker eds. Durham, NC: Duke University Press, 1988.

Fox J, Krummel J, Yarnasarn S, Ekasingh M, Podger N. Land use and landscape dynamics in Northern Thailand: assessing change in three upland watersheds. *Ambio* 1995; 24: 328-334.

Fukui H. *Food and population in a Northeast Thai Village*. Monographs of the Center for Southeast Asian Studies, Kyoto University, English-Language Series, No. 19, Honolulu: University of Hawaii Press, 1993.

Ghassemi F, Jakeman AJ, Nix AH. *Salinisation of Land and Water Resources: Human Causes, Extent, Management and Case Studies*. Sydney: University of New South Wales Press Ltd, 1995.

Hirsch P, Lohmann L. Contemporary politics of environment in Thailand. *Asian Survey* 1989; 29: 439-451.

Jensen JR. *Introductory Digital Image Processing: A Remote Sensing Perspective*, 2nd Edition. Englewood Cliffs, New Jersey: Prentice-Hall, 1996.

Kaida Y, Surarerks V. Climate and Agricultural Land Use in Thailand. In *Climate and Agricultural Land Use in Monsoon Asia*. MM Yoshino ed. Tokyo: University of Tokyo Press, 1984.

McGarigal K, Marks BJ. *FRAGSTATS: Spatial pattern analysis program for quantifying landscape structure*. Forest Science Department, Oregon State University, Corvallis, Oregon, 1993.

Parnwell MJG. Rural poverty, development and the environment: the case of North-East Thailand. *Journal of Biogeography* 1988; 15: 199-313.

Polthanee A, Marten, GG. "Rain-fed Cropping Systems in Northeast Thailand." In *Traditional Agriculture in Southeast Asia: A Human Ecology Perspective*. GG Marten ed. Boulder: Westview Press, 1986.

Rigg J. Homogeneity and heterogeneity: an analysis of the nature of variation in northeastern Thailand. *Malaysian Journal of Tropical Geography*, 1991; 22: 63-72.

Rindfuss RR, Walsh SJ, Entwisle B. *Land Use, Competition, and Migration*. Paper presented at the Population Association of America, New Orleans, LA 1996.

Rouse JW, Haas RH, Schell JA, Deering DW. Monitoring vegetation systems in the Great Plains with third ERTS. *ERTS Symposium*, NASA No. SP-351, 309-317, 1973.

Rundel PW, Boonpragop K. "Dry forest ecosystems of Thailand." In *Seasonally Dry Tropical Forests*, SH Bullock, HA Mooney, E Medina eds. New York: Cambridge University Press, 1995.

Walsh SJ. Deforestation and agricultural extensification in Northeast Thailand: a remote sensing and GIS study of landscape structure and scale. *Proceedings, Applied Geography Conference*, (FA Schoolmaster ed.) 1999; 22: 223-232.

Walsh SJ, Evans TP, Welsh WF, Entwisle B, Rindfuss RR. Scale-dependent relationships between population and environment in northeastern Thailand. *Photogrammetric Engineering and Remote Sensing* 1999; 65: 97-105.

Chapter 7

BARRIERS AND SPECIES PERSISTENCE IN A SIMULATED GRASSLAND COMMUNITY

David M. Cairns
Department of Geography, Texas A&M University, College Station, TX 77843-3147
cairns@tamu.edu

Keywords: cellular automata, fractal dimension, grasslands.

Abstract The fragmentation of landscapes that results from habitat destruction can produce barriers to species interactions across a landscape. In this study, a stochastic cellular automata model was used to investigate the effects of barriers on the persistence of five grassland species. By varying the number and pattern of barriers on the landscape, it was determined that regardless of the initial arrangement of species, the placement of barriers on the landscape slows the loss of species due to invasion by superior competitors. The pattern of barriers (as measured by their fractal dimension) was more important when the barriers accounted for smaller portions of the landscape.

INTRODUCTION

The diversity of many biological systems is maintained as a shifting mosaic of species or seral stages within a landscape over time and space. Although underlying environmental heterogeneity may foster diversity, high diversity communities can occur in relatively homogeneous environments (Hastings 1980, Tilman 1994). The maintenance of diversity in these homogeneous environments has been addressed both within the framework of metapopulation dynamics (Gilpin and Hanski 1991, Tilman 1994) and with the use of cellular automata models (Silvertown et al. 1992).

Human activities often result in habitat fragmentation that can lead to a reduction in biodiversity by decreasing available habitat and increasing insularization (cf. MacArthur and Wilson 1967, Andren 1994). Analytical studies that do not consider habitat destruction indicate that there is no limit to the number of species that can coexist in a spatially subdivided habitat, if they have the correct balance of competitive ability, dispersal and longevity characteristics (Tilman 1994). However, when fragmentation occurs as habitat destruction, diversity is lost in a predictable manner (Tilman et al. 1994, 1997). The amount and pattern of fragmentation are known to affect species persistence in communities where there is an inverse relation between competitive ability and dispersal (e.g., Dytham 1995).

In some cases, however, landscape fragmentation can lead to either retaining the status quo of diversity or may even increase biodiversity in an area. At the species and community levels, it is commonly known that forest fragmentation increases the relative amount of edge habitat. The destruction of interior habitat and its replacement with edge habitat can, at least temporarily, increase the β diversity of an area (Noss 1983). The increase in diversity does, however, come at the expense of losing some interior species, and few would argue that we should fragment the landscape, further to increase the number of edge species on the landscape since the interior species are those that are most often in need of conservation.

Landscapes are not only structured by the presence of competing species. Non-living portions of the landscape are not resource competitors, but instead shape the environment in which competition occurs by structuring the movement of species and their interaction across the landscape. At coarse spatial scales, barriers have been hypothesized to delay the arrival of potential competitors into some environments subsequent to climate change. For example, based on pollen data, Davis et al. (1986) hypothesized that the late arrival of beech (*Fagus*) into the Upper Great Lakes region, USA after the last glacial maximum was due to the dispersal barrier posed by the Great Lakes. At a finer spatial scale, Malanson and Cairns (1997), using a spatially explicit forest gap model, found that increases in the number of barriers on a landscape decreased the migration rates of tree species. They found little difference in the magnitude of migration rates between structured landscapes (i.e., those with regular patterns of fragmentation) and landscapes that were randomly fragmented. Also, in the Grand Canyon, USA barriers to dispersal allowed some species to survive on benches on the canyon walls long after they had been competitively excluded from other more easily accessible locations (Cole 1985).

Here, the objective was to examine how the number of barriers and their spatial pattern influence species richness in a simulated grassland landscape using a stochastic cellular model. If the extinction debt concept holds true under the conditions simulated here, a loss of diversity was expected with increasing amounts of uninhabitable area on the grid.

METHODOLOGY

The Model

For this research, the model developed by Silvertown et al. (1992) was used. The simulated landscape was composed of a 40 x 40 lattice of square cells with edges that act as absorbing barriers. Each cell was inhabited by only one of five species that differ in their ability to invade and be invaded by their neighbors. At each iteration, the status of each cell was determined by the species that currently inhabits the cell and the species in the four directly adjacent cells (i.e., horizontal and vertical). Replacement rates (Table 1) are the same as those used by Silvertown et al. (1992) for a grassland community and are based on empirical studies (Thórhallsdóttir 1990). Although the model was developed to represent a grassland community, the model is non-mechanistic and therefore can represent any type of vegetation community with a similar set of replacement rates. In Table 1, the replacement rates are weighted by the number and species present in the neighboring cells to determine transition probabilities (Figure 1). The status of the cell is then determined as a function of the transition probabilities and a randomly generated number between 0 and 1. By restricting the interaction of cells to only the four closest neighbors, local dispersal was assumed more important than global dispersal. This model formulation differs from metapopulation models in which global dispersal were assumed. Because a stochastic component was included in the model, it was not considered a strict cellular automata model, but rather a stochastic cellular model.

Experiments

To test the hypotheses, a set of simulation runs or experiments were conducted in which the initial pattern of species on the landscape, the proportion of the landscape occupied by barriers, and the fractal dimension of the barriers on the landscape were varied.

Species Pattern

Two very simple patterns of initial species arrangement were evaluated in this research. First, a random pattern was used as a control in which each species occupied 20% of the landscape and every cell in the landscape had an equal probability of being assigned any species (Figure 2a). Second, an aggregated pattern was created such that the landscape consisted of 5 monospecific bands each comprised of 20% of the landscape (Figure 2b).

Table 1. Pairwise invasion rates for the five species used in the stochastic cellular model. Calculation of the transition probabilities was based on these values and was described in Figure 1. Data from Silvertown et al. (1992).

Invader	Native species				
	Species 1	Species 2	Species 3	Species 4	Species 5
Species 1	-	0.09	0.23	0.37	0.32
Species 2	0.08	-	0.06	0.09	0.16
Species 3	0.02	0.06	-	0.03	0.05
Species 4	0.02	0.03	0.03	-	0.05
Species 5	0.06	0.06	0.44	0.11	-

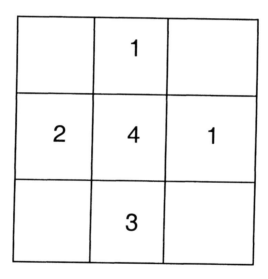

Figure 1. Calculation of transition probabilities based on the replacement rates reported in Silvertown et al. (1992). Transition probabilities were only calculated using the four horizontally and vertically adjacent cells. In this case, the cell of interest was in the center. The cell was occupied by Species 4 and had two neighboring cells occupied by Species 1, and one cell each of Species 2 and Species 3. The probability that the cell would be invaded by Species 1 was (2/4 x 0.37) = 0.185. The probability of being invaded by Species 2 was 1/4 x 0.09, and by Species 3 was 1/4 x 0.03. There was no probability that Species 5 would invade the cell since no adjacent cell was occupied by that species. The probability that the cell remained occupied by Species 4 was 0.785.

Number of Barriers

In the model, barriers were cells that could not support any species growth and were essentially dead spaces on the grid. The number of barriers varied between no barriers present on the landscape and 65% of the landscape consisting of barriers. Simulations at 5% intervals of fragmentation between 0 and 65% were completed (i.e., no barriers, 5%, 10%, etc.).

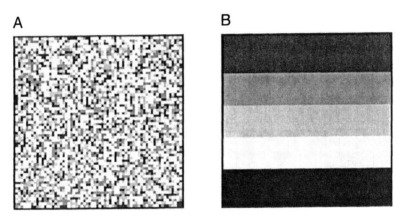

Figure 2. Initial species arrangements. Each shade of gray represents a different one of the 5 possible species in the simulation: A) random arrangement of species on the landscape, B) aggregated arrangement of species on the landscape.

Pattern of Barriers

The landscape pattern was varied by constraining the fractal dimension of the barrier portion of the landscape to certain values. Fragmented landscapes were created with fractal dimensions ranging from 2.1 to 2.9 at 0.1 intervals (Figure 3). A random barrier pattern was also simulated to allow for comparison. Fractal landscapes were created using the random midpoint displacement method (Saupe 1988).

Realizations

For each combination of initial species pattern, barrier pattern and barrier proportion, 30 realizations of the model were run. Each realization ran for 600 iterations. For the sake of simplicity, each iteration was referred to as a year. During each year, every "live" cell on the grid (i.e., not a barrier) was tested to see if it would remain the same species or be converted to an invading species. The 30 realizations for each combination of treatments were required because this model was stochastic and any single realization was not likely to represent the true performance of the model. Therefore, all results presented were the average results over the 30 realizations performed for each combination of species pattern, barrier pattern, and barrier amount. A total of 8,400 model realizations were performed

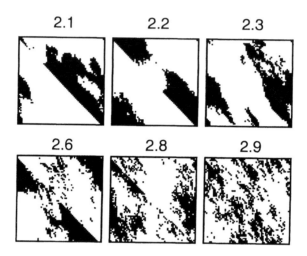

Figure 3. Fractal barrier patterns. In each landscape 30% of the landscape was occupied by barriers (black). The fractal dimension varied among the landscapes. Fractal dimensions 2.1, 2.2, 2.3, 2.6, 2.8 and 2.9 are shown. Increases in fractal dimension resulted in more smaller barriers than larger ones and a much more fragmented landscape.

RESULTS AND DISCUSSION

Initial Pattern of Species

The importance of initial species pattern on the performance of the model in the absence of barriers has been published previously (Silvertown et al. 1992). The results obtained here are qualitatively and quantitatively similar to those of Silvertown et al. (1992). In the random control pattern, the most competitive species quickly eliminated the lesser competitors and by year 150 only two species were left on the landscape. The best competitor completely excluded all other species by the end of the 600-year simulation. When the species were aggregated into monospecific bands, the same process occurred, but much more slowly. Typically by the end of the simulation only the two weakest competitors were entirely eliminated. The changes in abundances of the species over time were strongly non-linear in the aggregated initial pattern simulations, when no barriers were on the landscape (see Silvertown et al. 1992).

The results of t-tests comparing the final richness (i.e., number of species remaining on the landscape at the end of the simulations) for the random and aggregated patterns indicated that the two patterns were significantly different (t=2.843, df = 838, p < 0.005). Similarly for the date of first species loss, the first species were lost significantly earlier (one-tailed t-test) in the random species arrangement versus the aggregated one (t= 32.507, df=329, p < 0.001).

NUMBER OF BARRIERS

Effects on Final Richness

The proportion of the landscape that was occupied by barriers had a significant effect on the species richness at the end of the simulation. Mean species richness at the end of the simulation increased with fragmentation, and when barriers occupied more than 30% of the landscape, all five species persisted (Figure 4a). The pattern of species loss relative to the percent of the landscape occupied by barriers was qualitatively similar for both the aggregated and random initial species arrangements. The species richness in year 600 was significantly greater (α=0.05) for the aggregated pattern at fragmentation levels where 0, 5, 10 and 20 percent of the landscape was occupied by barriers. When 20% or more of the landscape was comprised of barriers, there was no difference in the final species richness of the simulations.

Effects on Date of First Loss

The speed at which the weakest competitor was lost from the simulation was evaluated by looking at the mean year of first species loss (Figure 4b). Since no species were lost from the initial pool when over 30% of the landscape was occupied by barriers, date of first loss information was only relevant for the cases where between 0 and 30% of the landscape was occupied by barriers. The date of first species loss increased with the number of barriers on the landscape for both the random and aggregated initial patterns. This indicated that placing barriers on the landscape not only increased the ultimate biodiversity, but also slowed the loss of diversity. The aggregated and random initial species patterns were significantly different (α=0.05) with regard to the date of first species loss for all tested levels of fragmentation.

116

Combined Effects on Final Richness and Date of First Species Loss

By plotting species richness against both the percent of landscape occupied by barriers and the number of years, temporal patterns in richness over the extent of the model realizations were illustrate (Figure 5). In both the case of random and aggregated initial species arrangements, it was clear that the placement of barriers on the landscape slowed the speed with which species were lost and also increased the species richness in the final year of the simulation. The random pattern of species on the landscape resulted in a loss of species considerably more quickly than the aggregated condition.

Pathways

The pathways that the species took toward their final abundances on the landscape were strongly non-linear when no fragmentation of the landscape was included. This has been discussed previously by Silvertown et al. (1992). The non-linearity of species response over time resulted from changes in the species composition of the landscape. For example, if two strong competitors were separated by a weak competitor, the two competitively dominant species would consume the space occupied by the weak competitor, both increasing their abundance on the landscape. However, when the weak competitor was eliminated from the simulation, the two strong competitors were in contact with each other and the stronger of the two would continued to increase in abundance, while the weaker declined in abundance (Figure 6). The non-linearity of species abundance over time was dampened by the placement of barriers on the landscape. The change in response from non-linear to linear was not present in the random species arrangement. For the random initial species pattern, the paths that the different species took toward their final abundance was not qualitatively different among the fragmentation amounts (Figure 6a). In contrast, for the aggregated initial species arrangement, there was a switch from a strongly non-linear to more linear paths toward the final abundance of species (Figure 6b). With increasing fragmentation, there was less chance that all of a weaker competitor would be removed from the landscape. Therefore, the best competitors were more likely to remain separated from each other, and not competitively eliminate the other.

A

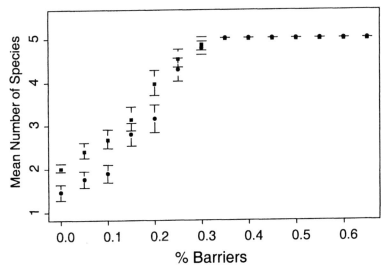

B

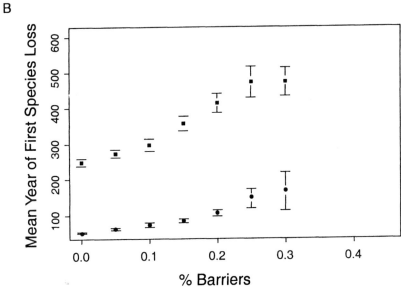

Figure 4. The effects of increasing the percent of the landscape occupied by barriers on A) the number of species present at the end of the simulation, and B) the year of first species loss from the simulation. Error bars represent the 95 % confidence level for the mean values presented in the graphs. The closed circles represent the mean values for the random initial species arrangement; the open boxes are the values for the aggregated initial species arrangements.

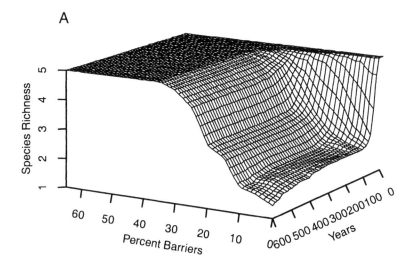

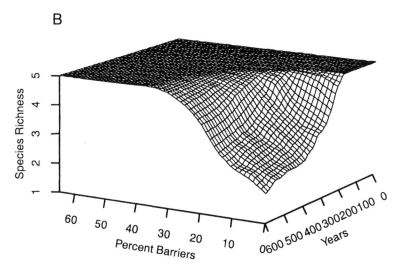

Figure 5. Response surfaces illustrating the effects of time (years in the simulation) and the percent of the landscape occupied by barriers on species richness for A) random, and B) aggregated initial species arrangements on the simulated landscapes.

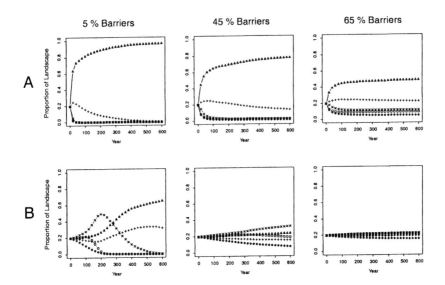

Figure 6. Proportion of the landscape occupied by each of the 5 species over time for the A) random, and B) aggregated initial species arrangements on the simulated landscapes. Results are presented for 5 %, 45 %, and 65 % of the landscape occupied by barriers. Each point is the average value obtained over 30 repetitions. Δ = Species 1, + = Species 2, o=Species 3, ◊ = Species 4, and x= Species 5.

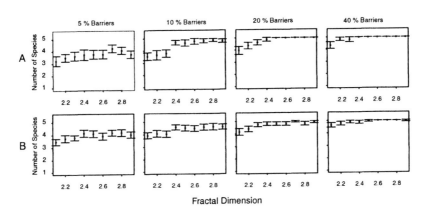

Figure 7. Mean species richness at the end of the 600-year simulations (n=30) for each fractal dimension tested (2.1 - 2.9). Error bars represent the 95 % confidence interval for the mean values. Results for 5 %, 10 %, 20% and 40% of the landscape are presented for both A) random, and B) aggregated initial species arrangements on the landscapes.

Pattern of Barriers

The effects of barrier pattern on the landscape was evaluated by varying the fractal dimension of simulated landscapes and investigating the ultimate species diversity, and the date of first loss of a species from the simulation. Increasingly complex patterns of barriers (i.e., higher fractal dimensions) increase the species richness at the end of the 600-year runs of the model for both initial patterns of species on the simulated landscape.

The effect of an increasingly complex pattern of barriers on the landscape was not dramatic. Higher fractal dimensions increased the number of species present at the end of a simulation more for cases where less of the landscape was occupied by barriers than in those cases where barriers occupied a greater proportion of the landscape (Figure 7). For example, there was no statistically significant difference ($\alpha = 0.05$) between any of the fractal dimensions for the aggregated species pattern when 40% of the landscape was occupied by barriers, whereas at 10% barriers, fractal dimensions less than or equal to 2.3, were significantly different than those greater than 2.3 (Figure 7). Similar results were obtained for the random arrangement of species (Figure 6). This indicated that pattern of barriers was more important when it occupied less of the landscape.

The results presented here are qualitatively similar to those achieved for other fractal landscapes (Palmer 1992). Although Palmer (1992) did not consider barriers in the same way, he did find that for landscapes comprised of as many as 10 species, landscape diversity (i.e., the number of species present at the end of a simulation) increased with fractal dimension. Although at the very highest fractal dimensions (> 2.75), the species richness decreased. Palmer (1992) attributed this pattern, in part, to the effects of habitat area. At low fractal dimensions, the landscape was more interconnected and the better competitors were able to invade more easily thereby reducing diversity. The same processes operated on the landscapes simulated here.

Although there was an observable increase in species richness with increasingly complicated fractal patterns, the same effect was not apparent for the year of first species loss. In nearly all cases for both initial species patterns there was no statistically significant difference among the different fractal dimensions for the different amounts of barriers on the landscape (Figure 8). This underscored the lesser importance of pattern of barriers (as measured by fractal dimension) relative to the absolute number of barriers on the landscape.

CONCLUSIONS

The results presented here provide some interesting conclusions that may at first be counter-intuitive. First, by placing barriers on the landscape biodiversity was actually augmented rather than reduced. This was the result of the barriers releasing species from competitive pressures that ultimately could result in their elimination from the landscape. The findings presented here differ from those of Tilman et al. (1994), because there was no inverse relationship in the model between the ability of a species to compete once established and disperse. It has yet to be conclusively proven that there is an inverse relationship between competitive ability and dispersability (Kareiva and Wennergren 1995), therefore, these results are simply different from previous models (e.g., Tilman et al. 1994, Dytham 1995); they are not necessarily more or less representative of reality. The results must be viewed with caution when applied to a conservation question, because the model assumed that the smallest simulated area (a single cell) contained a genetically viable population. This assumption may not hold in many ecological scenarios.

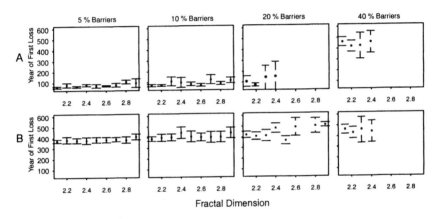

Figure 8. Mean date of first species loss (n=30) for each fractal dimension tested (2.1 - 2.9). Error bars represent the 95 % confidence interval for the mean values. Results for 5 %, 10 %, 20% and 40% of the landscape are presented for both A) random, and B) aggregated initial species arrangements on the landscapes.

This study has shown the utility of using a fairly simple model to study the complex pathways and patterns that may be present in ecological systems. Additional research is necessary to determine if effects of pattern measurable in ways other than fractal dimension are important in maintaining species diversity or in reducing the speed with which diversity is lost.

REFERENCES

Andren H. Effects of habitat fragmentation on birds and mammals in landscapes with different proportions of suitable habitat: a review. Oikos. 1994; 71:355-366.

Cole K. Past rates of change, species richness, and a model of vegetational inertia in the Grand Canyon, Arizona. American Naturalist. 1985; 125:289-303.

Davis M B, Woods DL, Webb SL, Futyama RP. Dispersal versus climate: expansion of *Fagus* and *Tsuga* into the Upper Great Lakes Region. Vegetatio. 1986; 67:93-104.

Dytham C. The effect of habitat destruction pattern on species persistence: a cellular model. Oikos. 1995; 74:340-344.

Gilpin M, Hanski I. eds. *Metapopulation Dynamics: Empirical and Theoretical Investigations.* London: Academic Press, 1991.

Hastings A. Disturbance, coexistence, history, and competition for space. Theoretical Population Biology. 1980; 18:363-373.

Kareiva P, Wennergren U. Connecting landscape patterns to ecosystem and population processes. Nature. 1995; 373:299-302

MacArthur RH, Wilson EO. *The Theory of Island Biogeography.* Princeton NJ: Princeton University Press, 1967.

Malanson GP, Cairns DM. Effects of dispersal, population delays, and forest fragmentation on tree migration rates. Plant Ecology. 1997; 131: 7-79.

Noss RF. A regional landscape approach to maintain diversity. BioScience. 1983; 33:700-706.

Palmer MW. The coexistence of species in fractal landscapes. American Naturalist. 1992; 139:375-397.

Saupe D. "Algorithms for random fractals." In *The Science of Fractal Images*, HO Peitgen, D Saupe. eds. New York: Springer-Verlag, 1988.

Silvertown J, Holtier S, Johnson J, Dale P. Cellular automaton models of interspecific competition for space—the effect of pattern on process. Journal of Ecology. 1992; 80:527-534.

Thórhallsdóttir TE. The dynamics of five grasses and white clover in a simulated mosaic sward. Journal of Ecology. 1990; 78:909-923.

Tilman D. Competition and biodiversity in spatially structured habitats. Ecology. 1994; 75:2-16.

Tilman D, Lehman CL, Yin C. Habitat destruction, dispersal, and deterministic extinction in competitive communities. American Naturalist. 1997; 149:407-435.

Tilman D, May RM, Lehman CL, Nowak MA. Habitat destruction and the extinction debt. Nature. 1994; 371:65-66.

Chapter 8

FEEDBACK AND PATTERN IN COMPUTER SIMULATIONS OF THE ALPINE TREELINE ECOTONE

Matthew F. Bekker[1], George P. Malanson[1], Kathryn J. Alftine[1] and David M. Cairns[2]
[1]*Department of Geography, University of Iowa, Iowa City, IA 52242 USA*
matthew-bekker@uiowa.edu, george-malanson@uiowa.edu
[2]*Department of Geography, Texas A&M University, College Station, TX 77842 USA*

Keywords: ecotone, treeline, computer simulation, feedback.

Abstract: The pattern of ecotones relative to gradients in the abiotic environment may be affected by spatial biotic positive feedback, i.e., where trees improve the conditions for other trees in their neighborhood. Computer simulation models were used to investigate the consequences of degrees of positive feedback. A hybrid model combining FOREST-BGC, ATE-BGC, and FORSKA was used. The BGC models created a surface of potential growing conditions at the alpine treeline. FORSKA was then used to compute the leaf area that was produced with varying levels of feedback strength. Feedback must be strong relative to the abiotic environment for growth to occur. Such feedback created an abrupt boundary on a smooth abiotic gradient and was higher than needed to generate more complex patterns. These results have implications for some of the major hypotheses about treeline and for studies combining modeling and field parameterization.

INTRODUCTION

The pattern of ecotones (transitional areas between adjacent ecological systems; Risser 1995) and their relation to the abiotic environment (primarily temperature, moisture, and substrate) may be mediated by biotic processes. Differentiating between ecotones and more gradual ecoclines, van der Maarel (1990) emphasized differences in the steepness of the gradient, i.e., the rate of change in the environment over distance, that led to different relative importance of processes such as competition. Wilson and

Agnew (1992) detailed the possibility that positive feedback, where plants improve the conditions for conspecifics, in the biological processes within the plant communities on either side of an ecotone could account for an abrupt ecotone. In these cases the plants modify the environment that they experience, so that what was a continuous abiotic gradient becomes discontinuous in terms of the potential carbon balance; e.g., Armand (1992) attributed alpine treeline ecotone patterns to species interactions. Timoney et al. (1993) hypothesized that a sigmoid wave model might be used to explain the spatial transitions of vegetation at the Arctic treeline and, more generally, at undisturbed biome transitions. The observed sigmoid wave patterns might be linked directly to positive feedback mechanisms. Malanson (1997) examined the one-dimensional pattern at a treeline as it responds to differences in the strength of positive feedback and in seed rain. Ecotones have recently been analyzed as phase changes, analogous to the physics of condensed matter (Milne et al. 1996, Loehle et al. 1996). In these studies, nonlinear relations of pattern and process have been demonstrated for tree/grassland ecotones. We will expand this concept by using a more biologically mechanistic model.

The objective of this research was to assess biological spatial processes as explanations of ecotone pattern. These processes could include competition, mediated herbivory, parasitism and disease, and negative or positive feedback. Here, we present an analysis of the feedback strength needed to maintain krummholz vegetation in a hybrid simulation that joins a physiologically mechanistic model with a mechanistic tree growth model.

This work is also germane to the idea of ecotones as indicators of the impacts of climatic change (Hansen et al. 1988). Because species are at the limit of their realized niche at an ecotone, changes in the abiotic environment might either expand or contract the spatial realization of the niche. The degree to which alpine ecotones can be indicators depends, however, on their rates of response and the multiplicity of important niche dimensions (Kupfer and Cairns 1996).

Observations of Pattern

A variety of patterns have been observed in the alpine treeline ecotone (Walsh et al. 1994a). Brown (1992, 1993, 1994b) performed an analysis of the spatial distribution of structurally defined vegetation types in Glacier National Park, Montana, USA using Landsat Thematic Mapper (TM). Working at a spatial resolution of 30 m, he was able to identify, and validate through field sampling, four major types of vegetation that straddled the treeline ecotone. The types were based primarily on the density and structure of the canopy vegetation. They were, in order of decreasing canopy density and dominance of tree species: closed-canopy forest, open-canopy forest, meadow/tundra, and unvegetated. The extent to which studies make a link between treeline pattern and process is that they show how site conditions vary among different treeline pattern types (Baker and Weisberg 1995, Baker et al. 1995; Allen and Walsh 1996). Allen and Walsh (1996) quantified spatial patterns on about 1,100 slopes; patterns were related to topographic types. Four types of spatial pattern tended to be dominant: a gradual change from tall trees, to short trees, to krummholz, to tundra (a textbook gradient); an abrupt change from trees to tundra; a gradient of trees to krummholz fingering upslope into tundra; and patches of krummholz scattered upslope of the trees into tundra (Figure 1).

Explanations of Pattern

Abrupt ecotones at a local scale have been observed where trees border wetlands, grasslands, and tundra. Stevens and Fox (1991) concluded that treeline was largely a phenomenon expressing the carbon balance of trees, so that trees ended where their carbon balance was zero (the carbon balance hypothesis). Explanations of abrupt treelines fall on a gradient from abiotic control, i.e. abrupt differences in resources/climate, to biotic interactions such as competition (e.g., Armand 1992).

Stevens and Fox (1991) proposed a theory to account for the locations of ecotones between species of different physiognomic types such as the forest-steppe and alpine treeline ecotones. They proposed that the boundary represents differences in nutrient availability. This hypothesis is based on the concept of nutrient averaging: in an environment that on average is nutrient poor, small patches that are nutrient rich exist, and small plants with their less expansive root systems are able to take advantage of these nutrient rich pockets. Larger growth forms, such as trees with larger root systems, must average nutrients over a larger area. Therefore, a tree seedling may establish in a nutrient rich pocket, but it quickly outgrows the pocket and encounters a nutrient poor environment. Consequently, the smaller plants out-compete the trees and become the dominant physiognomic type. The heterogeneity of the abiotic resource leads to spatial patterns in the vegetation.

While soil fertility may be a factor influencing the alpine treeline ecotone (cf. Bamberg and Major 1968, Holtmeier and Broll 1992), other abiotic factors potentially affecting the ecotone pattern include slope, dominant geomorphic processes, and late-lying snowpatches. In a way similar to the nutrient averaging hypothesis, krummholz and trees may also suffer more from periglacial disturbance of the substrate because of their more extensive roots. Slopes of lower angle may have had solifluction processes active in the ecotone during the Holocene. Active solifluction would probably prevent tree establishment (Hansen-Bristow and Ives, 1985). Another spatially variable resource could be snow (Kullman 1990, Scott et al. 1993, Walsh et al. 1994b). Broad changes in aspect probably affect treeline elevation but not pattern.

Abiotic control of carbon balance alone may not, however, explain an abrupt treeline on a gradual abiotic gradient; perhaps trees modify their environment, making the establishment of additional trees more likely (Gosz and Sharpe 1989, Armand 1992, Timoney et al. 1993a). Wilson and Agnew (1992) detailed the possibility that positive feedback in the biological processes within the plant communities could account for an abrupt ecotone. In these cases the plants modify the environment that they experience, so that what was a continuous abiotic gradient becomes discontinuous in terms of the potential carbon balance. The processes that might lead to a positive feedback include microclimate, i.e. albedo and temperature in canopy and soil (e.g. Bonan 1992, Chalita and Le Treut 1994) to physical effects such as resistance to erosion (Zonneveld 1995). These feedbacks compound the effects of competition (Armand 1992). Although the possibility of a positive feedback, in which the plants modify the abiotic environment so that establishment and growth are increased, is obvious, the pattern that might be produced for a feedback of a given strength has not been investigated

Korner (1998) has proposed an alternative hypothesis for the general elevation of treeline. He suggested that cooler temperatures in the canopy, due to its stature, and in the soil, due to shading, would limit growth per se, rather than reducing carbon production. This is a negative feedback hypothesis.

Tree Growth Modeling

Noble (1993) modeled the response of a treeline to climatic change, where the climatic change was hypothesized to act as response to disturbance. In the primary model the advance is only by contagion. A positive feedback was included at a landscape scale; the treeline becomes smoother when this feedback was strengthened. Feedback from ecotone shape smoothed the boundary and reduced the variability of response to simulated climatic change. In a modification, Noble (1993) included the possibility that individual trees could establish by dispersal rather than direct contagion with extant trees. Dispersal increased the rate of advancement, but also increased the variability of the treeline so that the noise was greater than the potential signal of response to climatic change. Noble (1993) did not discuss other implications of dispersal, and he only examined the binomial occupancy of sites without regard for abundance.

Using a spatially explicit version of a JABOWA-FORET model, Malanson (1997) studied an ecotone as might occur at a mountain treeline. Seed rain and seedling survival modified the dominant patterns determined by the strength of feedback and the steepness of the abiotic gradient. The feedbacks were spatially autocorrelated and so created waves of mortality and regeneration as have been observed on mountain slopes. These dynamics mean that the pattern at the ecotone at any point. in time were ephemeral and may respond differently to environmental change. In a major revision of this type of model, Leemans (1989) developed improved, more mechanistic growth functions in a model called FORSKA.

Cairns and Malanson (1997) used a modified version of FOREST-BGC that incorporated treeline-specific hydrological and physiological processes (ATE-BGC; Cairns 1994, 1995) to test the carbon balance hypothesis at the alpine treeline ecotone (ATE). They found that carbon balance was in dynamic equilibrium with climate, but the ATE was not. They concluded that the carbon balance hypothesis was useful for predicting the potential, but not actual location of the ATE. Cairns and Malanson (1998) examined the effects of several environmental variables on the carbon balance of krummholz. They concluded that temperature was the most important factor in determining the potential location of the ecotone, but the complex patterns of the treeline were determined primarily by moisture related variables, including soil depth. Cairns (1998) concluded that disturbance plays a small role in determining the pattern of the ecotone, except in areas with particularly steep slopes, where geomorphic activity was most likely. In developing the model ATE-BGC for use at the ATE, Cairns (1994, 1995) found that factors related to feedbacks were potentially important. Brown et al. (1994) suggested that a combination of empirical and physical models was needed to properly describe vegetation patterns at the treeline ecotone because of the multi-scale influences on the patterns.

METHODS

We examined the levels of feedback needed to maintain krummholz vegetation at the alpine treeline ecotone by using two simulation models. For a treeline site in Glacier National Park, Montana (GNP), we simulated a surface of potential carbon balance using ATE-BGC. We then relativized this surface to the carbon balance simulated by FOREST-BGC for a productive site. The relative surface was then used as the site quality in a multi-cell implementation of FORSKA. FORSKA then computed a surface of leaf area. We compared the amount of feedback strength needed here to that used by Malanson (1997) and to the amount needed to produce realistic patterns in a cellular automaton. Although the model components were similar, the integration differed from that of Friend et al. (1993).Carbon balance values for a treeline site were predicted using ATE-BGC (Cairns

Figure 1A-B. Common patterns at the alpine treeline ecotone: (A) gradual gradient from trees to tundra, (B) abrupt transition from trees to tundra.

Figure 1C-D. Common patterns at the alpine treeline ecotone: (C)patches of krummholz, (D) fingers of trees and/or krummholz.

and Malanson 1997). ATE-BGC is a version of FOREST-BGC (Running and Coughlan 1988) that were modified to include physiological processes important at treeline locations. The model predicts carbon balance as the sum of photosynthetic inputs to the vegetation system balanced by carbon outflows due to respiration and tissue loss (senescence and injury). A full description of the model is presented elsewhere (Cairns and Malanson 1997, 1998); it is important to note that ATE-BGC explicitly considers the effects of winter injury (frost drought) and the effects of extreme low temperatures on the photosynthetic mechanism.

We simulated pattern for a single site in Glacier National Park; it is site 20 in Cairns (1995). The site is comprised of 400 points evenly spaced 30 m apart. For each point, we used ATE-BGC to predict the average carbon balance for 1981-1991. ATE-BGC requires information describing the site and vegetation characteristics at every point to predict carbon balance values (Table 1). All parameters in Table 1 varied from point to point with the exception of soil depth, leaf area index, and woody biomass. The vegetation parameters used here correspond to the average conditions found at sparse krummholz locations (moderate leaf area and low woody biomass). Soil depth was set to the average measured soil depth for 27 treeline sites in Glacier National Park (Cairns 1995). Site-specific parameters were calculated using a digital elevation model in a GIS.

Table 1. Sample parameters used in ATE-BGC carbon balance simulations for treeline and valley simulations.

Parameter Name	Treeline Values	Valley Site Values
Latitude (°)	48.722	48.699
Elevation (m)	2217	1524
Slope (°)	19.93	8.20
Meteorological Base Station	Many Glacier	Many Glacier
East horizon angle (°)	18.77	1.25
West horizon angle (°)	17.57	14.15
Site average precipitation (mm)	1774	783
Snow potential index	0.001	0.027
Soil Type	Babb Association	Loberg Association
Soil Depth (cm)	8.5	100
Total leaf area index	11.07	6.8
woody biomass (kg)	253.59	10260

The simulation of a non-treeline site was calculated in the same manner as above, but was based on the average values for 10 randomly located low elevation sites within the St. Mary River catchment of GNP. These sites are representative of full-sized sub-alpine forest in GNP. For the low elevation runs, the treeline specific processes in ATE-BGC (photosynthetic maturation and winter injury) were disabled. In this configuration there was no difference between ATE-BGC and FOREST-BGC.

The relative performance of treeline sites in comparison to low elevation mature forest sites was calculated as the ratio of site-specific carbon balance predictions and the average low elevation carbon balance prediction.

FORSKA. The fundamentals of how individual tree growth is simulated have been explained in detail elsewhere (Leemans 1989). The fundamental growth equation is:

$$d/dt(D^2H) = (1 - W_{tot}/W_{max}) \int_B^H S_L(\gamma P_z - \delta z) dz * f(\text{environment}) \qquad 1)$$

where D is dbh, H is tree height, W is biomass (total and maximum), B is bole height, S_L is the vertical density of leaf area, P_z is the proportion of maximum possible annual assimilation achieved by leaves at depth z in the canopy, γ is the species-specific growth-scaling factor, and δ is a species-specific factor for maintaining the tree's actual size. Instead of the *f*(environment) based on input data, we used the relativized surface produced by ATE-BGC. All simulations were run with a cell size of 30 m.

Positive Feedback

The site quality of a given cell was increased if a neighboring cell was occupied. We increased site quality as a function of neighbor basal area:

$$Q = d*S*B/60 \qquad 2)$$

where S is feedback strength, B is basal area in cm^2, and d is 1 for cells sharing a side or .707 (difference in center to center distance) for cells sharing only a corner. We used feedback strength levels of 0, 0.25, 0.5, 1.0, 2.0, and 4.0. At levels >0, a cell neighboring at least one occupied cell could possibly have its site quality increased by 0.18, up to a maximum of 1.0. Twenty simulations were run for 500 years for each feedback level. Seed rain was held constant in these simulations.

RESULTS

Relativized carbon balance (site quality) values from ATE-BGC varied gradually across the site (Figure 2). Overall relative potential carbon balance was very low. When FORSKA was run using the relative carbon balance figures from Figure 2 as input, the results varied with feedback strength. Mean LAI values produced by FORSKA increased abruptly between no feedback and low feedback, then more gradually with higher feedback levels (Table 2). There was an increase in both the minimum number of cells with trees, and leaf area index (LAI) per cell. Notably, it was necessary to increase feedback strength by an order of magnitude to get much vegetation on the site. Actual LAI for krummholz patches on the site was 13.35, but the patches occupied less than 25% of any 30 m pixel area. Tree growth was restricted to cells with the highest site quality values, even at the highest feedback level (Figure 3).

DISCUSSION

To produce leaf area in the ranges observed at alpine treeline, it was necessary to use high feedback strength. The feedback strength needed produced very abrupt boundaries in the simulations on smooth unidimensiona gradients simulated by Malanson (1997) (Figure 4). This contrast indicated that the patterns at treeline, other than the most abrupt transition from trees to tundra, resulted from a combination of pattern in the abiotic environment, such as soil depth, and feedback. Another way in which some patterns can be generated, however, is through directional feedback. Directional feedbacks can produce patterns without abrupt.

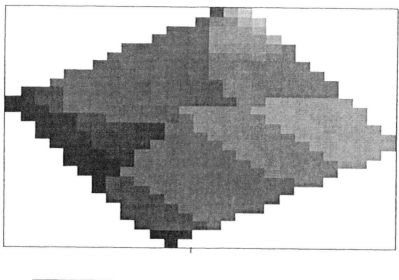

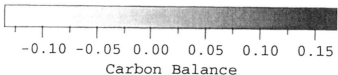

Figure 2. Pattern of relativized carbon balance (site quality) values.

transitions even at relatively high feedback strength. For a simple stochastic cellular model, the surrounding trees increased in probability of a cell being occupied by trees. Occupancy in one direction (e.g., upwind) can have a stronger effect. This model produced an advancing treeline on a smooth gradient that had some of the elements of pattern observed (Figure 5).

Table 2. Leaf area index (LAI) projected for a site for varying levels of feedback strength.

Feedback	Result	Mean	Variance	Min	Max
0	#cells>0	11.1	5.253	6	15
	LAI, cells>0	0.001	0.000	0.001	0.005
	LAI, all cells	0.000	0.000	0.000	0.005
0.25	#cells>0	17.6	1.516	14	18
	LAI, cells>0	0.494	0.090	0.001	1.047
	LAI, all cells	0.022	0.015	0.000	1.047
0.5	#cells>0	17.95	0.050	17	18
	LAI, cells>0	0.594	0.052	0.001	1.236
	LAI, all cells	0.027	0.018	0.000	1.236
1.0	#cells>0	17.95	0.050	17	18
	LAI, cells>0	0.594	0.046	0.001	1.203
	LAI, all cells	0.027	0.018	0.000	1.203
2.0	#cells>0	19.5	3.737	18	22
	LAI, cells>0	0.606	0.035	0.001	1.182
	LAI, all cells	0.029	0.019	0.000	1.182
4.0	#cells>0	20.7	3.484	18	22
	LAI, cells>0	0.591	0.033	0.003	1.245
	LAI, all cells	0.031	0.019	0.000	1.245

These results raise questions, which can be considered as new hypotheses, for the major hypotheses about ecotone pattern. The questions are: Can the resource averaging hypothesis account for the variations in spatial pattern at treelines? Can positive feedback alone produce complicated spatial patterns on a smooth environmental gradient? How strong is the negative feedback of cooler soils relative to positive feedbacks? The new hypotheses are:

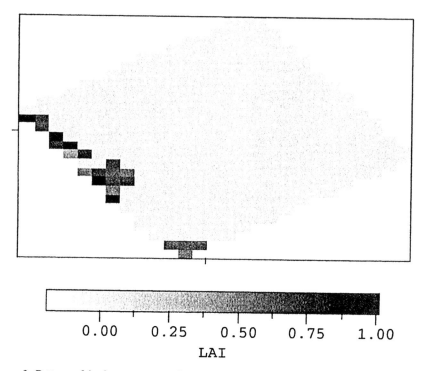

Figure 3. Pattern of leaf area generated on the site quality values of Figure 2 using the highest feedback strength (lightest gray is zero).

➢ Resource averaging hypothesis: patterns at treeline can be generated by varying distribution of soil resources only if the gradient is irregular and if positive feedback exists. Resource averaging alone, without feedback effects, cannot produce the patterns seen at fine resolution.

➢ Positive feedback: while positive feedback is necessary, it does not produce the detailed patterns observed unless, perhaps, it has unusual spatial dimensions. Variation in the spatial distribution of the resources or stresses is necessary.

➢ Growth limitation: the negative feedback of cooler soil and canopy limits on growth may not outweigh positive feedbacks, but a more thorough evaluation needs to incorporate this effect in the mechanistic models, guided by accurate data.

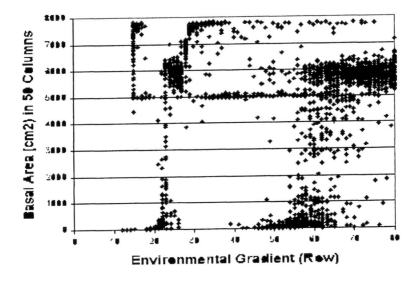

Figure 4. Pattern of abundance for high feedback from Malanson (1997).

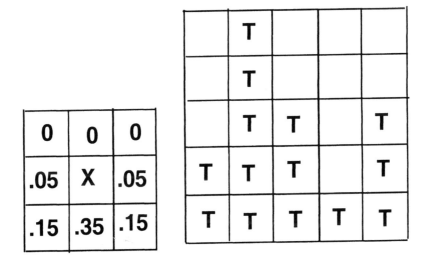

Weighting Filter **T - Tree Locations**

Figure 5. Neighborhood contributions to site quality and pattern produced in a simple stochastic cellular model.

The most fruitful path for future modeling of the spatial patterns at the alpine treeline ecotone would seem to be to incorporate elements from the above hypotheses to determine the processes that are both necessary and sufficient to produce the observed patterns. First, the growth limitation hypothesis is based on specific mechanistic response to temperature. The degree to which leaf area alters temperature can be determined and the effects of temperature on growth processes modeled. This step could be included in the FORSKA portion of our model. Second, the spatial aspects of positive feedbacks need to be quantified and modeled: it is the pattern that determines the feedback that produces the pattern. The direction of wind and slope need to be included and can be on a grid model. Whether or not the simple moving window approach with additive process is adequate needs to be addressed; we suggest that using genetic algorithms for rule discovery may be appropriate. A finer resolution model, perhaps at 1 m, might be more appropriate. Third, the variation in the underlying resource gradient needs to be explored systematically. How variance in the gradient interacts with spatial feedbacks will produce a vast number of potential patterns. Because the resource gradient in field situations has already been affected by feedbacks, it may be necessary to design more careful field campaigns before we can understand what models tell us.

ACKNOWLEDGMENTS

The research was supported by grant SBR-9714347 National Science Foundation.

REFERENCES

Allen TR, Walsh SJ. Spatial and compositional pattern of alpine treeline, Glacier National Park, Montana. Photogrammetric Engineering and Remote Sensing 1996; 62: 1261-1268.

Armand, A. "Sharp and gradual mountain timberlines as a result of species interaction." In *Landscape Boundaries: Consequences for Biotic Diversity and Ecological Flows*, AJ Hansen, F di Castri, eds. New York: Springer-Verlag, 1992; 367-377.

Baker WL, Weisberg PJ. Landscape analysis of the forest-tundra ecotone in Rocky Mountain National Park, Colorado. Professional Geographer 1995; 47: 361-375.

Baker WL, Honaker JJ, Weisberg PJ. Using aerial photography and GIS to map the forest-tundra ecotone in Rocky Mountain National Park, Colorado, for global change research. Photogrammetric Engineering and Remote Sensing 1995; 61:313-320.

Bamberg SA, Major J. Ecology of the vegetation and soils associated with calcareous parent materials in three alpine regions of Montana. Ecological Monographs 1968; 38: 127-167.

Bonan GB. "A simulation analysis of environmental factors and ecological processes in North American boreal forests." In *A Systems Analysis of the Global Boreal Forest*, HH Shugart, R. Leemans, GB Bonan, eds. Cambridge UK: Cambridge University Press, 1992; 404-427

Brown DG. Comparison of vegetation-topography relationships at the alpine treeline ecotone. Physical Geography 1994a; 15: 125-145.

Brown DG. Predicting vegetation types at treeline using topography and biophysical disturbance. Journal of Vegetation Science 1994b; 5: 641-656.

Brown DG, Cairns D.M, Malanson GP, Walsh SJ, Butler DR. "Remote sensing and GIS techniques for spatial and biophysical analyses of alpine treeline through process and empirical models." In *Environmental Information Management and Analysis: Ecosystems to Global Scales*, WK Michener, JW Brunt, SG Stafford, eds. London: Taylor and Francis, 1994; 453-481.

Cairns DM. Development of a physiologically mechanistic model for use at the alpine treeline ecotone. Physical Geography 1994; 15: 104-124.

Cairns DM. Carbon Balance Modeling at the Alpine Treeline Ecotone, Glacier National Park, Montana. Ph.D. Dissertation, University of Iowa, Iowa City, 1995.

Cairns DM. Modeling controls on pattern at alpine treeline. Geographical and Environmental Modelling 1998; 2: 43-63.

Cairns DM, Malanson GP. Examination of the carbon balance hypothesis of alpine treeline location, Glacier National Park, Montana. Physical Geography 1997; 18: 125-145.

Cairns DM, Malanson GP. Environmental variables influencing carbon balance at the alpine treeline ecotone: a modeling approach. Journal of Vegetation Science 1998; 9: 679-692.

Chalita S, Le Treut H. The albedo of temperate and boreal forest and the Northern Hemisphere climate: a sensitivity experiment using the LMD GCM. Climate Dynamics. 1994; 10: 231-240.

Friend AD, Schugart HH, Running SW. A physiology-based gap model of forest dynamics. Ecology 1993; 74: 792-797.

Gosz JR, Sharpe PJH. Broad-scale concepts for interactions of climate, topography, and biota at biome transitions. Landscape Ecology 1989; 3: 229-223.

Hansen AJ, di Castri F, Risser PG. A new SCOPE project. Ecotones in a changing environment: the theory and management of landscape boundaries. Biology International Special Issue 1988; 17: 137-163.

Hansen-Bristow K J, Ives JD. Changes in the forest-alpine tundra ecotone: Colorado Front Range. Physical Geography 1984; 5:186-19

Holtmeier F-K, Broll G. The influence of tree islands and microtopography on pedoecological conditions in the forest-alpine tundra ecotone on Niwot Ridge, Colorado Front Range, U.S.A. Arctic and Alpine Research 1992; 24: 216-228.

Korner C. A reassessment of high elevation treeline positions and their explanation. Oecologia 1998; 115: 445-459.

Kullman L. Dynamics of altitudinal tree-limits in Sweden: a review. Norsk Geografisk Tidsskrift 1990; 44:103-116.

Kupfer JA, Cairns DM. The suitability of montane ecotones as indicators of global climatic change. Progress in Physical Geography 1996; 20: 253-272.

Leemans R. Description and simulation of stand structure and dynamics in some Swedish forests. Acta Universitatis Upsaliensis 1989; 221.

Loehle C, Li B-L, Sundell RC. Forest spread and phase transitions at forest-prairie ecotones in Kansas, U.S.A. Landscape Ecology 1996; 11: 225-235.

Malanson GP. Effects of feedbacks and seed rain on ecotone patterns. Landscape Ecology 1997; 12: 27-38.

Malanson GP, Butler DR. Competitive hierarchies, soil fertility gradients, and the elevation of treeline in Glacier National Park, Montana. Physical Geography. 1994; 15: 166-180.

Milne BT, Johnson AR, Keitt TH, Hatfield CA, David J, Hraber PT. Detection of critical densities associated with pinon-juniper woodland ecotones. Ecology 1996; 77: 805-821.

Noble IR. A model of the responses of ecotones to climate change. Ecological Applications 1993; 3: 396-403.

Risser PG. The status of the science examining ecotones. BioScience 1995; 45: 318-325.

Running SW, Coughlan JC. A general model of forest ecosystem processes for regional applications. 1. Production processes. Ecological Modelling 1988; 42: 125-154

Stevens GC, Fox JF. The causes of treeline. Annual Review of Ecology and Systematics 1991; 22:177-191.

Timoney KP, La Roi GH, Dale MRT. Subarctic forest-tundra vegetation gradients: the sigmoid wave hypothesis. Journal of Vegetation Science 1993; 4: 387-394.

van der Maarel E. Ecotones and ecoclines are different. Journal of Vegetation Science 1990; 1:135-138.

Walsh SJ, Butler DR, Brown DG, Bian L. "Form and pattern of alpine environments: an integrative approach to spatial modeling and analysis." In *Mountain Environments and GIS*, DI Heywood, MF Price, eds. London: Taylor and Francis, 1994a.

Walsh SJ, Butler DR, Allen T, Malanson GP. Effects of snow patches and snow avalanches on the alpine treeline ecotone. Journal of Vegetation Science 1994b; 5: 657-672.

Wilson JB, Agnew ADQ. Positive-feedback switches in plant communities. Advances in Ecological Research 1992; 23: 263-336.

Zonneveld IS. Vicinism and mass effect. Journal of Vegetation Science 1995; 5: 441-444.

Chapter 9

SPATIAL PATTERN AND DYNAMICS OF AN ANNUAL WOODLAND HERB

Lucy Bastin[1]* and Chris D. Thomas[2].
[1]*School of Biological Sciences, University of Birmingham, Birmingham, UK*
lb26@le.ac.uk
[2]*School of Biology, and Centre for Biodiversity & Conservation, University of Leeds, Leeds LS2 9JT, UK*
c.d.thomas@leeds.ac.uk
current address & correspondence: Department of Geography, University of Nottingham, University Park, Nottingham NG7 2RD

Keywords:	Conservation, introduction, scale, metapopulation, source-sink dynamics.
Abstract	We examined the effect of spatial scale on the distribution and probable dynamics of an annual woodland plant. The natural distribution of *Melampyrum pratense* was mapped within woods in Birmingham, UK. Its distribution was recorded within five defined habitat types (Holly, Heath, Birch, Glade and Bracken), and experimental transplants into each of these habitat types were monitored over two growing seasons. Populations were observed to persist naturally in habitat types where experimental sowings were incapable of sustained population replacement (i.e. "sinks"). These populations were all within close range (< 16 m) of natural populations in 'ideal' woodland glade habitat, and appeared to be maintained, in the long term, by immigration from these 'source' patches. Deduced metapopulation dynamics, involving local colonisations and extinctions in woodland glades, appeared to be occurring at medium "source-sink" dynamics mentioned above were restricted to a range of 0 to 16m. Natural dispersal events to separate woodland units (300 – 1600 m away), where *M. pratense* was absent from habitat proved to be favourable by the success of transplants, would appear to be rare. As a result, colonisation or re-colonisation of other suitable woods some kilometres away was probably an extremely rare occurrence. *Melampyrum pratense* was dependent on the characteristic spatio-temporal dynamics of mature woodland, or of managed coppice woodland, and its presence within any wood was indicative of past, as well as present, conditions within a wood. Declining rates of dispersal with increasing distance mean that source-sink dynamics, metapopulation dynamics and community history assume differing importance at different spatial scales. Thus, conclusions about the relative roles of habitat and dispersal in species distributions are critically dependent on the spatial scale of study.

The capacity of GIS to integrate mapped data from a variety of sources can make analysis of plant and animal population patterns across a variety of scales more manageable. Therefore, GIS tools are likely to be extremely valuable in such multi-scale analyses, and in identifying the dominant influences on species distributions at each spatial scale.

INTRODUCTION

The spatial resolution at which the distribution of a particular plant or animal is mapped may fundamentally affect the biological interpretation of which factors determine that distribution (e.g. Wiens 1989 1997, Thomas and Kunin 1999) and the conservation priorities attributed to different species (e.g., Kunin 1998, Cowley et al. 1999). GIS can provide the tools to examine species distributions easily at multiple scales. We provide a case study of how the spatial resolution of analysis affects the interpretation of biological processes in the plant *Melampyrum pratense* (common cow-wheat). This species is a hemiparasitic annual herb associated particularly with clearings and banks in ancient woodland.

Most animal and plant species have patchy distributions, both naturally and as a result of habitat fragmentation (Gilpin and Hanski 1991). There are two major causes of patchiness: a) limiting habitat requirements (including competition, microhabitat requirements, and the presence of associated species); and b) failure to colonise suitable habitats, usually because propagules are not reaching these habitats at all, or in sufficient numbers. However, a species may sometimes also be found in fundamentally unsuitable habitats, where it is maintained by the dispersal of propagules into "sink" habitat.

In general, places without a particular species may either be unsuitable non-habitat, or potential habitat waiting to be colonised (reviewed by Thomas and Kunin 1999). Identifying the difference is a key element in any study of spatial dynamics. However, places which currently support a species, and thus appear to be suitable as habitat, may not all be fundamentally suitable for local reproduction and population maintenance. Some locations (sinks) may be populated only because of the flow of individuals/propagules into them from population sources elsewhere. Therefore, one could possibly observe source-sink-type population dynamics at one spatial scale (involving frequent, small-scale movements of individuals or propagules) and metapopulation dynamics at another (involving colonisations which are mediated by infrequent, larger-scale movements between suitable breeding areas). It is crucial to the development of conservation programmes, to understand which processes are most important at different spatial scales.

This study attempted to distinguish between the effects of habitat suitability and seed dispersal on the distribution of a woodland annual plant, *Melampyrum pratense*, by considering its habitat mosaic at various resolutions of analysis. The major hypothesis tested within this study was that dispersal limits the species' distribution across large (regional) distances, but is still high enough to generate sink populations over shorter distances.

Melampyrum Pratense and its Distribution in Birmingham

Melampyrum pratense is a hemiparasitic member of the Scrophulariaceae, with spikes of tubular yellow flowers and characteristic "beake'" seed pods. The plant has a maximum of four seeds per pod, and their large weight (*ca.* 20 mg) is thought to be an adaptation to hemiparasitism in a leaf-litter environment, where there may be some distance between the germinating seed and the host roots to which it must attach (ter Borg 1985).

Within the study area, the species occurred only in or around woodland older than 300 years, often growing at locally high densities between May and September. *Melampyrum pratense* is usually found in areas of relatively open ground (e.g., woodland glades and banks, rides, wood margins) on acid ground (Streeter and Garrard 1983). The species is also recorded from more open heathy habitats, where it is commonly associated with *Calluna vulgaris, Vaccinium myrtillus* and the grasses *Deschampsia flexuosa* and *Agrostis tenuis*. A mix of *C. vulgaris* and *M. pratense* may indicate previously burnt ground, and characterise a transitional community (Ingrouille 1995).

Melampyrum pratense extracts water, minerals and organic compounds from the roots of its hosts through haustoria which form xylem-to-xylem connections. All *Melampyrum* species require a host to complete the life cycle (ter Borg 1985), and *M. pratense* appears to be able to utilise a variety of hosts, but to "prefer" some over others. The Scrophulariaceae as a whole are noted as parasites of grasses and grain crops, and many flora state that *M. pratense* parasitises grasses. However, Lars Ericson (pers. comm.) believes that *M. pratense* is reliant on the roots of birch and pine in Sweden. Other authors believe that *M. pratense* parasitises woody plants, including *Betula, Pinus* and Ericaceae (Bitz 1970), *Picea abies* (Barsukova and Pyatkovskaya 1977), *Quercus robur,* and *Corylus avellana* (Smith 1963), and that the apparent association with grasses arises because seeds in dense grass are better protected from small mammal predators (Masselink 1980, cited in ter Borg 1985).

The other requirements of *M. pratense* appear to be coincident with the stable conditions of older woodland; it does not survive any great degree of trampling, it thrives in the transient stronger light conditions produced by fallen trees and branches, and it can form carpets in the first year or two of new glades formed by coppicing (cutting). In Sutton Park, dense patches of *M. pratense* are typically seen around trees or large branches that fell three or four years before. Subsequently, these patches apparently become overgrown by bracken and taller brambles, but individual plants of *M. pratense* persist in reduced numbers underneath this shade.

Seeds produced by selfing in *M. pratense* are often large and may lack dormancy, in contrast to the smaller products of cross-fertilisation (Lars Ericson, pers. comm.). Ericson also indicates that selfing is relatively frequent. *Melampyrum pratense* has a substantial, persistent seed bank (Masselink 1980, ter Borg 1985) and around 50% of seeds have the capacity for dormancy (L. Ericson, pers. comm). Seed dispersal is largely reliant on ants (the seeds have an attractive oil gland), but seeds may be carried longer distances by voles, whose seed caches may be important in the plant's colonisation of new glades formed by fresh tree-falls.

Melampyrum pratense is characteristic of ancient woodland with mixed age structure, mostly on acid soils. As a poor coloniser (Tasker 1990), *M. pratense*'s spatial dynamics are likely to be highly dependent on the spatial and temporal dynamics of its habitat, and its presence or absence might therefore be a useful indicator of the history of the woodland in which the species occurs. In this paper, we investigated the extents to which *M. pratense*'s spatial dynamics were determined by the nature of its habitat, and by its own colonisation abilities.

Within the UK, *M. pratense* occurs widely from north to south, having been recorded in over half of the UK's 10 km grid squares (Figure 1a). At a regional scale, the species is currently present in only a few locations in and around Birmingham (Figure 1b), all of which consist at least partly of ancient woodland with some oak (*Quercus* spp.) (Amphlett and Rea 1909, Lee 1867, Readett 1971a, Readett et al. 1971b, Edee 1972). It is probable that it was also present, but unrecorded, in a wider range of woodlands in the past. Within the Birmingham boundary, *M. pratense* was present only within Sutton Park, and the species was recorded in only two of the Park's ten woods (Figure 1c), and not in the open heathland which makes up the majority of the park. The species' distribution within woods is very patchy, restricted primarily to relatively open glades between dense, mature holly trees (Figure 1d). Within the glades where they are present, individual plants tend to be more evenly spread (Figure 1e)

Melampyrum pratense has a seed bank (L. Ericson, pers. comm.), but can be expected to go through a number of generations within the study period (3 years). Populations in patches were observed to be increasing in some areas, declining in others (Table 1) and empty patches could well have been colonised within the time-scale of the project. These fast spatial dynamics mean that *M. pratense* is a plant species to which metapopulation dynamics might be expected to apply on timescales relevant to conservation management (i.e., decades).

Patch ID (assigned when mapped)	Canopy type	Number (7/8/93)	Number (7/8/96)
15a	Clearing - birch / oak	634	489
7	Clearing - fallen oak	435	357
4	Old birch / oak	212	201
3	Bracken	46	37
12	Oak	207	197
8	Rowan / oak	415	429
10	Oak	346	385
14	Oak (tree fall in 1994)	125	286

Table 1. Numbers of individuals in each of 8 mapped natural patches within a single woodland, counted in early August 1993 and in early August 1996.

Hypotheses on dispersal and habitat suitability were tested by artificial distribution of seed into "empt" habitat of various types. This paper investigated the role of dispersal *versus* habitat in limiting the distribution at the two intermediate scales (Figures 1c and 1d); woods within Sutton Park, and glades and alternative habitats within one wood.

The Study Site

Sutton Park, (UK National Grid coordinates SP 100 970, or 410000, 297000), consists of a complex of habitat types: heath, bracken, acid grassland, older woodland of predominantly oak (*Quercus robur / petraea*), rowan, (*Sorbus aucuparia*) and holly (*Ilex aquifolium*), and younger stands of regenerating birch (*Betula spp.*), in previously open areas that were maintained by rabbits (pre-myxomatosis) and occasional fires. The park is around 859 ha, of which about 180 ha is mature, deciduous woodland (Figure 2). Pollen cores indicate that woods 1, 3, 4, 5, 6, and 7 (Figure 2) have been wooded areas since *at least* the 16th century (Barlow, 1988). Wood 8 is also ancient (Cobham Resource Consultants, 1991). Wood 2 is a deciduous plantation, 80 years old. Woods 9 and 10 date from around 1850, and are predominantly oak. Substantial plantings of oak were made at this time in the six older woods (Peterken 1970), which means that all the older woods now have a broadly similar species composition. A mix of oak, holly and rowan characterises all of the woods older than 200 years, with some mature birch also present. The areas of young birch have regenerating oak and rowan in the understorey, and could be expected to mature into high oak forest very similar to the older woods, although this process would take decades or centuries (Pike 1975).

Wood 7 (Figures 1, 2) is a diverse, patchy complex of mature oak and holly, bracken, heath, young and mature birch, and rowan. It contains *M. pratense*. This wood was the subject of intensive mapping (Figures 1c, 1d); experimental sowings were made within and near it in 1994. Woods 5 and 6 (treated as one unit), wood 9 and wood 1 were also used for experimental sowings, and are referred to as unoccupied woods 1-3 (UNOCC 1-3, Figure 2). Although these woods contain some areas that are too shady for *M. pratense*, they do also contain apparently suitable glades and banks with a more open canopy and a ground flora of grasses and bilberry (*Vaccinium myrtillus*).

METHODS

Mapping

At the outset of this work, the occurrence of *M. pratense* was mapped within a small area of woodland, using a N-S gridding system of tent pegs at 10m intervals. Unoccupied and occupied glades between the stands of dense holly were measured and mapped (see Figure 1d). To supplement this initial work, individual plants were mapped and counted within a single glade, using a fine grid of 50 cm (Figure 1e).

Experimental Sowings

Sowings of *M. pratense* were made in 1994 and 1995 (Table 3), to test whether the absence of the species from particular glades, woods and other habitat types was due to dispersal failure or to inadequate habitat. Host availiability was also an important consideration, but in the study area it appeared unlikely that *M. pratense* was restricted to a single host. Above ground it was associated with oak, birch, bilberry, the grasses *Deschampsia flexuosa* and *Agrostis tenuis*, and holly.

Melampyrum pratense was only found in two, ancient woods within Sutton Park (Figure 1b). Initial observations suggested that (1) other woods in the Park also contained potentially suitable habitats, and (2) other locations within the woods that did support *M. pratense* might also be suitable for the plant, even though at the time of survey it was absent from these locations. Table 2 shows the areas where these different sowings were made, as well as listing the ages, areas and tree-species composition of all woods in Sutton Park.

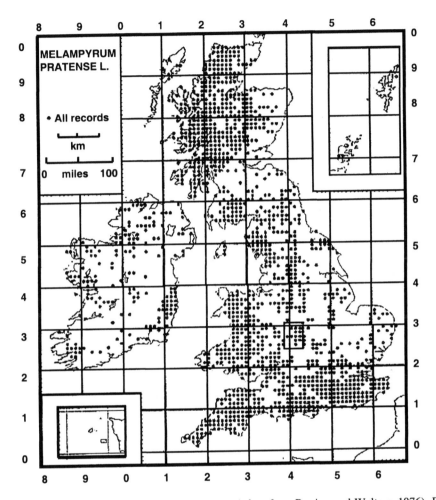

Figure 1. (a) UK distribution of *M. pratense* (taken from Perring and Walters, 1976). Black dots indicate 10 km grid squares in which the species had been recorded at this time. On each of these figures, a rectangle shows the approximate extent covered by the subsequent figure.

Two woods nearby contained areas which seemed ideally suited to *M. pratense*, but the species was absent. (This apparently suitable habitat type is referred to from here on as GLADE). Artificial sowings were made in these woods in 1994, to test whether the habitat was as suitable as it appeared. If the species succeeded in these plots, it implies that the reason for its absence was an inability to reach these other areas of woodland, 325 m (UNOCC1) and 1400 m (UNOCC2) from the nearest natural patches of *M. pratense*. In 1995, a further thirty-two sowings were made in a wood (UNOCC3), 1600 m from the closest occurrences of *M. pratense*.

b) Occurrences of *M. pratense* in and around Birmingham district boundary. Shaded sites are currently occupied. ⌀ symbol indicates a site where the species is now absent, but was recorded in Bagnall's flora of 1891. Sutton Park is the most northerly site shown.

c) Occupied and unoccupied woods within the boundary of Sutton Park. Narrow belts of suitable woodland which stretch between the separate woods are indicated on the map by dotted lines.

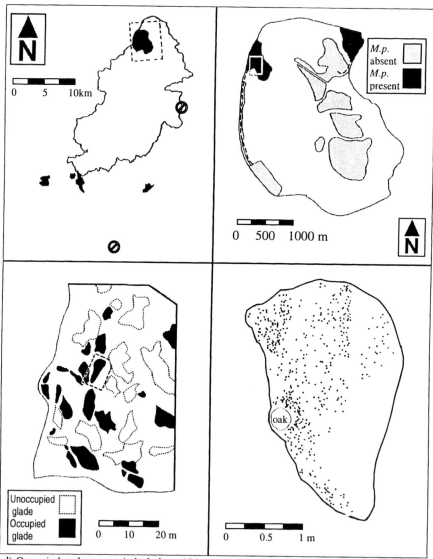

d) Occupied and unoccupied glades within a section of oak/holly woodland.

e) Distribution of individual plants of M. pratense within a single glade.

A number of sowings were made in unoccupied glades within wood 7, where *M. pratense* was already present in other glades. Again, the habitat appeared ideal, but *M. pratense* could have been absent from these patches either because they were actually unsuitable (e.g. because of competition, microhabitat or other factors), or because the distances were too far for seed dispersal and colonisation, (even though all these sowings were within 2.5 to 40 m from existing patches of *M. pratense*). For analysis, the sown plots were divided into two separate groups of 16 plots each; one group was less than 10 m from existing patches of *M. pratense* (SWCLOSE); the other more than 12 m away (SWFAR). In fact, the only successful plots in the SWFAR group were >= 17.5 m from existing populations, increasing the disparity between the two groups of plots.

Table 2. The main woods of Sutton Park (shown in Figure 2).

No.	Name	Area (ha)	Age (years)	Tree-species composition
1	Holly Hurst	36.2	140	Holly / planted pines / little Oak
2	Hill Hurst	4.8	80	Deciduous mixed plantation
3	Upper Nut Hurst	16.5	>250	Old hollies plus deciduous plantings from around 1850
4	Lower Nut Hurst	22.0	>250	Old hollies plus deciduous plantings from around 1850
5	Pool Hollies	22.1	250	Mixed deciduous plus planted conifers.
6	Darnel Hurst	14.7	250	Mixed deciduous - natural and planted.
7	Streetley Wood	14.6	>250	Oak / Holly, Oak / Rowan / Birch, Birch.
8	Gum Slade	24.0	>300	Oak / Holly - old wood pasture
9	Westwood Coppice	16.5	140	Oak / Holly
10	Wardens Belt	8.9	100	Mixed conifer / deciduous

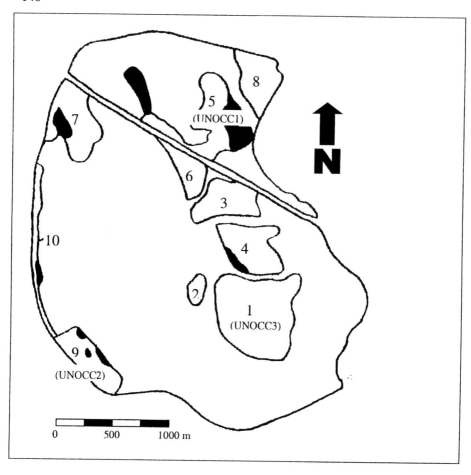

Figure 2. The main woods of Sutton Park. Shaded segments indicate areas of young birch. "UNOCC1", "UNOCC2" and "UNOCC3" are woods where the species is absent, and where introductions were carried out (see section 2.2).

Table 3. The plots of *M. pratense* planted out in 1994 and 1995.

Habitat type		Code	Number of plots	
			1994	1995
Glade habitat in woods where *M. pratense* is absent		UNOCC1	16	-
		UNOCC2	16	-
		UNOCC3	-	32
Unoccupied glade habitat in wood 7, where *M. pratense* is present		SWFAR	16	-
		SWCLOSE	16	-
Other habitats in or- adjacent to wood 7	Young birch	BIRCH	16	-
	- Bracken	BRACKEN	16	-
	- Dense holly	HOLLY	16	-
	- Open heath	HEATH	16	-

Sowings were also made in four other habitats in and around wood 7, as follows:
- BIRCH. Birch woodland contains the same two species of grass, *D. flexuosa* and *A. tenuis*, with which *M. pratense* is associated in the older woods, and birch trees themselves may also be hosts for *M. pratense* (see above). Few naturally-growing individuals of *M. pratense* are found in these areas, which may be a result of the young age of the woods (i.e., a lack of colonisation), unsuitable habitat in some way (e.g., no oak), or the more open canopy structure. The successional birch woods were between 30 and 60 years old. The experimental area was immediately adjacent to wood 7, where *M. pratense* is present (Figure 2). Birch woodland also occurs elsewhere in Sutton Park (Figure 2).
- HEATH. Open heath contains patches of grass and bilberry, which could possibly form suitable habitat for *M. pratense*, and this plant does sometimes grow in scrub and heath adjacent to wood 7. The heath patches to which the species were introduced were all over 18 metres from woodland edges, and they all contained some bilberry and fine grasses, as well as *C. vulgaris*.
- HOLLY. *Melampyrum pratense* is sometimes found in dense shade under holly, but usually at low density. Grasses and bilberry are usually absent. It was unknown whether this habitat could sustain persistent populations.
- BRACKEN. Sowings were also made in areas of dense bracken, to examine the ability of *M. pratense* to persist under its shade, since bracken tends to invade the woodland clearings where *M. pratense* initially thrives.

In experimental sowings, the propagules consisted of twenty fresh, ripe pods of *M. pratense*, wired together and fixed to a tent peg which was sunk at each site. Based on observed average seed count per pod, each plot had an estimated 64 seeds (mean seed number per pod = 3.2, standard deviation = 0.8, $n = 1168$ pods). 128 plots were planted out by this method in the autumn of 1994, and a further 32 in autumn 1995. The sowings were timed in order that the seed could germinate *in situ* in late autumn. These plots are listed in Table 3. The permission of English Nature and the site warden was obtained before this and other experimental work was carried out.

Monitoring Plots and Natural Patches

For each of the above major habitat types (i.e., heath, woodland glade, young birch woodland, holly shade and bracken), naturally-occurring densities of *M. pratense* were measured in twenty random quadrats of 1 m² each (Table 4).

Table 4. Occupancy of random quadrats (n=20, for each habitat type) by M. pratense, in and around woodland 7.

Habitat Type	Average density of *M. pratense* ($\pm$ s.d) in occupied quadrats (plants m^{-2}).	Proportion of quadrats occupied	Maximum distance of *M. pratense* from mixed deciduous woodland.
GLADE	185.3 $\pm$ 103.8	0.55	n/a
HOLLY	36.0 $\pm$ 15.6	0.10	4 metres
HEATH	78.3 $\pm$ 27.9	0.20	15 metres
BIRCH	82.0 $\pm$ 39.6	0.15	6.5 metres
BRACKEN	67.8 $\pm$ 29.7	0.30	8 metres

The progress of all sown plots was followed in 1995 and 1996, and for each, the following were recorded: (1) number of plants, (2) mean number of seed pods per plant, and total number per plot, (3) estimated number of seeds per original seed (i.e., level of replacement or recruitment), and (4) survivorship of seedlings to maturity.

Pod numbers were monitored by counting the number of ripe pods (with visible seed swellings) on each occasion, and ignoring dehisced or unripe pods. Based on field observations, it was assumed that pods which were ripe on one occasion would probably discharge their seed before the next twenty-day census, so that the total number of observed pods in each plot over one year would give a good estimate of the total pod number produced in that plot for that year. This assumption was checked by counting the number of empty seed pods on a visit, and comparing it to the number of pods last recorded for that plot, and the two were closely correlated.

In 1995 and 1996, the plots were surveyed at approximately twenty-day intervals starting from late May. The dates of survey are referred to as follows throughout the text; (1) "late May": 30th May - 1st June, (2) "mid June": 19th June - 21st June, (3) "early July": 10th July - 12th July, (4) "late July": 31st July - 2nd August, (5) "mid August": 19th August - 21st August and (6) "early September": 11th September - 15th September.

On each of these occasions, in 1996, data from forty random plants in natural populations were also recorded.

RESULTS

Mapping

Mapping across a number of glades (Figure 1d) demonstrated the presence of unoccupied glades with apparently suitable habitat within woods where the species was present. This phenomenon formed the basis of much of the experimental work here described. Within glades, distribution was more even (Figure 1e). The clustering and aggregation which does exist at this scale may be generated by localised seed production, balanced against intraspecific competition.

Comparison of Natural Occurrences in the Different Habitat Types

Melampyrum pratense was found naturally in holly, heath, young birch and bracken habitats, but at lower densities than in the mature mixed woodland (Table 4). The proportion of occupied quadrats shows a similar pattern. One-way ANOVA revealed significant differences between the densities of *M. pratense* in the different habitat types ($F = 8.88$, $p < 0.0001$, 4 and 95 d.f.). Plants were significantly more abundant in glade habitat than in heath ($F = 9.299$, $p = 0.004$, 1 and 38 d.f.), holly ($F = 13.127$, $p = 0.0008$, 1 and 38 d.f.), bracken ($F = 8.404$, $p = 0.0062$, 1 and 38 d.f.) and birch ($F = 10.26$, $p = 0.0027$, 1 and 38 d.f.). All remain significant at $p<0.05$ with Bonferroni correction. Chi-squared tests (with Yates' correction) on quadrat occupancy showed significantly more occurrences of *M. pratense* in glade quadrats than in birch ($\chi^2 = 11.36$, $p = 0.005$, 1 d.f.), holly ($\chi^2 = 14.59$, $p = 0.005$, 1 d.f.), bracken ($\chi^2 = 4.09$, $p = 0.05$, 1 d.f.) and heath ($\chi^2 = 8.53$, $p = 0.005$, 1 d.f.). All except glade versus bracken remain significant at $p<0.05$ with Bonferroni correction.

All the quadrats from holly, heath, birch and bracken habitats were grouped together, and their distance from oak/holly canopy was found to have a marked effect on their occupancy by *M. pratense* (Figure 3a) and on the mean density of *M. pratense* found across occupied and unoccupied quadrats (Figure 3b). When only those quadrats from all habitats, which had *M. pratense* present were considered, a linear regression of quadrat density against their distance from oak/holly woods showed a negative slope. However it was not significant. Nonetheless, the occurrence of any *M. pratense* in quadrats significantly declined with increasing distance (logistic regression, slope = -0.199, intercept = 0.3645, $p = 0.0043$, $n = 80$).

Furthermore, heathland quadrats where *M. pratense* was found were qualitatively different from the majority of the open heath in the area. These quadrats contained many young birch and oak trees, as well as tall bilberry, and could be described as scrubby or "successional" heath, as opposed to the low-growing mix of heather and grasses in most other areas of the park.

Regeneration from Dormant Seed

Pollination-bag experiments in the summer of 1993 indicated that *M. pratense* was successfully selfing. Because outcrossed seeds often have the capacity for dormancy (see above) it was necessary to investigate the species' seed bank.

In order to identify any dormant seed bank, all *M. pratense* plants were removed from an abundant natural patch before seed set in 1993, being pulled up by the roots. This procedure was repeated in 1994, 1995 and 1996. The implication is that any individuals found in 1994-6 were the products of dormant seed in the soil, rather than seed produced in 1993, although low levels of immigration are possible. In 1994, 232 plants were removed from the patch, and in 1995, 55 plants were found and removed. In 1996, a further 29 plants were removed.

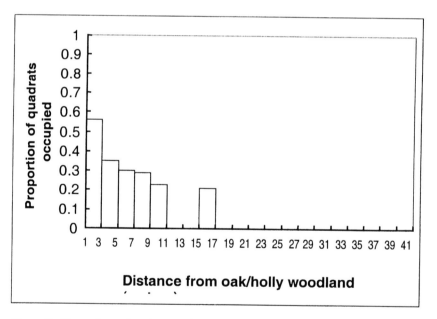

Figure 3a. Proportions of random quadrats occupied by M. pratense, at various distances from occupied woodland.

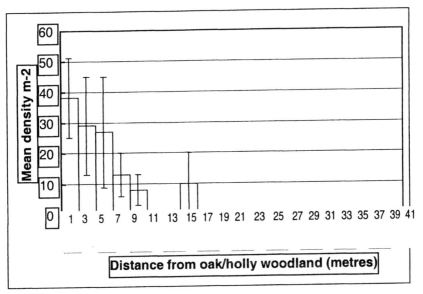

Figure 3b. Mean densities (± s.e.) of M. pratense in random quadrats (see Table 4).

These numbers indicate that some seeds of *M. pratense* remain dormant for three years or more. However, calculating dormancy proportions from these results would be complex, since, for example, the plants in 1994 arise from seed set in 1992, 1991, 1990 and possibly earlier, when local densities of *M. pratense* were unknown.

Experimental Sowings

The initial success of sown plots in early 1995 was good, but extremely hot and dry weather over the summer meant that even in well-established, shady plots, many plants died before seed set. However, there was survival of plants to maturity in a number of the artificial plots, and some seed was produced (Table 5). Occupancy of the planted plots was similar in 1996 to plot occupancy at the end of the 1995 season, and seedlings appeared (though they did not survive long enough to seed themselves) in two plots which had been unoccupied in 1995, demonstrating that a proportion of *M. pratense* seeds remain dormant at least until the second year.

There was variation in success within and between habitat types (Table 5). Habitat averages may be misleading because some of the sowings may have been in unsuitable microhabitats within generally suitable habitat types (*M. pratense* is locally distributed even within single woods; Figure 1d). So, it is instructive to consider the fate of individual experimental plots (Table

5). Of the surviving plots, some were highly productive, while others in the same habitat type might produce only a few pods in both years. The general decline in the drought year ($r<0$) was followed by a general increase ($r>0$) in woodland habitats, little change in bracken, and further declines in holly and heath.

Once unsuccessful plots had been selected out in the first, drought year, there was an overall tendency towards population increase in the unoccupied woods, the occupied wood and the young birch woods (i.e., positive values are seen in the "$r96/95$" column in 24/31 plots, or 77%).

Even considering the severe drought year, half (16/31) of the plots showed an increase from the beginning of the sowing (96/94 column). Only in bracken and holly habitat was a general population decline observed in this second year.

The experimental plots were visited again four years after sowing (July 27th 1998), to check for persisting populations (Table 5). At this date, three populations survived in wood UNOCC1, four in wood UNOCC2, and three in SWFAR, while all plots in SWCLOSE, bracken, birch, holly and heath habitat had disappeared. In addition, populations survived in four plots in wood UNOCC3 in 1998, three years after sowing. However, as can be seen from Table 5, surviving populations were mostly in locations where initial population increase had been observed ($r > 0$ for 96/94) while other sowings in the same woods had failed (generally where $r < 0$ for 96/94). It appears that the early population increase seen in young birch woods was insufficient to maintain any populations over the four year period. Numbers of individuals in surviving populations ranged from 1 to 34, and most populations contained healthy flowering plants at this date.

The implication of these results is that there are a number of glade sites within Sutton Park, within which *M. pratense* could survive and increase, and from which the species is currently absent. The artificial introductions allowed us to calculate r in the absence of immigration, with a known propagule size. Emigration from plots was also negligible; the areas around the sowings were searched for individuals of *M. pratense* which had colonised from these seed sources, but the maximum distance spread over both years was small (112 cm). Even after four years, the maximum distance spread was 181 cm (Table 5).

DISCUSSION

The spatial extent and resolution of analysis strongly influenced our interpretation of the relative importance of different factors in the distribution of *M. pratense*. At a national scale (Figure 1a), climate, land use and soil series are major determinants of the distribution. At a regional scale

(Figure 1b), woodland area and soil series were most important. At the scale of woods in one large park (Figure 1c), woodland history was the key determinant of the distribution. At the scale of glades within a single wood (Figure 1d), metapopulation dynamics of colonisation and extinction, and habitat quality became the most important influences. At the finest study scale of a few metres, rates of local birth and death combined with dispersal to generate source-sink dynamics. Thus, the resolution of analysis is crucial in population biology. GIS tools have a particularly valuable contribution to make to multi-scale analysis of species distributions, because of their capacity to handle and combine point and polygon records at a variety of scales, and to record the effects when areal units are modified.

Metapopulation Dynamics within Woods

Glades and other open habitats within oak/rowan/holly woodland appear to be ideal for *M. pratense*. Many of these habitats are temporary, becoming shaded by canopy, and dominated by bracken and brambles, some 5-15 years after the initial disturbance. These conditions are found in mature woodlands, where branch- and tree-falls are regular occurrences (as in wood 7 in this study). Given the maximum observed spread of 112 cm in two years, and 181 cm in four years, it is reasonable to regard each glade as containing a separate local population with respect to seed production (gene flow by pollen transfer may be another matter). Thus, each local population in a single glade has some likelihood of becoming extinct, as succession takes place. Persistence in such woods occurs through dispersal; in time, through a seed bank (which may be relatively short-lived), and in space, via seed dispersal. The occurrence of unoccupied glade habitats within wood 7 (Figure 1d), where *M. pratense* occurs naturally, combined with successful experimental seeding into these glades and very low rates of dispersal away from points of introduction, suggests that dispersal between patches is slow. Thus, any reduction in the rate at which new clearings were formed could threaten metapopulation persistence.

Table 5. Numbers of pods for 1995 and 1996, and r-values for 1994-1996 and 1995-1996, in each of the plots which survived to 1996. Between 1994 and 1995, all plots declined, giving negative r-values. 'Y' in the 1998 column denotes that this population survived at June 27th 1998, and the final column shows the spread by the time of the 1998 survey.

Patch	pods 95	pods 96	r 96/95	r 96/94	1998	98 spread (cm)
UNOCC1.1	15	133	2.18	1.89	Y	38
UNOCC1.3	29	33	0.13	0.50	Y	181
UNOCC1.4	2	20	2.30	0.00	Y	32

UNOCC1.6	1	5	1.61	-1.39		
UNOCC1.8	13	8	-0.49	-0.92		
UNOCC1.13	1	8	2.08	-0.92		
UNOCC1.15	1	12	2.48	-0.51		
UNOCC2.1	4	28	1.95	0.34		
UNOCC2.2	89	127	0.36	1.85		
UNOCC2.4	39	29	-0.30	0.37	Y	49
UNOCC2.6	2	18	2.20	-0.11		
UNOCC2.8	2	33	2.80	0.50	Y	30
UNOCC2.9	27	16	-0.52	-0.22	Y	50
UNOCC2.10	1	6	1.79	-1.20		
UNOCC2.14	5	1	-1.61	-3.00		
UNOCC2.15	26	30	0.14	0.41	Y	18
SWFAR.10	2	5	0.92	-1.39		
SWFAR.14	8	1	-2.08	-3.00		
SWFAR.17	7	1	-1.95	-3.00		
SWFAR.27	4	21	1.66	0.05	Y	55
SWFAR.28	23	47	0.71	0.85	Y	52
SWFAR.23	29	177	1.81	2.18	Y	74
SWCLOSE.6	9	37	1.41	0.62		
SWCLOSE.8	9	41	1.52	0.72		
SWCLOSE.19	5	11	0.79	-0.60		
BIRCH.2	1	5	1.61	-1.39		
BIRCH.4	3	17	1.73	-0.16		
BIRCH.5	7	57	2.10	1.05		
BIRCH.12	6	28	1.54	0.34		
BIRCH.15	4	30	2.01	0.41		
BIRCH.16	1	0	-3.00	-3.00		
BRACKEN.5	6	5	-0.18	-1.39		
BRACKEN.6	9	8	-0.12	-0.92		
BRACKEN.7	8	7	-0.13	-1.05		
BRACKEN.8	15	14	-0.07	-0.36		
BRACKEN.15	5	8	0.47	-0.92		
HOLLY.7	0	0	-3.00	-3.00		
HOLLY.10	0	0	-3.00	-3.00		

Source-Sink Dynamics

In and around wood 7, *Melampyrum pratense* occurred at highest density within glades, and could be successfully introduced to empty habitats. It occurred at lower frequency in heathland, holly and bracken, and sowings into these habitat types had low success (bracken) or failed (heath, holly); (i.e. $r <= 0$). In these habitats, frequency of natural occurrences declined with increasing distance from prime *M. pratense* habitat (Figure 3a), suggesting that immigration may be responsible for establishing low-density "sink" populations near to these population sources. Thus *M. pratense* occurs in relatively poor habitats close to population sources, but fails to colonise more suitable, but distant habitats, where successful introductions were made and where positive rates of population increase were seen.

Dispersal appears to be a major determinant of this species' distribution. We deduce that one wood with productive 'source' populations supplies seed to "sink" habitat, such as holly edges and heath, a few metres away, but seed dispersal is not always sufficient for good quality habitat to be colonised 17.5 to 40 m away. "Good" habitat above a certain distance from the natural populations is unoccupied, while small populations can be found in "poor" habitat such as holly shade and bracken, provided that it is close to existing seed sources. Therefore, source-sink dynamics appear to occur at roughly one tenth of the linear scale of the regular metapopulation dynamics (appearance and disappearance of glade populations). Thus, the limited dispersal ability of *M. pratense* is a very strong determinant of its current distribution within Sutton Park.

More Isolated Habitats

While much unsuitable heathland and grassland habitat lies between the different woods of Sutton Park, belts of suitable habitat do in fact exist between existing patches and the woods in which 1995 sowings were made (Figure 1c). It is possible to walk from natural patches of *M. pratense* to both Woods 9 and 5 without having to move out of the mature oak/holly/birch woodland in which the species survives, and there are a number of banks with suitable grasses and bilberry at small intervals along the way. However, these strips of habitat are narrow (usually between 1 and 13 m), and the distance to travel from existing patches of *M. pratense* is over 325 m to one wood, and 1400 m to the other.

Successful sowings (in the short term) of *M. pratense* into 30-60 year-old birch woodland, immediately adjacent to the oak / rowan / holly wood where substantial natural populations were present, emphasise the low rates of colonisation, as do the successful sowings into unpopulated glades in wood 7; spread at a median rate of about 0.5m in three years. Therefore, "normal" dispersal is very unlikely to result in the arrival of *M. pratense* seeds in the unoccupied woods of Sutton Park, 300 to 1400 m away, where suitable habitat patches do exist (Table 5). Of course, occasional seeds might be transferred further, in mud attached to large vertebrates (mainly humans, dogs, foxes and cattle), but the absence of *M. pratense* from nearby oak woods is testament to the rarity of such events. They may be very important on a glacial time scale, accounting for large changes in the species' distribution, but insignificant relative to the time scale of recent habitat fragmentation and land management.

Conservation of an Indicator Species

Melampyrum pratense is often used as an indicator of ancient woodlands on acid soils. Its spatial dynamics reveal why it is so restricted, and illustrate the importance of woodland dynamics to the persistence of this focal species. *Melampyrum pratense* requires intermediate levels of shading or open conditions in forested areas. Such conditions are characteristically temporary within the woodlands it inhabits. Furthermore, *M. pratense* is very poor at colonising new habitats across space, and probably does not have a very persistent seed bank (Lars Ericson, pers. comm.). Therefore, it survives under two sets of conditions within the study region. In natural, ancient woods, tree and branch falls are frequent, generating small areas of increased light on the forest floor, suitable for *M. pratense*. Although these habitats are temporary, clearings are close together, and the species can persist by colonising across space. The second set of conditions occurs in traditionally coppiced woodlands, where regular clearing on a 5-20 year cycle creates a shifting patchwork of openings, enabling the plant to colonise new habitats, through time, via its seed bank.

Since coppicing is now uneconomic, *M. pratense* has begun to decline in this class of woodlands. A case in point is Rough Hill Wood, near Redditch (30 km to the South of this study site). *Melampyrum pratense* was recorded in this woodland, and in others around it, in Bagnall's flora of 1891 (Figure 1b). However, the species is now absent, and this is almost certainly due to the cessation of coppicing over the past 40 or 50 years; 20-40 years after coppicing has ceased, the woodland floor is much darker than in an ancient wood with regular treefalls. The ground flora is now dominated by bracken and brambles, although some areas of bilberry and heather also survive. *Melampyrum pratense* appears to have become extinct during a period when open habitats were not available.

The dynamics of this species are such that its presence indicates moving mosaics of partially-open conditions, persisting over long periods on the forest floor of mature woodlands. Thus, the presence of *M. pratense* can be used to infer the presence not only of static parameters, such as suitable soil type, but also of the habitat patterns, over time and space, which indicate ancient woodland, usually regarded as being of high conservation value.

CONCLUSIONS

The spatial population dynamics of animals and plants are often characterised as belonging to one or another theoretical population type, such as metapopulations or source-sink populations (Hanski and Gilpin 1997). The patterns described here, and the results of experimental sowings, indicate that interpretation of population structure for *M. pratense* is scale dependent (Thomas and Kunin 1999). Over very small distances, (<1 m) within habitat patches, normal birth and death processes dominate. This was seen in the experimentally sown plots, which were isolated from other seed sources. Within established populations in the middles of glades, rates of immigration and emigration of seeds will be low, but probably with a slight excess of emigration (because more seed is produced within the glade). At a slightly larger scale (0-16 m), *Melampyrum pratense* is found at reduced densities and frequencies in sub-optimal habitats, which appear not to sustain persistent populations, and which are maintained by immigration from ideal glade habitats: source-sink dynamics take place at this scale. Within woods, populations in temporary light conditions in glades are likely to decline and eventually become extinct as the habitat becomes too shady. Thus persistence depends on the occasional colonisation of new empty glades, up to 40 m away. At this scale, metapopulation dynamics (colonisation and extinction) are predominant. Finally, suitable habitats exist 300 to 1600 m away, in areas where new habitats have been created, or where the plant may have been eliminated when conditions were temporarily unsuitable. At this scale, vegetation history is the key issue.

The actual spatial ranges over which which these processes dominate are species or system dependent, but their order of importance with increasing spatial scale is likely to be very general. Source-sink dynamics are likely to take place across relatively small distances, where the flow of individuals can maintain populations in fundamentally unsuitable habitats. Metapopulations may exist across larger areas, and then historical factors determining distributions will become influential at broader scales still. This means that population ecologists and conservation biologists must be very careful about scales of study when interpreting spatial patterns and implementing conservation management.

ACKNOWLEDGEMENTS

This work was supported by a Departmental Studentship to Lucy Bastin from the University of Birmingham. Thanks to Lars Ericson for valuable information about the biology of *M. pratense*, and to Jeff Bale, who co-supervised this work.

REFERENCES

Amphlett J, Rea C. *The Botany of Worcestershire*. Birmingham, UK: Cornish Bros (for Birmingham University), 1909.

Bagnall JE. *The flora of Warwickshire: the flowering plants, ferns, mosses and lichens*. Birmingham, UK: Cornish Bros, 1891.

Barlow SH. Sutton Park - A history of its landuse. Unpublished BSc. dissertation, Combined Studies, Worcester College of Higher Education.

Cobham Resource Consultants. A management plan for Sutton Park, (prepared for Birmingham City Council Department of Recreation and Community Services), 1990.

Cowley MJR, Thomas CD, Thomas J, Warren MS. Flight areas of British butterflies: assessing species status and decline. Proceedings of the Royal Society B, 1999; 266: 1587-1592.

Edees ES. *Flora of Staffordshire: flowering plants and ferns*. Newton Abbot, Devon: David & Charles, 1972.

Gilpin ME, Hanski IA. "Metapopulation dynamics: brief history and conceptual domain." In *Metapopulation dynamics: empirical and theoretical investigations*, Gilpin ME, Hanski, IA eds. London: Academic Press, 1991.

Hanski IA, Gilpin ME. *Metapopulation Biology: Ecology, Genetics and Evolution*. San Diego: Academic Press, 1997.

Ingrouille M. *Historical ecology of the British flora*. London: Chapman & Hall, 1995.

Kunin WE. Extrapolating species abundance across spatial scales. Science 1988; 281: 1513-1515.

Lees E. *The Botany of Worcestershire*. Worcestershire Naturalists' Club, 1867.

Nature Conservancy Council. Sutton Park - a woodland management review. Arlington, Virginia: The Nature Conservancy Council, 1980.

Peterken G. Sutton Park, Warwickshire: Recommendations for the future management of woodlands. Monks Wood, Cambs: Institute of Terrestrial Ecology, 1970.

Pike GV. Abandoned cultivated land of Sutton Park. Unpublished B.A. dissertation, Department of Geography, University of Birmingham, 1975.

Readett RC. *A flora of Sutton Park, Warwickshire*. Birmingham UK: Birmingham Natural History Society, 1971.

Readett RC, Cadbury DA, Hawkes JG. *A computer-mapped flora: A study of the county of Warwickshire*. Birmingham UK: Birmingham Natural History Society, 1971.

Smith AJE. Variation in Melampyrum pratense L. Watsonia 1963; 5: 336-367.

Streeter D, Garrard I. *The Wild Flowers of the British Isles*. London: Macmillan, 1983.

Tasker A. ed. The nature of Warwickshire. Buckingham: Barracuda Press, Buckingham, Report for the Warwickshire Nature Conservation Trust & Royal Society for Nature Conservation, 1990.

Thomas CD, Kunin WE. The spatial structure of populations. Journal of Animal Ecology 1999; 68: 647-657.

ter Borg SJ. "Population biology and habitat relations of some hemiparasitic Scrophulariaceae." In *The Population Structure of Vegetation*, White J. ed., Lancaster UK: Dr W. Junk Publishers, 1985.

Wiens JA. Spatial scaling in ecology. Functional Ecology 1989; 3: 385-397.

Wiens JA. "Metapopulation dynamics and landscape ecology." In *Metapopulation Biology: Ecology, Genetics and Evolution*, Hanski IA and Gilpin ME. eds. San Diego: Academic Press, 1997.

Chapter 10

SPATIAL ANALYSIS OF MICRO-ENVIRONMENTAL CHANGE AND FOREST COMPOSITION IN BELIZE

Peter A. Furley[1], Malcolm Penn[2], Neil M. Bird[3] and Malcolm R. Murray[1]
(with a contribution from Doug R. Lewis[4])
[1]*Department of Geography, The University of Edinburgh, Drummond Street, Edinburgh, EH8 9XP, UK. paf@geo.ed.ac.uk, mm@geo.ed.ac.uk*
[2]*Natural History Museum, Cromwell Road, London SW7 5WD, UK. mp@nhm.ac.uk*
[3]*Natural Resources Institute, Medway University Campus, Central Avenue, Chatham Marine, Kent, UK. ME4 4TB*
[4]*Scottish Agricultural College, West Mains Road, Edinburgh, EH9 3JG, UK. d.lewis@ed.sac.ac.uk*

Keywords:	forest composition, micro-environments, scale, geostatistics, soil properties, GIS, Belize.
Abstract	The results are presented for a pilot study of micro-environmental variation and plant distribution in a semi-deciduous rain forest at Las Cuevas in the Chiquibul Forest Reserve, Maya Mountains. The aim was to produce data at various micro-scales to understand the spatial distribution of commercially important tree species and to contribute to an understanding of forest dynamics. The project utilised the detailed tree surveys of the Belize Forest Department and projected the plant species, topographic survey data, soil physical and chemical properties and observations on ecological variation onto a common geo-referenced grid. The initial survey compared the distributions in a control (unlogged) plot with an adjacent selectively logged plot. The project considered environmental variation at a number of scales from 10 cm to 1 ha, to account for the dynamic processes which occur within the forest at each of these levels. The results show that the two plots are distinct, despite their initial selection on the assumption of a uniform vegetation class. The principal control appears to be topography, which was shown to be closely related to a number of physical and chemical soil properties. Significant differences occur between and within plots and these can be related to tree distribution.

INTRODUCTION

In recent years, the conservation of tropical forests has received worldwide publicity whereas effective forest management, particularly for timber extraction, has attracted little attention and gained some notoriety. The overall aim of the present paper was to examine how environmental micro-variation in the Chiquibul Forest Reserve of Belize can influence species distribution and thereby inform management strategy. The paper deals first with the background to forest management in Belize, then considers the methodology used in the present study and finally assesses the preliminary results.

The specific objectives are: (1) to assess the effects of changing scale on the variability of selected individual soil properties in forest plots within the same vegetation class; and (2) to examine the variation in soil properties and tree species distribution, and to integrate environmental and ecological data over a range of scales.

BACKGROUND

Whereas the global and regional distribution of tropical forests is broadly governed by climatic and altitudinal variation, individual forest tracts need to consider a range of other, locally important factors to explain species distribution and change. With very high species diversity, tropical forests present a major challenge in the attempt to unravel controlling factors in distribution and growth (Swaine et al. 1987). Research that attempts to explain diversity has looked at species distribution according to a range of factors, with a general recognition that soil fertility plays a significant if ill-defined role (Swaine 1996). These approaches have tended to ignore micro-environmental influences and most notably variations in soil properties although, recently, the possibility that meso-and micro-scale soil variation is important has been emphasised. For example, Clark et al. (1998) studied the effect of soil variation over short environmental gradients in upland tropical rain forest in Costa Rica. This meso-scale approach discovered a highly significant relationship between tree growth and soil type related mainly to parent material, slope angle and slope position. They concluded that relatively short edaphic gradients play an important role in the structure of tropical forest, a feature illustrated by earlier work (e.g. Furley and Newey 1979, Baillie and Ashton 1983, Johnston 1992). Nutrient-rich patches have also been shown to affect plant growth response through root proliferation and/or increased rates of nutrient ion uptake (Farley and Fitter 1999, Furley 1996, Ratter et al. 1978). Luizao *et al.* (1998) examined the effect of artificial gap size upon soil and litter processes in Brazil, but conclude that there was no consistent relationship at the scales considered (40 m^2 to 2500 m^2). Clark et al. (1995) also highlight the importance of past anthropogenic disturbances but comment that soils may be responsible for perpetuating such effects. Research on canopy and sub-canopy palms (Clark et al. 1995) reflected similar patterns and influences (see also Furley (1975) on human disturbance and impact on the cohune palm in Belize; Svenning (1999) on palm clustering in Amazonian Ecuador; or Whitmore (1982) on patterns and processes). The dynamic nature of tropical forests has also been explored (e.g., Newbery et al. 1998) and this brief overview highlights both the spatial and temporal character of tropical forests.

The present research focuses explicitly on plant-soil relationships at a range of scales from 10 cm to 1 ha plots. Micro-scale is defined in this work as all scales below 1 ha. Individual plant species differ in their tolerances, requirements, germination, and dispersal mechanisms. Local soil conditions may impact upon these factors both directly and indirectly. Differences in soil properties act to limit or encourage the pool of potential species able to inhabit any given site and thus contribute to forest diversity. The distribution of plants is the result of a range of processes that are likely to operate at different scales (Levin 1992) as well as over different periods of time. For example, reproduction and the short-term survival of seedlings may be dependent on very small-scale processes (for example, symbiotic or parasitic relationships with fungal mycelium, the transport of seeds by insects). As trees grow, larger scale processes become important, for example, where adventitious roots seek out more distant supplies of water and nutrients. Herbaceous plants exploit a different root volume from that of trees and this is also likely to change over the life cycle of plants at a given location. Local topography and position relative to the other plants affects light levels and wind damage. We argue that a range of multi-scale factors need to be considered if we are to understand the process which contribute to an individual tree's chance of long-term success.

Several authors have commented on the lack of knowledge concerning spatial and temporal variation, particularly at scales applicable to the root systems of individual plants. They point to the importance of micro-habitats in future research (*viz.* Svenning 1999, Farley and Fitter 1999). Such work often remains problematic because the spatial distribution of the roots is, to a large extent, unknown.

FOREST MANAGEMENT IN BELIZE

Timber harvesting in Belize has been largely uncontrolled and characterised by short-term timber licences (Mitchell 1997). Timber production was dependent on natural forests with negligible supplies from plantations and with little enrichment planting. Illicit logging has also been a problem, notably for mahogany, and there have been periodic disastrous fires and hurricanes (Friesner 1993). The Tropical Forest Action Plan (ODA 1989) recommended major changes in forest management. As a consequence, the Forest Management and Planning Project (FPMP) was initiated in 1992, with the remit of designing fresh forest management systems.

Very little information existed on the dynamic processes affecting tree growth and mortality in natural broadleaf forest. Therefore, the FPMP instituted 30 permanent sampling plots across a range of forest types throughout the country, thus establishing a comprehensive data set on tree diversity and forest structure. The present project (Penn and Furley 1998) builds on the plots set up in the Chiquibul Forest in the Maya Mountains (a location map is given in Figure 1 – on the CD-ROM). They are situated within a semi-deciduous forest according to the classic surveys of Wright et al. (1959) and cover sharply accentuated, karstic limestone relief (Bateson and Hall 1977). Although a productive forest reserve today with light selective logging and located in a relatively inaccessible part of the country, the area was once heavily utilised by the Maya. The general nature of their impact has become increasingly understood since the pioneering work of Lundell (1937), Furley (1998), and Brokaw (in press). Despite surveys in the 1970s (Johnson and Chaffey 1973, Johnson 1975), little quantitative research had been conducted prior to the FPMP plot experiments (Bird 1994). A framework was established for collecting, integrating and analysing the data over a hierarchy of scales using geostatistical, geographical information systems and remote sensing techniques.

METHODOLOGY AND SAMPLING

The approach adopted in this research is summarised diagrammatically in Figure 2 (on the CD-ROM).

Remote Sensing and Spatial Analysis

The research employed remote sensing and GIS techniques. Remote sensing imagery was used as an exploratory tool and enabled a visual representation of the Maya Mountain forests to be assembled with a digital terrain model of specific project areas (for example the Raspaculo catchment, Sutton and Penn, (pers. comm., Natural History Museum, London)). For the purposes of the present pilot study, the authors used ArcView 3.1 GIS and the S+ 2000 spatial statistical package. All data were geo-referenced and interpolated into spatial layers. This was intended to enable direct visual and statistical comparisons between environmental variables in a common format.

All data layers (including tree species, soil chemical and physical properties, and topographic data) were transformed into Universal Transverse Mercator (UTM: WGS84) co-ordinates and re-projected to the local North American Datum: NAD 27. This permitted direct spatial comparison with previously published vegetation maps and remotely sensed satellite images.

To help with the visualisation of the raw data sets it was necessary to transform and interpolate the data within the GIS. Topography and soil depth data have been interpolated using kriging techniques, following detailed statistical analysis. For other soil properties, a preliminary picture has been obtained using the inverse distance weighting interpolator (IDW). This interpolator is based on a moving window technique, which estimates non-sampled points by summarising sample points within the area of the window (Berry 1995). The average or summary is weighted so that the immediately surrounding values have a higher influence on the non-sampled data point than data further away. One drawback of this method compared with kriging, is that it is impossible for interpolated data points to have higher or lower values than the input data range (Burrough and Donnell 1998).

The first step was to compare plots. Basic statistical measures were used to assess the normality of the data and t-tests were undertaken to determine whether there was a significant difference between the samples sets.

The topographic data from the field survey were collected using a Leica "total station" instrument. These were integrated into the GIS to produce elevation and contour data for all the forest plots. The resulting digital elevation models are shown in Figure 3 (on the CD-ROM)

The Plot System

The Las Cuevas plots analysed in this paper form part of a network of four forest experimental sites established by the Belize Forest Department within the Chiquibul Forest Reserve. Each site consists of a pair of plots that extends over 18 ha, i.e. 300 x 600 m (Figure 4a – on the CD-ROM). Each pair is further subdivided into (a) two 9 ha (300 x 300 m) treatment areas: one selectively logged in 1995, the other left as a control; and (b) a central 1-hectare zone within each 9 ha area, acting as the permanent plots and the main focus for tree species sampling. In this central core zone, all trees above 10 cm dbh were identified, measured, tagged and spatially located. The surrounding 4 ha forms an internal buffer, where vegetation sampling is more limited with only trees above 40 cm dbh being measured and located.

The remaining outer 4 ha zone is unmeasured, acting as an external buffer zone. In addition, an extra 100 m wide zone has been created outside the unlogged sub-plot where timber extraction is not permitted. Figure 4a shows this nested arrangement for the Las Cuevas plots. The selective logging procedures and sampling protocols for this procedure have been detailed elsewhere (Bird 1994).

In addition, an ecological assessment of each 10 x10 m zone was undertaken using a derivative of Holdridge et al.'s (1971) classification. UTM co-ordinates were added to the database, locating the plots, sampling points and individual trees. In parallel with this work, two separate surveys of saplings (Brokaw pers. comm.) and seedlings (Lyle pers. comm.) were also underway. Together with the forest incremental program, scheduled to run for 40 years (the period of time between successive timber harvests), the data should provide an invaluable source for time series analysis. Finally, bulk density and soil moisture assessments were made at selected zones within the plots. Preliminary observations are made on soil water content and hydraulic properties.

Each central 1 ha plot was demarcated into 20 m squares in the original tree inventory. Subsequently, each 20 x 20 m zone was further subdivided into 10 x 10 m blocks. Within the central 60 x 60 m zone, thirty-six 10 x 10 m squares were marked out for detailed study. A further 20 x 5 m zone was selected for micro sampling at a 1 x 1 m scale. This left a 20 m perimeter zone within each permanent plot which was not sampled and acted as an internal buffer zone. This further site demarcation is shown in Figure 4b (on the CD-ROM).

Soil Sampling and Scale Variation

Soil properties were analysed at two depths (0-10 cm and 10-20 cm) in the centres of each 10x10 m plots, at the centre of each 1 m square in the "micro" plot, and from the soil profiles surrounding all the logged and control plots. At each site, any cover of litter was removed before sampling and stored for separate analysis. At each location, the following soil properties were determined: % moisture loss; % weight loss on ignition; pH (in H_2O); pH (in $CaCl_2$); % $CaCO_3$; exchangeable calcium, magnesium, potassium and sodium; total exchangeable bases (TEB); the cation exchange capacity (CEC); readily available phosphorus; % total nitrogen; % organic carbon and texture.

Soil depth (measured by percussion drilling to bedrock) was surveyed at the centre of each 10 m square within the 60 x 60 m core area, and the centre of each 1 m square in the "micro" plot. Further detailed soil depth measurements were taken in the pits formed from logging the soil profiles, here sampling every 10 cm. Full details of this work are given in Murray (1999). The soil pits were located outside the 1 ha perimeter to minimise physical disturbance to the standing vegetation in the plots. For the purposes of the depth study, it would have been better if these profiles were within the 60x60 m area, maintaining the integrity of the nested sampling strategy. The variogram analysis considered later shows that soil depth displays spatial continuity over a distance of 40 to 50 m. Thus the profiles are sufficiently near to the measurement points in the main plots to make combining these data valid. Soil depth was studied in greater detail in order to validate the sampling methodology.

The studies of the spatial variability of soil moisture required methods which are relatively non-destructive and rapidly implemented. The equipment consisted of:

➢ A Guelph permeameter – this measures saturated hydraulic conductivity at a variety of depths.
➢ A tension infiltrometer – this measures surface hydraulic conductivity at a variety of soil water tensions.
➢ A Delta-T thetaprobe – this measures volumetric water content.
➢ Bulk density – which was measured using standard steel cups.

Meteorological data is available from the Global Historic Climate Network. This consists of daily measures of past rainfall, air temperature, dew point temperature and wind speed. Currently, data are only available for four sites in Belize: Belize International Airport, Belize/Landivar, Half Moon Cay and Hunting Cay. These sites are marked on Figure 1. Some are significant distances from the study area (the furthest, Half Moon Cay is approximately 150 km away). At the Las Cuevas site today, a new regime of sub-daily measurements of air temperature, relative humidity, wind speed and rainfall has been implemented, although as yet, the record remains incomplete.

PRELIMINARY RESULTS

Geostatistical Analysis

In addition to simple comparative statistics, emphasis was placed on variogram modelling to give a more accurate idea of changing patterns with scale. Soil depth is used as an example, in view of its suspected importance in tree root strategies. Variogram modelling can be used to explain the degree of spatial autocorrelation amongst the data (Isaaks and Strivistava 1994). It allows a visual separation of the regional, local and random contributions of variation in the data (Matheron 1971). Geostatistical methods have been successfully applied to the estimation of soil properties (e.g.,. Webster and Oliver 1990, Burrough, 1995).

The process of creating a continuous surface of properties from the sample measurements will be considered using two contrasting examples: elevation and soil depth. The elevation data are scattered irregularly over the site, following features such as small ridges and hollows. It is reasonably likely that these measurements include the extremes – the highest and lowest points of the site. This is because elevation to a large extent can be measured directly. The resulting digital elevation model can be compared visually with the site, enabling any significant errors to be detected and corrected. In contrast, the soil depth measurements are from a series of nested regular grids. These use different sampling intervals. Because the soil cover is continuous across the site, the pattern of variation is hidden. It is likely that there are areas of soil cover both deeper and more shallow than those sampled despite the intensity of the sampling network.

Results from the Topographic Survey

Variograms were calculated following the procedure outlined in Isaaks and Strivistava (1994) and Pannatier (1996). Shown in Figure 5 (on the CD-ROM), they all reveal a generally linear relationship. This indicates that the samples nearby are most similar, with dissimilarity increasing uniformly with distance (at least up to 50 m away). Superimposed upon this pattern is a degree of directionality (anisotropy). Spatial continuity is greater along certain "preferred" directions. In the present case, this can be explained by the presence of a small hill at the eastern edge of the site. Incorporating this directionality in the model (by using several directional, rather than one omnidirectional variogram model) improves the accuracy of the estimates obtained (Cressie 1991).

Using linear variograms, a topographic surface was interpolated, using a search radius limited to 50 m. The results, presented as a digital elevation model are given in Figure 3 (on the CD-ROM). In the case of site elevation (topography), where the data are readily measured, the choice of interpolation algorithm is not usually critical. Estimates of topography produced using other approaches (cubic splining and binomial interpolation) produced very similar models.

Analysis of Soil Depth

Unlike the topographic survey, the soil depth sampling was carried out in three phases. Thirty six depth measurements were taken in the centre of the 10 m sample squares, to provide a general picture of depth variation across the logged and control plots. Secondly, 100 depth measurements were taken every metre in the "micro" plots, providing information about more local change. Finally, 268 samples 10 cm apart were taken from three exposed soil pits (located just off-site). These provide a picture of very small scale changes.

Variograms (shown in Figure 6a – on the CD-ROM) were plotted both for the complete dataset (374 measurements) and the individual phases. The first three models show slight differences in the pattern of spatial dependence. These differences are best combined by building a fourth variogram model which incorporates both the very fine scale variation from the soil pits, and the larger 1 m and 10 m variation. The interpolation is again performed using kriging. This time, the choice of interpolator is more critical. Kriging allows point estimates of soil depth to be produced which are larger or smaller than any actual measurement. This is a function of the incorporation of the variogram in the estimation process (Isaaks and Strivastava 1989). Other interpolation methods are likely to strongly underestimate the degree of variation in soil depth, producing a falsely smoothed model.

Draping this surface over a digital elevation model allows an exploration of the underlying processes. Figure 6b (on the CD-ROM) shows that there is a degree of correspondence between the depth of soil and the shape of the land; not surprisingly deeper soils seem more common on the lowest slopes. This is confirmed by a simple aspatial correlation. There is a weak (but statistically significant) negative correlation between soil depth and elevation. This is important in a geostatistical sense, as the existence of a measurable relationship between soil depth and topography means that using a co-kriging approach might improve future estimates of soil depth. Soil depth appears to be an important feature in understanding the patterns of tree growth and species clustering.

Future analyses will produce interpolated surfaces for the remaining chemical and physical soil parameters and compare the control with the logged plots at Las Cuevas and at the neighbouring San Pastor plots. The use of a spatial model of variation in the interpolations means that the gaps are filled in an intelligent manner, recognising that the pattern of change may be neither linear or uniform. As such, these surfaces provide the best estimates for use in the final ordination phase. Because some of the variograms produced differ according to the sample interval used, this shows that separate, scale-specific interpolations may be needed for later ordination analyses at different scales.

Spatial Variation of Soil Properties and Tree Species

Soil Patterns

At this preliminary stage, only the micro-scale chemical properties of the soil have been compared across plots and between depths. The soil data appear to be normally distributed, hence the more powerful parametric t-tests were used to analyse differences between the sample populations.

With reference to Table 1, it is clear that there is a significant variation of data between plots and within plots. The levels of total nitrogen and organic carbon vary across both the control and logged plots. More predictably, they also vary within each plot at different depths. Exchangeable potassium shows a similar set of t-values. Hence it is justified to suggest that at least statistically, the soil chemical variables shown vary considerably across what were originally considered to be quite uniform plots, both of which were allocated the same vegetation class by Wright et al. (1959).

Table 1. A summary of the differences between soil properties at the Las Cuevas logged plot and control plot, at different depths.

Variable	Control plot		Logged plot	
	0-10 cm	10-20cm	0-10 cm	10-20 cm
% $CaCO_3$	2.97		3.76	
Exch. Ca			3.50	5.42
Exch. Mg			8.50	7.44
Exch. K	3.10	4.60	10.20	12.96
Exch. Na	4.52		3.91	7.66
CEC			5.7	7.94
% Total N	3.90	5.14	7.91	12.29
% Organic C	2.90	5.10	8.85	9.64

Notes to Table 1.
Exchangeable cations and CEC values are expressed in $cmol_c$ kg-1.
All variables with differences greater than 2.1 show a difference in the class means considered significant at a 5% level.

Figure 7a (on the CD-ROM) shows the topography of the Las Cuevas logged and control plots, covered by drapes mapping the patterns of exchangeable sodium and potassium, measured at depths of 0-10 cm. There is visual overlap in data values for these exchangeable cations but also, and more significantly, the data ranges are different supporting the t-test results. As well as cross-plot variation, within-plot variation is also highlighted by the t-test results. Figure 7b (on the CD-ROM) illustrates the variation in organic carbon levels at two different depths for the control plot, confirming field observations that surface organic horizons are clearly distinct from sub-surface ones in these limestone soils.

Variations in the soil water content and hydraulic conductivity are illustrated for the control plot. Figure 8a (on the CD-ROM) shows the results of measurements taken on the 11th September 1999 using the Delta-T thetaprobe. Prior to this date there was a three day period of relatively little rainfall, which resulted in the relatively dry surface soils seen on the hill top. This spherical distribution is also evident in Figure 8b (on the CD-ROM). This shows measures of saturated hydraulic conductivity made using the Guelph permeameter.

Soil-water relationships have been measured which allows the modelling of water resources at the plot scale. To complete this process, realistic water fluxes between the land surface and the atmosphere need to be ascertained. A water SVAT (Soil Vegetation Atmosphere Transfer) scheme is used to represent these fluxes. This predicts the effects of changes in vegetation or climate upon the water resource. The physically based model SOIL has been designed to provide estimates of daily evapotranspiration and soil moisture deficit for forest soils, using averaged soil parameters and daily synoptic weather data as input. Figure 8c (on the CD-ROM) shows the model's results. Patterns in soil moisture can be ascertained throughout the year. Low values indicate periods of potential plant stress. The extent of plant stress can be related in particular to sapling mortality rates, and these have been measured by staff from the Natural History Museum in both plots. These results reinforce the trends identified in the soil chemistry data and highlight the powerful influence of slope on soil moisture characteristics.

Tree Species Distribution

The influence of a variety of active and passive factors affect species abundance and the spatial distribution of individual plant species. They are extremely difficulty to quantify. What can be inferred at this pilot stage, is the spatial variation of individual tree species across the plots. These distributions are presented in Figures 9 a and b (on the CD-ROM) for two species abundant in the semi-deciduous vegetation class of Wright *et al.* (1959). These are *Dialium guianense* (Ironwood) and *Sebastiana tuerckheimiana* (White Poisonwood). The clustering of tree species is well illustrated for the logged plot at Las Cuevas. (Figure 9c – on the CD-ROM).

The first significant difference of note is that the distributions and number of individuals differ significantly between the plots. *S. tuerckheimiana* is abundant in the logged plot but has few individuals in the control plot. The reverse distribution occurs for *D. guianense* which has higher abundance in the control plot than the logged plot. This relationship of similar species occurrence, but with different numbers of individuals appears to be the norm for most of the tree species present at Las Cuevas. It raises the question as to why a single vegetation class should show such a significant variation in tree composition. He et al. (1996) suggest that unidentified causes were responsible for small scale changes in tree species composition/numbers in what was perceived to be a uniform vegetation class in Malaysia. The initial pilot study here seems to point to the "unidentified causes" as being changes in the chemistry and soil moisture levels of the soil coupled with local topographic effects though, as Newbery et al. (1998) indicate, a number of other factors and dynamic processes may operate locally.

SUMMARY AND CONCLUSIONS

The paper details the methodology and preliminary results from a pilot study of micro-environmental change and species distribution in a semi-deciduous tropical forest overlying limestone soils. The geo-referencing of topographic data, soil properties and tree species identifications has permitted the first stage in an analysis of the causes of species distribution. These findings will eventually be applied to models of forest growth.

The results suggest a number of characteristics associated with the Las Cuevas plots:

- The 36 10 x 10 m subplots, which comprise the 60 m x 60 m sample area, provide an appropriate scale to pick up changes resulting from local changes in the topography (e.g. topslope vs. backslope) and at the level of individual trees (e.g. differences in rooting depth, nutrient and moisture-rich patches).
- Geostatistical analysis shows the existence of different soil patterns at different scales. From this it can be inferred that the dynamic processes affecting plant growth may also vary similarly.
- Topographic variation over each plot is shown to be influential in determining soil depth and a range of soil properties including those relating to soil moisture content and hydraulic conductivity.
- A physically based model (SOIL) has been designed to provide estimates of daily evapotranspiration and soil moisture deficit. The model identifies periods of potential plant stress.
- Significant differences in soil chemical characteristics occur both between plots, and within plots.
- Tree species distribution varies between the logged and unlogged plots, but the total pattern of species distribution and the pattern of individual species show distinct spatial identities.

During the design stage of this study, replicates were located to minimise within-plot variance, recognising the fact that a major constraint of field research in the tropics is the heterogeneity of large plots (van der Hout 1999). Despite this planning, soil sampling has shown that significant, large, within-plot variance still remains. This confirms the view that the inclusion of all soil sampling points within a single vegetation class may result in gross over-simplification. It strengthens the need for detailed studies of micro-variation, together with the analysis of the causes of spatial variation and plant growth.

ACKNOWLEDGEMENTS

The pilot study was carried out with a research grant from the Natural History Museum in London, supplemented by a grant from the Moray Fund of the University of Edinburgh. The assistance of staff from the Museum, notably Professor Steve Blackmore, Dr David Sutton and Dr Alex Munro has been much appreciated. We are equally grateful for the help and encouragement received from Forestry Department and the field staff at Las Cuevas (particularly Mr Nicodemus Bol (Chapal) and Mr Chris Minty). Discussions with Dr Nick Brokaw (Manomet Centre for Conservation Studies) have proved of great value.

REFERENCES

Baillie IC, Ashton PS. "Some soil aspects of the nutrient cycle in the mixed dipterocarp forest of Sarawak, East Malaysia." In *Tropical forest ecology and management*, SL Sutton, TC Whitmore, AC Chadwick eds. *Tropical forest ecology and management*. Oxford: Blackwell, 1983.
Bateson JH, Hall IHS. *The geology of the Maya Mountains, Belize*. Institute of Geological Sciences, Report No 30. London: HMSO, 1977.
Berry JK. *Spatial reasoning for effective GIS*. GIS World Inc.: Fort Collins CO, 1995.
Bird NM. *Experimental design and background information*. Silvicultural Research Paper No. 1: The Forest Planning and Management Project. Belmopan, Belize: Ministry of Natural Resources, 1994.
Bird NM. *Sustaining the Yield: Improved timber harvesting practises in Belize 1992-1998*. Chatham: Natural Resources Institute, 1998.
Brokaw, N. in press "A history of plant ecology in Belize." In *Ecological and environmental research in Belize, Volume 1*, PA Furley, J Iyo eds. Belize City.
Burrough PA. Spatial aspects of ecological data. In *Data analysis in community and landscape ecology*. RGH Jongman, CJF ter Braak, OFR van Tongeren eds. Cambridge: Cambridge University Press, 1995.
Burrough PA, McDonnell RA. *Principles of Geographical Information Systems*. Oxford: Oxford University Press, 1988
Clark DA, Clark DB, Sandoval MR, Castro CMV. Edaphic and human effects on landscape scale distributions of tropical rain forest palms. Ecology 1995; 76, 2581-2594.

Clark DA, Clark DB, Read JM. Edaphic variation and the mesoscale distribution of tree species in a neotropical rain forest. Journal of Ecology 1998; 86, 101-112.

Cressie NA. *Statistics for spatial data*. New York: Wiley Interscience, 1991.

van der Hout P. *Reduced impact logging in the tropical rain forest of Guyana. Ecological, economic and silvicultural consequences*. Georgetown, Guyana: Tropenbos Guyana, 1999

Farley RA. Fitter AH. Temporal and spatial variation in soil resources in a deciduous woodland. Journal of Ecology 1999; 87, 688-696.

Friesner J. *Hurricanes and the forests of Belize*. Occasional Series No. 1: The Forest Planning and Management Project. Belmopan, Belize: Ministry of Natural Resources, 1993.

Furley PA. The significance of the cohune palm, *Orbignya cohune (Mart.) Dahlgren*, on the nature and in the development of the soil profile. Biotropica 1975; 7, 32-36.

Furley PA. "History and destiny of Middle American forests: the inheritors of the Mayan landscape." In *Human activities and the tropical rain forest*. BK Maloney ed. Dordrecht, Netherlands: Kluwer, 1998.

Furley PA, Newey WW. Variations in plant communities with topography over tropical limestone soils. Journal of Biogeography 1979; 6 ,1-15

He F, Legendre P, LaFrankie JV. Spatial pattern of diversity in tropical rain forest in Malaysia. Journal of Biogeography 1996; 23, 57-74

Holdridge LR, Grenke WC, Hatheway WH, Liange T, Tosi JA. *Forest environments in tropical life zones*. Oxford: Pergamon Press, 1971.

Isaaks EH, Strivastava RM. *An introduction to applied geostatistics*. Oxford: Oxford University Press, 1989.

Johnson MS. Chiquibul Forest Reserve Working Plan (Draft). Internal Report. Belize City: Forest Department, 1975.

Johnston MH. Soil-vegetation relationships in a tabonoco forest community in the Luquillo Mountains of Puerto Rico. Journal of Tropical Ecology 1992; 8, 253-263.

Johnson MS, Chaffrey DR. 1973 *An inventory of the Chiquibul forest reserve, Belize*. Land Resource Study No 14. Tolworth UK: Land Resources Division, Overseas Development Administration, 1973.

Levin SA. The problem of pattern and scale in ecology. Ecology 1992; 73, 1943-1967.

Luizao FJ, Proctor, J, Thompson J, Luizao RCC, Marrs RH, Scott DA. Viana V. Rain forest on Maraca Island, Roraima, Brazil: soil and litter process response to artificial gaps Forest Ecology and Management 1998; 102, 291-303.

Lundell CL. *The vegetation of Peten*. Washington: Carnegie Institute of Washington, 1937.

Matheron G. *The theory of regionalized variables and its application*. Les Cahiers du Centre de Morphologie Mathematique de Fountainbleau. Paris: Ecole Nationale Superiere des Mines de Paris, 1971.

Mitchell A. *Review of the timber licenses of Belize*. Occasional Series No. 13: The Forest Planning and Management Project. Belmopan, Belize: Ministry of Natural Resources, 1997.

Murray MR. Kriging soil depth: the benefits of prior statistical investigation. Proceedings of the Geostats-UK 1999 Conference. PM Atkinson, AE Riding, NJ Tate eds. Leicester: University of Leicester, 1999.

Newbery DM, Brown NDL, Prins HHT. eds. *Dynamics of tropical communities*. Blackwell Scientific: Oxford, 1998.

Overseas Development Administration (ODA) *Belize Tropical Forestry Action Plan*. London: HMSO, 1989.

Pannatier Y. *VARIOWIN: Software for spatial analysis in 2D*. New York: Springer Verlag, 1996.

Ratter JA, Askew A, Montgomery RF, Gifford DR. Observations on forests of some mesotrophic soils in central Brazil. Revista Brasiliera de Botanica 1978; 1, 47-58.

Svenning JC. Microhabitat specialisation in a species-rich palm community in Amazonian Ecuador. Journal of Ecology 1999; 87, 55-65.

Swaine MD. Rainfall and soil fertility as factors limiting forest species distributions in Ghana. Journal of Tropical Ecology 1996; 84, 419-428.

Swaine MD, Lieberman D, Putz FE. The dynamics of tree populations in tropical forest: a review. Journal of Tropical Ecology 1987; 3, 359-366.

Webster,R, Oliver MA. *Statistical methods in soil and land resource survey.* Oxford: Oxford University Press, 1990.

Whitmore TC. "On pattern and process in forests." In *The plant community as a working mechanism,* EI Newman ed. Blackwell: Oxford, 1982.

Wright ACS, Romney DH, Arbuckle RH, Vial VE. *Land in British Honduras.* Colonial Research Publications No.24 London: HMSO, 1959.

Chapter 11

THE RADIATE CAPITULUM MORPH OF SENECIO VULGARIS L. WITHIN SUSSEX: THE USE OF GIS IN ESTABLISHING ORIGINS

Stephen Waite and Niall Burnside
Earth & Environmental Science Research Unit, University of Brighton, Brighton BN2 4GJ, UK
s.waite@brighton.ac.uk

Keywords: *Senecio vulgaris, Senecio vulgaris* var. *hibernicus*, hydridisation, introgression, species distribution.

Abstract: The geographical distribution of the radiate morph of Groundsel (*Senecio vulgaris*), *Senecio vulgaris* var. *hibernicus* was examined in the county of Sussex in southeast England using records from 1980 and 2000. In addition the geographical distribution of Oxford Ragwort (*Senecio squalidus*) was also investigated for the same area. The spatial distribution of *S. squalidus* was strongly correlated with the past and present railway network, as is that of *S. vulgaris* var. *hibernicus*.
The number of 2 x 2 km squares containing the radiate morph of *S. vulgaris* have declined since 1980, though those with *S. Squalidus* have remained more or less the same.
The spatial evidence presented supports the hypothesis that the radiate morph of *S. vulgaris* occurs because of hydridisation and introgression betweeb the native non-radiate form of *S. vulgaris* and *S. squalidus*.

INTRODUCTION

Senecio vulgaris L. (Groundsel) is a common native annual species of cultivated ground, disturbed and wasteland sites. It is abundant throughout the UK, reaching an altitude of 530 m in Scotland. *Senecio vulgaris* produces large numbers of small seeds, which possess a pappus and are readily wind dispersed (Clapham et al. 1987, Hodgson et al. 1995, Stace 1997). It is able to spread rapidly, colonising disturbed sites on a wide range of soil types. Harper (1977) describes *S. vulgaris* as a 'precociously reproducing annual' which may be found flowering during all months of the year in Britain. The flowers of *S. vulgaris* are polymorphic for capitulum type. Two principal morphs occur: the common non-radiate flower morph, where the capitula consist solely of hermaphroditic disc florets; and a rarer radiate morph that bears an additional outer ring of approximately 8 pistillate ray florets (Abbott 1986). Two forms of the rarer radiate morph have been described. One, *S. vulgaris* var. *denticulatus* (Mueller, Sell), is readily distinguishable from the normal form of *S. vulgaris*, being typically unbranched or weakly branched, having less deeply lobed leaves, producing fewer flowers and being densely pubescent (Trow 1912, Allen 1967). *Senecio vulgaris* var. *denticulatus* is confined to sea cliffs and sand-dunes, occurring locally in the Channel Isles and along the southwest and west coasts of Britain as far north as the Isle of Man (Allen 1967, Clapham et al. 1987). The second radiate form, *S. vulgaris* var. *hibernicus*, which apart from the production of rayed flowers is indistinguishable from the normal non-radiate form of *S. vulgaris*, occurs more widely. *S. vulgaris* var. *hibernicus* was first described in the British Isles in 1866 and has since spread extensively throughout Britain (Stace 1977). The application of GIS as an aid to establishing the possible origin and spread of this rayed morph of *S. vulgaris* throughout Sussex forms the subject of the present study.

S. vulgaris var. *hibernicus* occurs throughout southern and central England, Wales and northwards to Durham, occurring locally in Scotland, especially near Edinburgh (Stace 1977, Clapham et al. 1987). It has been present in the British Isles from at least 1866, when it was recorded in Cork, Eire (Syme 1875). The form may have occurred in Britain before this date. Crisp (1972) identified the same variant in the herbarium at Liverpool University. The specimen was recorded as having been collected from Oxford in 1832. Trow (1912) reports observing the form near Cardiff and Penarth around 1890, and states that it had recently spread widely in the counties of Glamorgan and Monmouth, and was also known to occur at sites near Swindon, Cork, Cambridge, Northwick in Cheshire, Freshfield in Lancashire and Bigbury in South Devon. By the 1930s the new morph had become considerably more widespread, reaching its current distribution by the 1950s (Stace 1977).

THE PATTERNS OF INHERITANCE OF THE RAYED FORM OF *S. VULGARIS*.

Using seeds and plants of the normal non-radiate morph of *S. vulgaris* and the radiate morph (*S. vulgaris* var. *hibernicus*) collected around Cardiff, Trow (1912, 1916) conducted a large number of garden and greenhouse breeding experiments. He was able to demonstrate that radiate florets are inherited as a single incompletely dominant gene (R). The rayed plants of *S.*

vulgaris var. *hibernicus* were homozygous for the gene *RR*, while the non-rayed form were homozygous for the non-radiate allele *r*. The flowers of the heterozygotes, *Rr*, were very variable, but always had much smaller and often fewer ray-florets than the homozygotes *RR*. Although uncommon, the heterozygote *Rr* form has been observed in both natural and experimental populations of *S. vulgaris* growing among plants of the other two forms (Trow 1912, Hull 1974, Oxford and Andrews 1977, Marshall and Abbott 1982, 1984). Trow (1912) recognised five distinct morphological varieties of *S. vulgaris*. By crossing these varieties with *S. vulgaris* var. *hibernicus* and breeding from the progeny, Trow was able to produce fully rayed forms *(RR)* of each variety. In these crosses the observed frequency of plants showing the heterozygous *(Rr)* and homozygous *(RR)* rayed phenotypes was consistently less than expected. Trow noted that there was a 'constant tendency for a slight excess' of the typical non-rayed form of *S. vulgaris*. Under normal circumstances both forms of the species are predominantly self-pollinating (Clapham et al. 1987, Warren 1988, Stace 1997). Trow estimated that among his populations, the frequency of out-crossing was approximately 1%. Hull (1974) also obtained similar levels of out-crossing for populations of the two forms from central Scotland and was able to confirm the pattern of Mendelian inheritance described by Trow.

POSSIBLE EVOLUTIONARY ORIGINS OF THE RADIATE MORPH

The simple Mendelian pattern of inheritance provides a mechanism accounting for the origin of the radiate morph. It has been suggested that the appearance of *S. vulgaris* var. *hibernicus* may result from a rare 'back mutation' from the normal non-radiate to the radiate form. Ancestral forms of the species would have been radiate. Among all species closely related to *S. vulgaris*, the radiate form is the common typical morph (Stace 1977). An alternative, and now generally accepted hypothesis (Abbot 1992), is that *S. vulgaris* var. *hibernicus* (2n=40) has arisen from the inter-hybridisation of the native *S. vulgaris* (2n=40) with the introduced *Senecio squalidus* L. (Oxford Ragwort, 2n=20). Subsequent back-crossing of the initial hybrid is thought to have resulted in the introgression of a radiate allele *(R)* into populations of *S. vulgaris* giving rise to the radiate form *S. vulgaris* var. *hibernicus* (2n=40). Regardless of the origin of the radiate allele, once established the generally low levels of out- and intra-morph crossing will help to maintain locally evolved forms of the species.

The evidence for both of these hypotheses has been summarised by Stace (1977). Stace points out that evidence for the hybridisation and introgression of *S. vulgaris* with *S. squalidus* was initially largely circumstantial, based on an apparent correlation between the spread of the introduced species and the subsequent appearance of the radiate morph of *S. vulgaris*. Although subsequent work summarised by Abbott (1992) strongly supports the introgression hypothesis, all of this work including studies of the genetic diversity of populations of *S. vulgaris*, *S. vulgaris* var. *hibernicus* and *S. squalidus* (Abbott *et al* 1992a, 1992b) has been conducted on populations from Central and Northern England, Wales and Central Scotland. Despite Ingram et al. (1980) stressing the need to establish the frequency of the radiate morph among the mixed populations *S. squalidus* and *S. vulgaris* in the south-east of England, these populations have not been studied.

It should be possible to distinguish between the two alternative hypotheses on the basis of the expected distribution and morphology of the radiate and non-radiate forms of *S. vulgaris*. If the first hypothesis is correct (that the radiate morph is the result of a rare back mutation from non-radiate to radiate) then the two morphs of *S. vulgaris* should be morphologically and ecologically very similar and the distribution of the radiate morph should be independent of *S. squalidus*. In contrast, if the second hypothesis is correct (that the rayed morph is the result of hybridisation and introgression between *S. vulgaris* and *S. squalidus*) then *S. vulgaris* var. *hibernicus* should have phenotypic and ecological characteristics in common with both *S. squalidus* and the non-radiate form of *S. vulgaris*. In addition, the appearance and distribution of *S. vulgaris* var. *hibernicus* should show a close spatial and temporal correlation with the distribution and spread of the non-native *S. squalidus*.

APPLICATION OF GEOGRAPHICAL INFORMATION SYSTEMS (GIS)

The study established a Geographical Information System (GIS) database of the County of Sussex. The GIS and cadastral database used secondary data sources and focused upon the county as a whole. The GIS database contained information on the distribution of *S. vulgaris* var. *hibernicus* and *S. squalidus* derived from the 1980 and 2000 Sussex Plant Surveys (Hall 1980, Harmes, pers. comm.). Plant species information was mapped on a 2 km^2 tetrad basis georeferenced to the National Grid. The analysis incorporated tetrad data from 1,022 individual tetrads in Sussex (Hall 1980). Environmental variables relating to the county boundaries and rail transportation network were also included from 1902 to present (Griffiths 1983). Using a cadastral system and overlaying plant species point/grid data and rail network line data the county was examined for correlations and similarities between all species distributions and environmental factors. The GIS package used was MapInfo 4.5.

The presence of associations between factors recorded in tetrads was tested for using the chi-square procedure. Where appropriate Yates correction factor was applied. When testing for the presence of an association Grieg-Smith (1983) has shown that chi-square procedures may be applied to non-random samples. Statistical analysis was performed using Minitab 12.1.

The Spread and Distribution of *Senecio squalidus*

The Oxford ragwort, *S. squalidus*, a native of southern Europe, is an introduced annual, occasionally perennial plant that has naturalised widely on wastelands, walls, waysides and disturbed ground. At some date before the 1770s *S. squalidus* had been introduced into the Oxford University Botanical Gardens, reputedly from material collected from the volcanic rocks and cinders of Mount Etna, Sicily, where it is native. In 1794, *S. squalidus* was recorded as established on several old walls in Oxford and was by 1810 widespread throughout the city on walls and waste ground (Druce 1886, Mabey 1996). By 1830, it was reported to be common about Oxford Railway Station and to be spreading along the line, colonising the granite-chippings, clinker and piles of cinders by the side of the railway line which provide conditions similar to those found in its native habitats (Kent 1964). In 1866, Druce noted it as being present 'at various places along the line', and having spread along the line to Littlemore and Reading. Around 1900, *S. squalidus* had reached London and was abundant there by 1940, becoming common elsewhere in the south east by 1950, approaching its present distribution in other parts of the country by the 1960s (Stace 1977). Typically found on stony, well-drained soils, the species is still extending its range and is now common in England, Wales and locally abundant in Scotland (Clapham et al. 1987, Stace 1997).

In Sussex, *S. squalidus* was first recorded on walls in Chichester in 1907 and later in 1931, on a rubbish tip in Hove (Wolley-Dod 1937). This last record is particularly important as it provides an approximate date from which both species, *S. vulgaris* and *S. squalidus* may have coexisted in Sussex, the native species *S. vulgaris* being ubiquitous. Based on data from the Sussex Plant Atlas (Hall 1980) the distribution of *S. squalidus*, circa 1980 is shown in Figure 1a (on the CD-ROM). The species is widespread throughout Sussex. No obvious spatial patterns are evident, although there does appear to be a concentration of sites towards the coast. If the current railway network system is overlaid (Figure 1b – on the CD-ROM), a clear association between the railway system and the distribution of *S. squalidus* becomes apparent. The association is statistically significant ($\chi^2 = 163.6$, $p<0.001$, df = 1), 58% of the tetrads containing *S. squalidus* are either traversed by railway lines or are immediately adjacent to tetrads that contain railway lines. The strength of the association is even stronger when the distribution of *S. squalidus* is compared to that of the railway system present in 1902 (Figure 1c – on the CD-ROM). Again the relationship is highly significant (χ^2 = 238.9, p<0.001, d.f. = 1), with 78% of all tetrads containing *S. squalidus* also containing railway lines or adjacent to tetrads traversed by railway lines. The railway system present in 1902 depicts the original routes of colonisation available to *S. squalidus* during the first half of the 20th C. as its range expanded (Griffiths 1983). The persistence of a relationship between the distribution of *S. squalidus* and past and present rail routes suggests that although *S. squalidus* has managed to spread widely throughout the county, it is still largely confined to sites and habitats close to those originally colonised along railway routes. Similar findings have been reported elsewhere in Britain. For example, Mabey (1996) reports that *S. squalidus* is abundant and the dominant ragwort present on waste ground within a quarter of a mile of the mainline railway from Abergavenny to Hertford. Further away from the line, *S. squalidus* becomes scarce and the common ragwort *S. jacobaea* is the dominant ragwort species present. The apparent coastal concentration of Sussex records for *S. squalidus* (Figure 1c – on the CD-ROM) probably reflects the geographical distribution of the major railway terminals such as Hastings, Eastbourne, Newhaven, Kemptown (no longer present), Brighton, Hove and the east-west coastal railway which provides an effective dispersal route linking the railways radiating from London (Figure 1b, c – on the CD-ROM).

The Spread and Distribution of the Radiate S. *vulgaris* var. *hibernicus*.

Following the first probable record of *S. vulgaris* var. *hibernicus* at Oxford in 1832 (Crisp 1972), Druce (1886) recorded the rayed form of *S. vulgaris* as occurring with the non-radiate form and *S. squalidus* on waste ground about Oxford. By the 1930s, it had become widespread and frequent, its range extending rapidly between 1950 and 1970. In many areas the spread of the radiate form of *S. vulgaris* is said to closely follow, in time and space, the spread of *S. squalidus* (Crisp and Jones 1970, Crisp 1972, Hull 1974, 1975, Richards 1975). However, Stace (1977) pointed out that despite *S. squalidus* being well established in London by 1940, only six records for the radiate form had been recorded by 1952, and although *S. squalidus* has been widespread throughout southern England since 1950s the radiate form of *S. vulgaris* remains rare. No radiate forms of *S. vulgaris* were recorded in the 1937 edition of the Flora of Sussex (Wolley-Dod 1937). By 1978 radiate forms were recorded from 23 tetrads in Sussex (Hall 1980). Between 1980 and 1998 the total number of tetrads containing records had declined to 7 (Harmes pers. comm.). The distribution of the radiate form in Sussex circa 1980 was centred on the coast around Newhaven, Brighton, Hove and Shoreham, all of which remain, or have in the recent past been major railway stations and junctions (Figure 2a – on the CD-ROM). When the distribution of the rayed form of *S. vulgaris* is compared with that of the 1902 railway network and the distribution of S. *squalidus*, all except two of the 23 tetrads containing the radiate form of *S. vulgaris* also contain or are adjacent to tetrads that contain sections of railways lines and *S. squalidus* (Figure 2b – on the CD-ROM). Within the lower central section of the map that encloses the area in which *S. vulgaris* var. *hibernicus* occurs (188 tetrads), both the association between the rayed morph and the presence of *S. squalidus* and the occurrence of railway lines are significant ($\chi^2 = 41.42$ and 161.09, respectively, p<0.001, d.f. = 1). Under the null hypothesis, if the distributions of the radiate form of *S. vulgaris*, *S. squalidus* and the occurrence of the railway are independent of each other, the expected number of tetrads where all three occur together is approximately two. The difference between this expected number, and the observed number of 22 tetrads suggests that the association between the three factors is very strong.

The distribution of the radiate form of *S. vulgaris* in Sussex has contracted considerably over the last twenty years (Figure 3a, b – on the CD-ROM). During this period, there is no evidence of a decrease in the distribution of *S. squalidus*. Unfortunately, data for the new edition of the Sussex County Plant Atlas has only been mapped at a scale of 10 km squares. Unless isolated single records occur it is not possible to resolve the location of the record below this scale. Thus, from Figure 3a it is not possible to comment on changes in the distribution and frequency of *S. vulgaris* var. *hibernicus* along the coast around Brighton and Hove. However, it is clear that with the exception of the record at Lewes, inland records north of Brighton and Hove present *circa* 1980 have all been lost. Two new inland records of *S. vulgaris* var. *hibernicus* have been made (Figure 3a – on the CD-ROM). One, on the approaches to Pulborough railway station, occurs in an area where *S. squalidus* is known to occur, and is clearly associated with the railway. The other, recorded on a sandstone cliff at Stedham, is neither associated with railway routes or with the presence of *S. squalidus* (Figure 3b – on the CD-ROM).

MORPHOLOGICAL AND ECOLOGICAL DIFFERENCES BETWEEN THE RADIATE AND NON-RADIATE S. VULGARIS.

Numerous studies have compared the various phenotypic and morphological characteristics of the radiate and non-radiate forms of *S. vulgaris*. While evidence for the greater phenotypic resemblance of the radiate *S. vulgaris* var. *hibernicus* to *S. squalidus* rather than to the normal non-radiate form of *S. vulgaris* is often equivocal (Stace 1977), numerous studies have shown that the two forms of *S. vulgaris* differ substantially from one another in many characteristics. For example, Abbot et al. (1988) confirmed earlier work by Richards (1975), who found that seed of the normal non-radiate morph germinate more rapidly and achieve higher levels of germination than seed from the radiate morph. Other workers have reported that the two forms differ in terms of their growth rates, the number of seed per capitula, number of capitula per plant, seed size, seed weight, degree of leaf dissection and, at the biochemical level, the allozymes of esterases and aspartate aminotransferase present (Richards 1975, Oxford and Andrews 1977, Kadereit and Briggs 1985, Abbott 1986, Marshall and Abbott 1987, Abbott et al. 1988, 1992a,b). The direction of many of these differences is not consistent between studies or populations (Oxford et al. 1995). However, consistent differences between the two forms have been found for seed germination, growth rates and the levels of outcrossing. Under a wide range of experimental conditions, germination and seedling growth rates for the normal non-radiate morph of *S. vulgaris* exceed those obtained for the radiate morph. The level of out-crossing differs between the two forms. The radiate form is more attractive to pollinators (syrphid flies), and generally cross-pollinates more frequently than the non-radiate morph. The level of cross-pollination for the radiate morph can approach 30%, while levels of around 10% out-crossing may be achieved by the non-radiate form (Campbell and Abbott 1976, Abbott and Irwin 1988, Warren 1988, Warren et al. 1988). Given these levels of out-crossing, the shorted-rayed heterozygote (Rr) might be expected to occur more frequently in the wild. Their apparent scarcity in natural mixed populations of *S. vulgaris* (Trow 1912, Hull 1974, Oxford and Andrews 1977) may indicate that natural levels of inter-morph out-crossing are lower than those obtained for experimental populations. Alternatively selection may reduce the abundance of the heterozygous form. Both mechanisms would favour the maintenance of the rayed polymorphism in populations.

Laboratory and greenhouse experiments conducted on radiate and non-radiate forms of *S. vulgaris* collected from the edge of a footpath in Brighton (TQ532106) confirm these results. The population sampled contained plants of the radiate and non-radiate morph growing alongside plants of intermediate character (presumed to represent the heterozygotes Rr). Measurements and seeds were only taken from fully radiate morphs and normal non-radiate plants. It is apparent that the two morphs differ substantially for a number of important morphological and ecological characters (Table 1). The differences between the radiate and non-radiate forms within the Brighton population conform to those previously reported. Higher seed viability and germination of the normal non-radiate seed was recorded (Abbott et al. 1988), along with greater growth rate and flower numbers within the non-radiate morph (Oxford et al. 1996).

Among selected members of the Asteraceae, Sheldon and Burrows (1973) showed that terminal velocity of seed falling in still air was closely related to the ratio of pappus to achene diameter. Their results show that *S. vulgaris* has a higher pappus diameter to achene diameter than *S. squalidus* and therefore has considerably greater dispersal potential. A crude relationship was also found to exist between seed terminal velocity and the ratio of achene to pappus weight. The increased ratio of seed (achene) weight to pappus weight for the radiate morph (Table 1) is likely to decrease dispersal ability of the radiate morph relative to that of the non-radiate morph.

DISCUSSION AND CONCLUSIONS

The results of the study support the hypothesis that in Sussex the radiate morph of *S. vulgaris* has arisen from the hybridisation and introgression of the native non-radiate form of *S. vulgaris* and the introduced species *S. squalidus*. The spatial and temporal distribution of *S. squalidus* and *S. vulgaris* var. *hibernicus* (Figures 1, 2, 3 – on the CD-ROM) are consistent with this hypothesis as are extent of phenotypic and ecological differences shown by the Brighton populations of the two forms of *S. vulgaris* (Table 1). Once established in Sussex the failure of *S. vulgaris* var. *hibernicus* to expand its distribution throughout the county has parallels with *S. squalidus*, and contrasts dramatically with the ubiquitous distribution of the normal non-radiate form of *S. vulgaris*. Unless the presence of rayed flowers increases the levels of out-crossing sufficiently to disrupt locally evolved gene combinations rapidly, it is difficult to accept that the morphological and ecological differences between the two forms can be explained by the presence of a single gene. Although the radiate form is more attractive to potential pollinators and relatively high rates of out-crossing have been demonstrated (e.g. Marshall and Abbott, 1984), among natural populations levels of inter-morph crosses appear to be very low. Oxford et al. (1996) concluded that gene exchange between the morphs may be very small, so that each acts as a semi-isolated gene pool. Oxford et al. (1996) suggest that the variation in characteristics observed between the two morphs may result from one of three causes: (1) that they result from the linkage between characters on the portion of introgressed chromosome from *S. squalidus* that carries the radiate gene, (2) that they are the result of founding effects which are maintained by the low levels of outcrossing between the two morphs, (3) that they are the result of selection and represent a co-adapted suite of characters. Oxford et al. (1996) found evidence to support the first two mechanisms, and while not able to completely reject the third, suggest that it is extremely unlikely to account for the diversity of traits that have been found to be associated with the radiate allele. Because *S. vulgaris* is an ephemeral weedy species and an early coloniser of open disturbed sites, it is likely that many populations will have been founded by very small numbers of propagules. This combined with the generally low levels of out-crossing shown by *S. vulgaris*, means that local populations are likely to differ considerably from each other. Because of this, the characters associated with the radiate morph are unlikely to be consistent, but will reflect the characteristics of the founding populations during hybridisation and introgression (Oxford et al. 1996).

The distribution of *S. vulgaris* var. *hibernicus* will depend on the frequency of successful formation events e.g., production of fertile hybrid seed, the dispersal ability of the rayed morph once formed and its fitness relative to the typical non-radiate morph. Although not impossible, the temporal and spatial distribution of the rayed morph in Sussex is inconsistent with its gradual spread and establishment along the major rail routes from London (Figure 2a, b – on the CD-ROM). The pattern of distribution is more easily explained if the morph has arisen, possibly several times, at a southern major railway station(s) and then spread slowly outward from the sites of origin. Larger railway stations, such as Brighton, support very extensive populations of *S. squalidus* and *S. vulgaris* associated with rail track ballast and marshalling areas. The population densities at such sites are considerably higher than those observed among populations scattered along the tracksides between major stations. It seems plausible that the initial hybridisation and the formation of the radiate morph may have been, and continues to be, more likely to occur at such sites where substantial populations of both *S. squalidus and S. vulgaris* are maintained. The concentration of *S. vulgaris* var. *hibernicus* records (*circa* 1980) along the coast either side of Brighton and northward along the principal railway corridor are consistent with the morph having arisen around the Brighton-Hove area, subsequently colonising suitable sites along the major rail routes. This speculation is supported by the early record in the county for *S. squalidus* at Hove in 1931 (Wolley-Dod 1937), suggesting that this is where the two species, *S. squalidus* and *S. vulgaris*, have coexisted for longest.

Table 1. Differences between the radiate and non-radiate plants of *Senecio vulgaris* (Groundsel) collected from Brighton, Sussex. Mean percentage germination values were obtained from four separate experiments, where the germination of seeds from the radiate and non-radiate morphs was tested under the same conditions on a range of soil types. In each case, the germination of the non-radiate form exceeded that of the radiate form. Viability testing was conducted on fresh seeds collected from the field. Other measurements were obtained from populations grown in an unheated polythene tunnel. All plants were grown during 1996, under the same conditions, for the same length of time (1-2 months).

Variables	Senecio vulgaris var. *hibernicus*	Senecio vulgaris var. *vulgaris*
Mean % Germination	52.4 ±0.092	84.4±0.079
% Seed Viability (Triphenyltetrazolium)	40±0.97	76±0.85
% Seed Viability (Indigocarmine)	56±1.02	76±0.85
No. of seeds/Flowerhead	36.6±1.17	44.95±1.15
No. Flowerheads	9.88±0.511	13.88±0.621
Flowerhead wgt (g)	0.00829±0.00047	0.00583±0.00060
Petal wgt (g)/flowerhead	0.00337±0.00039	0
Root wgt (g)	0.0707±0.0104	0.0892±0.0087
Leaf and Stem material wgt (g)	0.2297±0.0683	0.3413±0.0758
Seed wgt (g)/Flowerhead	0.0058±0.0683	0.0056±0.00122
Pappus wgt (g)/Flowerhead	0.00289±0.00061	0.00376±0.000963
Total Plant wgt (g)	0.502±0.0786	0.641±0.0850

The reasons for the reduction in the distribution of *S. vulgaris* var. *hibernicus* in recent years are unknown. Drawing an analogy with the appearance of a radiate morph of *Aster tripolium* L. (Sea Aster), another member of the Asteraceae which increased greatly in frequency along the east coast of Britain from the 1930s but subsequently declined during the late 1960s, Stace (1977) suggests that the abundance of *S. vulgaris* var. *hibernicus* might be expected to decrease as a balanced polymorphism becomes established. Further work on this is required. To date, most growth studies have simply formed part of breeding and crossing experiments designed to establish the possible mechanisms of origin. In order to understand the factors determining the abundance of the two morphs in the field, there is a need to establish the relative competitive ability of *S. vulgaris* var. *hibernicus*, the non-radiate morph of *S. vulgaris* and *S.*

squalidus under a range of ecologically relevant conditions. *Seneciovulgaris* is able to establish a large and persistent soil seed bank (Roberts 1974, Thompson et al. 1997). The role of the seed bank in maintaining the radiated morph is unknown. Where the radiate allele (R) has been lost from the above ground population it may still persist within the seed banks. Recruitment from the seed bank may subsequently re-establish the morph above ground. Vavrek et al. (1991) have shown that seed banks can preserve genetic diversity and re-introduce maladapted genes into the above ground population.

While the introgression and hybridisation hypothesis provides an acceptable explanation for the majority of Sussex records, it is difficult to see how it can account for the recent record of *S. vulgaris* var. *hibernicus* on a sandstone cliff at Stedham, which is neither associated with railway routes or the presence of *S. squalidus* (Figure 3b – on the CD-ROM). There seems no *a priori* reason to believe that only one mechanism explains all appearances of the rayed form. Although early workers such as Druce (1886) and Trow (1912) report that the radiate form often occurs in association with *S. squalidus*, Stace (1977) points out that neither suggested there was a close association between the two species and Trow (1912) stated that the radiate form of *S. vulgaris* "flourishes in localities where *S. squalidus* does not occur." Thus, it is possible that the recent appearance of the radiate morph inland at Stedham is the result of a rare back mutation from the normal non-radiate to radiate form.

ACKNOWLEDGEMENTS

Special thanks to Nikki Barker, Kings Garden Centre, Hassocks for the use of facilities and equipment. The help and assistance provided by Paul Harmes, County Recorder, Sussex Plant Atlas is also gratefully acknowledged.

Data sources: Plant point data - Sussex Plant Atlas (1980) & Paul Harmes (pers. comm.); Sussex County Boundary information - Manchester Information Datasets and Associated Services (MIDAS); Rail Network information – The Geographical Editorial Committee (University of Sussex).

REFERENCES

Abbott RJ. Life history variation associated with the polymorphism for capitulum type and outcrossing rate in *Senecio vulgaris* L. Heredity 1986; 56, 381-391.

Abbott RJ. Plant invasions, interspecific hybridisation and the evolution of new plant taxa. Trends in Ecology and Evolution 1992; 7, 401-405.

Abbott RJ, Horrill JC, Noble DG. Germination behaviour of the radiate and non-radiate morphs of groundsel *Senecio vulgaris* L. Heredity 1988; 60, 15-20.
Abbott RJ, Irwin JA. Pollinator movements and the polymorphism for outcrossing rate of the ray floret in Groundsel, *Senecio vulgaris* L. Heredity 1988; 60, 295-298.
Abbott RJ, Ashton PA, Forbes DG. Introgressive origin of the radiate groundsel, *Senecio vulgaris* L. var. *hibernicus* Syme: Aat-3 evidence. Heredity 1992a; 68 425-435.
Abbott RJ, Irwin JA, Ashton PA. Genetic diversity for esterases in the recently evolved stabilised introgressant *Senecio vulgaris* L. var. *hibernicus* Syme, and its parental taxa *S. vulgaris* L. var. *vulgaris* L. and *S. squalidus*. L. Heredity 1992b; 68, 547-556.
Allen DE. The taxonomy and nomenclature of the radiate variants of *Senecio vulgaris* L. Watsonia 1967; 6, 280-282.
Campbell JM, Abbott RJ. Variability of outcrossing frequency in *S. vulgaris*. Heredity. 1976; 36, 267-274.
Clapham AR,Tutin TG, Moore DM. *Flora of the British Isles*. 3rd edition. Cambridge: Cambridge University Press, 1987.
Crisp PC. Cytotaxonomic studies in the section Annui of *Senecio*. Ph.D. Thesis, University of London, 1972.
Crisp PC, Jones BMG *Senecio squalidus* L., *S. vulgaris* L. and *S. cambrensis* Rosser. Watsonia 1972; 8, 47-48.
Druce CG. *The Flora of Oxfordshire*. Oxford: Parker and Co., 1886.
Grieg-Smith P. *Quantitative Plant Ecology*. Oxford: Blackwell, 1983.
Griffiths ILl. "Road and Rail in Sussex ." In *Sussex: Environment, Landscape & Society*. The Geographical Editorial Committee eds. Cheltenham: Alan Sutton Publishing Limited & University of Sussex, 1983.
Hall PC. *Sussex Plant Atlas*. Brighton: Brighton Borough Council, Booth Museum of Natural History, 1980.
Harper JL. *Population Biology of Plants*. London: Academic Press, 1977.
Hodgson JG, Grime JP, Hunt, R, Thompson K *The Electronic Comparative Plant Ecology*. London: Chapman & Hall, 1995.
Hull P. Self fertilisation and the distribution of the radiate form of *Senecio vulgaris* L. in Central Scotland. Watsonia 1974: 10, 69-75
Hull P. Selection and hybridisation as possible causes of changes in the frequency of alleles controlling capitulum-type in *Senecio vulgaris* L. Watsonia 1975; 10, 395-402
Kadereit JW, Briggs D. Speed of developments of radiate and non-radiate plants of *Senecio vulgaris* L. from habitats subject to different degrees of weeding pressure. New Phytologist 1985: 99, 155-169.
Kent DH. *Senecio squalidus* L. in the British Isles – 4, Southern England (1940-). Proc. BSBI. 1964; 5, 210-213.
Mabey R *Flora Britannica*. London: Sinclair-Stevenson, 1996.
Marshall DF, Abbott RJ. Polymorphism for outcrossing frequency at the ray floret locus in *Senecio vulgaris* L. I. Evidence. Heredity 1982; 48, 227-235.
Marshall DF, Abbott RJ. Polymorphism for outcrossing frequency at the ray floret locus in *Senecio vulgaris* L. I I. Confirmation. Heredity 1984; 52, 331-336.
Marshall DF, Abbott RJ. Morph differences in seed output and the maintenance of the polymorphism for capitulum type and outcrossing rate in *Senecio vulgaris* L. Transactions of the Botanical Society of Edinburgh 1987; 45, 107-119.
Oxford GS, Andrews T. Variation in characters affecting fitness between radiate and non-radiate morphs of groundsel (*Senecio vulgaris* L.). Heredity 1977; 38, 367-371.
Oxford GS, Crawford TJ, Pernyes K. Why are capitulum morphs associated with other characters in natural populations of *Senecio vulgaris* (groundsel)? Heredity 1996; 76, 192-197.

Richards AJ. The inheritance and behaviour of the rayed gene complex in *Senecio vulgaris*. Heredity 1975; 34, 95-104.

Roberts HA. Emergence and longevity in cultivated soil of seeds of some annual weeds. Weed Research 1964; 4, 296-307.

Sheldon JC. Burrows FM. The dispersal effectiveness of the achenepappus units of selected Compositae in steady winds with convection. New Phytologist 1973; 72, 665-675.

Stace CA. The origin of radiate *Senecio vulgaris* L. Heredity 1977; 39, 383-388.

Stace CA. *New Flora of the British Isles*. 2nd edition, Cambridge: Cambridge University Press, 1997.

Syme JTB. *Senecio vulgaris* L. var. *hibernica mihi. Botl. Exch. Club, Rep. Curators, 1872-74*, 27-28, 1875: cited in Stace (1977).

Thompson K, Bakker JP, Bekker RM *The soil seed banks of North West Europe*. Cambridge: Cambridge University Press. 1997.

Trow AH. On the inheritance of certain characters in the common groundsel- *Senecio vulgaris*, Linn-and its segregates. Journal of Genetics 1912; 2, 239-276.

Trow AH. On the number of nodes and their distribution along the main axis in *Senecio vulgaris* and its segregates. Journal of Genetics 1916; 6, 1-63.

Warren JM. Outcrossing frequencies within- and between-capitulum morphs in groundsel *Senecio vulgaris* L. Heredity 1988; 61, 161-166.

Warren JM, Crawford TJ. Oxford GS. Inhibition of self pollen germination in *Senecio vulgaris* L. Heredity 1988; 60, 33-38.

Wolley-Dod AH. *Flora of Sussex*. Bristol: The Chatford House Press, 1937.

Vavrek MC, McCgraw JB, Bennington CC.Ecological genetic variation in seed banks. III. Phenotypic and genetic differences between young and old seed populations of *Carex bigelowii*. Journal of Ecology 1991; 79, 645-662.

Chapter 12

A GEOGRAPHICAL INFORMATION SCIENCE (GISC) APPROACH TO EXPLORING VARIATION IN THE BUSH CRICKET *EPHIPPIGER ePHIPPIGER*

David M. Kidd[1] and Michael G. Ritchie[2].
[1]*Department of Geography, Faculty of the Environment, University of Portsmouth, Portsmouth, UK. david.kidd@port.ac.uk.*
[2]*School of Biology, Division of Environmental and Evolutionary Biology, University of St. Andrews, St. Andrews, UK. mgr@st-and.ac.uk*

Keywords: Geographical Information Science, Principal Component Analysis, interpolation, barriers, Fst, intraspecific variation.

Abstract A GISc analysis of *E. ephippiger* presents an alternative methodology for the investigation of intraspecific variation. The analyses identified potential patterns to the population structure that were expected in the organism. The population groups detected by the analysis of interpolated trait surfaces produce clear predictions. Hierarchical Fst analysis with populations nested within the putative refugial groups was confirmed if most of the variation in genetic variation reflected these patterns. Mantel tests comparing genetic variation with geographic distance, morphological or behavioural variation, or the presumed refugial origins of forms provided an alternative approach.

INTRODUCTION

Commercial off-the-shelf geographical information systems (GIS) do not provide solutions to all geographical problems. Hence, Goodchild (1992) suggested the use of Geographical Information Science (GISc) as a wider definition of geographical information technology which employs stand-alone or coupled specialist software, e.g., spatial databases, statistics packages, scientific data visualisation, remote sensing, modeling, web servers and data browsers, as well as GIS.

The potential of GIS for the analysis of intraspecific geographic variation, and its potential use in distinguishing the effects of selection, drift and population history was recognised by Young (1995). However, there have been few published GIS applications in this area. These few demonstrate a variety of techniques that aid in unravelling complex trait patterns and evolutionary processes (Cavalli-Sforza 1994, Cesaroni et al. 1997, Kidd and Ritchie 2000). This chapter describes our GISc approach that uses a combination of GIS, statistics packages, and custom spatial techniques to investigate spatial trait variation in *E. ephippiger* (Saddle-backed Bushcricket). Topics discussed include trait and environmental data capture, surface interpolation, covariate and multivariate pattern analysis, and significance testing of evolutionary hypothesis. In addition, future system enhancements and research avenues are discussed. System enhancements include the addition of data at the individual as well as population level, data on other species exhibiting variation in the same geographical area, and palaeo-environmental data to further examine evolutionary and migratory hypotheses. Future research will investigate linking the GISc results with more established geographical population genetics analysis techniques such as hierarchical Fst analysis and Mantel tests.

BACKGROUND

E. ephippiger is a large, flightless tettigoniid which has a fairly widespread but patchy distribution in Europe, occurring from the Low Countries, France and Northern Spain eastwards across Central Europe and Northern Italy to the Balkans, although in recent decades it has disappeared from the northern part of this range. The species is highly geographically variable for a number of traits including morphology (Oudman et al. 1990, Hartley and Bugren 1986, Hartley and Warne 1984, Grandcolas 1987), male calling song and mating preferences (Duijm 1990, Ritchie 1996), allozymes (Oudman et al. 1989, 1990) and various DNA markers (Ritchie et al. 1997, Hockham 1999). Variation is especially high in Southeast France between the Pyrenees and the Alps. This variability has contributed to the variety of taxonomic classifications and evolutionary interpretations of the history of the species which have been proposed.

Oudman et al. (1990), using traditional cartographic and non-spatial statistical techniques, observed that patterns were generally non-coincident and inter-population genetic distances (based on allozyme variation) low. From this information they concluded that local forms were mostly of primary origin (i.e. a response to geographically variable contemporary selection or drift). More recently, Kidd and Ritchie (2000) captured trait data from a number of published and unpublished sources within a GIS. Subsequent analysis looked for univariate and multivariate trait clines and related them to present topography and climate. In contrast to Oudman et al.

(1990), they suggested that patterns in individual and groups of traits are of both primary and secondary origin. They suggested variation in body size was mostly the consequence of primary selection due to modern climate, whereas a multi-trait cline extending from the Mediterranean coast near Narbonne and Béziers to Andorra in the Pyrenees was more likely to be of secondary origin, reflecting expansion from glacial refugia (Hewitt 1996). In addition, they identified other patterns along the coast, and suggested a number of potential explanations for these.

ORGANISM DATA

An essential function of GIS is the storage of georeferenced data. Species are hierarchical entities that are composed of individuals grouped into local populations that in turn can be grouped into linked metapopulations. The ideal situation would be to have accurate spatial and attribute data for individuals. A more common situation, especially when collating data from published documents, is to have attribute data on individuals assigned to a nominal population or simply to have population statistics alone.

The *Ephippiger* trait data originates from a number of published and unpublished sources (Oudman et al. 1990, Duijm 1990, Ritchie et al. 1996, Ritchie unpublished). The majority of the data is population statistics; however, individual morphological measurements are available for some populations. As no information is available on the spatial extent of populations, they are georeferenced as point entities in the GIS with latitude-longitude coordinates. Trait data is stored in an MS Access database, which also stores metadata on data origin and the data capture process (Figure 1).

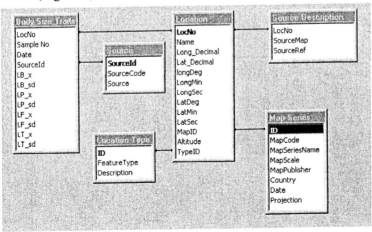

Figure 1. Microsoft Access database for E. ephippiger (only one trait table shown).

Traits sampled at a common set of locations are stored in a single trait class table (Figure 1 shows only the table for the body size class, LB – length of body, LP – length of pronotum, LF – length of hindfemur, LT – length of foretibia). Additional trait class tables include body weight, genitalic traits such as cercal and epiproct dimensions, and titillator size and form, mating song, allozyme frequency, and RFLPs. The "source," "source description," "location type," and "map series" tables store the associated metadata. The database now contains trait data for 125 locations (Figure 2).

Figure 2. Distribution of sample locations.

ENVIRONMENTAL DATA

Hypotheses suggested for the origin of variation patterns in *E. ephippger* include primary contemporary selection or drift (Oudman et al. 1990, Kidd and Ritchie 2000) and secondary contact of differentiated populations (Kidd and Ritchie 2000). In addition, Kidd and Ritchie suggest three possible historical scenarios that could apply to the ambiguous coastal populations. Study of primary hypotheses was limited to contemporary or very recent populations, i.e., those in the majority of France, Northern Iberia and the Low Countries (Figure 2). In contrast, investigation of secondary contact hypotheses requires consideration of all areas the organism has occupied since the last glacial maximum. The area of interest for historical (palaeo-)

data is thus determined by the location of all potential refugia. For *E. ephippiger* there was almost certainly one or more refugia in southern Iberia. In contrast, the locations of eastern refugia are very speculative and thus the area of interest should perhaps extend as far east as the Black Sea (Cooper et al. 1995). The different requirements for modern and palaeo-data mean modern data is required at a regional scale, whereas the palaeo-data is required at the continental scale. The differences in scale, coverage and data quality will have implications if modern and palaeo-data are compared.

A variety of contemporary regional and continental scale environmental data are now available in digital format from organisations such as the United Nations Environment Program (UNEP), United States Geological Survey (USGS), National Oceanic and Atmospheric Administration (NOAA), and European Union (EU). A range of modern environmental data sets from these and other sources have been, or are being, collated (Table 1).

GEOGRAPHICAL ANALYSIS

With the trait and environmental data sets integrated within the database, we can begin to interrelate them via the GISc techniques of visualisation, statistical analysis and modeling to identify patterns and examine evolutionary hypotheses. Here we discuss a number of techniques that would have been impossible to implement outside of the GISc framework. Techniques discussed include the interpolation of trait surfaces with real barriers, the use of surface scatter plots to identify local covariation relationships, and principal component analysis to identify multivariate trait patterns.

Trait Interpolation with Real Barriers

The usual situation in self-contained studies is to have one or more traits sampled at a number of locations enabling direct statistical comparison between sites (e.g. Thorpe and Baez 1987, Oudman et al. 1990, Brown and Thorpe 1991a,b). However, with the collated *E. ephippiger* data the exclusion of sites without all traits sampled would reduce the number of locations available for analysis drastically. Hence, to use all available data we must use surface interpolation to estimate trait values at unsampled locations.

Interpolation is "half-art-half-science" in that the operator decides which surface generation algorithm to use and sets algorithm parameters on the basis of their understanding of the processes involved in creating the real-world patterns and the mathematics of the chosen algorithm. In addition, there may be a need to preprocess data into a suitable format before interpolation, for example, where there are multiple samples at a location but the algorithm only accepts a single value per location. The *Ephippiger* data has multiple samples of the same trait at some locations, varying sample sizes for all traits at all locations, and known barriers to dispersal. Kidd and Ritchie (2000) generated trait surfaces with an inverse-distance weight (IDW) function; a location average weighted by sample size was calculated when there were multiple samples. This method does not take account of variation in sample size between locations or the known barriers to dispersal.

Variation in sample size individual averages towards the average of population averages; the magnitude of the shrinking factor depending on sample size (Equation 1). Sets of access queries were written to calculate *JSz* for the morphological traits.

Barriers are clearly of great potential importance in shaping patterns of geographic variation in organisms, yet most analyses (of phenotypes or genotypes) incorp'orate them poorly, if at all. Both modeling and empirical studies suggest barriers can create patterns by reducing gene flow across them, or act as 'traps' where differences accumulated during allopatric divergence come to lie (Barton 1979, Nichols 1988). Barriers can be defined from obvious geographic features such as mountain ranges or major rivers, or simply from unoccupied patches (where accurate distribution data are available). The Mediterranean Sea and mountainous areas above ~2050 m are barriers to the dispersal of *Ephippiger*.

The James-Stein estimator: 1)

(a) The estimator

$$JSz = \bar{y} + c(y - \bar{y})$$

Where $\bar{y}$ is the average of population averages,
c is the 'shrinking factor', and
y is the average of a single population.
(b) Individual population shrinking factor.

$$c = 1 - \frac{(k-3)\sigma^2}{\Sigma(y-\bar{y})^2}$$

Where, k is the number of populations,
σ^2 is the square of the standard deviation of the sample, and
$\Sigma (y - \bar{y})^2$ is the sum of the squared deviations of the individual averages from the grand average $\bar{y}$.

Table 1. Modern environmental data sets.

Data Set	Source	Scale/ resolution	Source Coverage
Base mapping (rivers, coasts, populated places, etc)	Environmental Systems Research Institute (1993)	1:1,000,000	Global
Digital Terrain Model	USGS (1996)	30 arc second	Global
Digital Terrain Model	NOAA, National Geophysical Data Center (1988)	5 arc minute	Global
Natural vegetation	Council of Europe and Commission of the European Communities (1987)	1:3,000,000	EC countries
Soil	Commission of the European Communities (1985)	1:1,000,000	EC countries and the Adriatic coast only of the former Yugoslavia and Albania
Solar irradiation April – August	Commission of the European Communities (1979)	~1:2,500,000	EC countries[1]
Annual precipitation	Ministère des Transportes (1980)	~1:5,000,000	France[1]
30yr monthly temperature statistics	NOAA (1995)	-	Global[2]
30yr monthly precipitation statistics	NOAA (1995)	-	Global[2]
River attributes	Not Captured	-	-

Notes. Coverage limited to 0°-7°E, 42°-45°N; 2. Local climate models under development.

A survey of over twenty interpolation algorithms revealed that only one, ARC Triangular Irregular Network (TIN, ESRI 1997), supports the inclusion of true barriers. The TIN surface model is the preferred technique to create elevation surfaces from topographic surveys (see Lane et al. 1998), although other techniques have been investigated (e.g., Hutchinson 1996a). In contrast, biologists, soil scientists, sociologists and climatologists have tended to use other techniques, including moving averages (IDW - Kidd and Ritchie 2000, a custom method based on Shepard's algorithm - Cavalli-Sforza et al. 1994), splines (Hutchinson 1996b, Kesteven and Hutchinson 1996), and kriging (Maurer 1994, Cessaroni et al. 1997). From herein we collectively refer to these non-TIN models as "field interpolators."

The rational for using the TIN model for topography is that the model maintains the geometry of the Earth's surface, which the field models do not (Dixon et al. 1998). In contrast, the phenomena under investigation by the other disciplines are seen as statistical rather than geometric entities. As an example, the TIN model uses only the values of the three triangle nodes (or two if the location to be interpolated lies on a triangle vertex) determined by the TIN algorithm (usually the Delauney algorithm) to interpolate values at unknown locations within the TIN facet. Other points that are in fact closer to the unknown location are ignored (Figure 3). For field-like phenomena, such as organism variation patterns, the limited use of local points by the TIN model is an arbitrary product of sample distribution and the triangulation algorithm, rather than being representative of true surface pattern.

A "hybrid" interpolation technique NetSURF, was developed which can incorporate both absolute and partial barriers within a distance-based (i.e. field-type) interpolation procedure. NetSURF is implemented via a combination of ARC/INFO commands and a custom IDL program (Research Systems Inc. 1999).

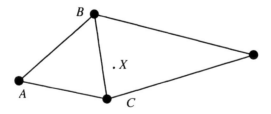

Figure 3. Limited use of node points in TIN interpolation. The interpolation of surface value at X depends on the values at B, C and D despite A being closer to X then D.

The NetSURF technique employs a user-designed network to calculate values at network nodes on the basis of network distance to the nearest x sampled locations. The *E. ephippiger* network was constructed from three data sets: the sample locations, mountain and sea barriers, and a set of 'space-filling' point which function both to increase network connectivity (making distance calculations more accurate) and subsequently as mass points when generating the final TIN surface. A TIN was generated from these data sets, which was subsequently converted to a line network (Figure 4). Subsequently network lines crossing the mountain or the sea barriers were deleted as these features are absolute barriers to *Ephippiger*; if they had been partial barriers the network lines would have been assigned a resistance cost relative to non-barrier lines instead of being deleted (Figure 5).

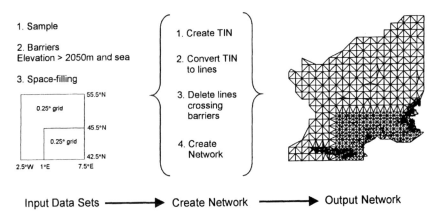

Figure 4. User-designed network for real-barrier interpolation of Ephippiger traits.

The ARC network tracing functions are used to calculate least-cost distances, around absolute and through partial barriers, between all pairs of sample locations and network nodes around absolute, and through partial barrier nodes. The IDL program is then used to calculate the estimated value at each network node based on a function of the nearest x sample locations weighted by network cost; for *Ephippiger* the six nearest sites were used weighted inversely by network distance. With values estimated at all network nodes a TIN surface is generated using the network node values as mass points. Comparison of equivalent NetSURF and normal distance weight surfaces shows the effects of barrier inclusion (Figure 6). The Pyrenean barrier has a considerable effect on the interpolated values south of the Pyrenees, whereas in contrast, including the maritime barrier appears to have little effect. One extremely useful by-product of NetSURF is the matrix of inter-sample site distances around barriers that can be used to test isolation-by-distance hypotheses via Mantel tests.

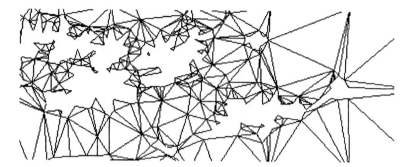

Figure 5. Detail of Ephippiger network in Eastern Pyrenees showing barrier holes.

Figure 6. Average male calling song syllable number interpolated with the six nearest locations weighted inversely with distance, (a) ArcView IDW without barriers, (b) NetSurf with barriers.

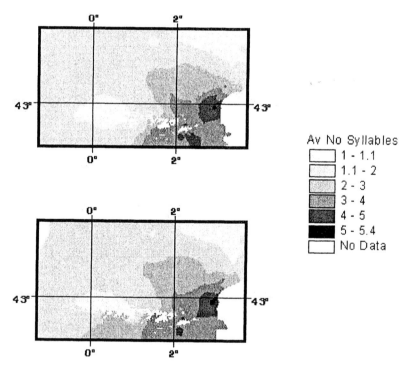

Pattern Analysis

Identifying evolutionary processes from observed patterns is a difficult task that relies on inferring the most likely scenario from a number of possible candidates. The most important spatial pattern forming processes are environmental selection and genetic drift in isolated or semi-isolated populations. These two processes occur within the context of historical environmental change, which acts at a multitude of spatial and temporal scales from local microclimates to ice ages and the effects of continental drift. Location trait values are the result of one or both processes superimposed on historical patterning.

The identification of causal evolutionary processes relies on the identification of covariance relationships between traits and the environment and between multiple traits. Partial correlation and Mantel tests have been suggested as suitable techniques for testing the significance of evolutionary hypotheses and covariation relationships (Smouse et al. 1986, Brown and Thorpe 1991a,b, Cessaroni et al. 1997). However, these statistics have been employed on entire sample sets (i.e., on the whole map) without consideration of spatial variation in the covariation relationships or barriers. Despite the complexity of identifying spatial congruence between traits, and traits and the environment, examination of modern variation patterns remains an important approach to identifying causal evolutionary processes.

Local Covariance

Surface scatter plots provide a simple visual means to identify at least some forms of spatial variation in covariation relationships (Kidd 1998). A surface scatter plot is a scatter plot with sampled locations displayed over the (interpolated) surfaces. Clusters in the surface scatter allow visual grouping of sample points; mapping the groups shows if they are spatially contiguous (Figure 7). The same technique can be employed to examine covariation relationships between traits and environmental variables. An extension of the technique is to use a 3D plot to show relationships between three parameters (Figure 8).

Multivariate Patterns

Covariance between trait patterns can arise from a variety of causes including sharing a common phylogenetic history, pleiotropy, gene linkage and selection. If selection is the dominant process we expect traits under direct selection and those linked to them to covary with the controlling environmental variable(s). Determining if a set of spatially covarying traits is likely to be subject to selection by the same environmental variable can only come from a thorough understanding of trait function. In contrast, if the dominant process were the consequence of historical population movements (secondary contact) or modern drift then we would expect to see covariation between unlinked traits. If the covariance is modern (primary) we should observe changes in trait covariance relationships across dispersal barriers. In contrast, if the covariance were the consequence of secondary contact we would look at evidence of historical range splits and covariation relationship changes congruent with potential historical barriers. Whatever the causal processes the first thing to do is to look for multivariate trait patterns; we used principal component analysis (PCA).

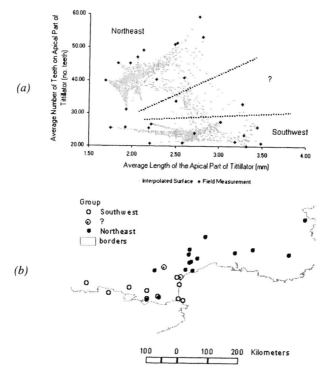

Figure 7. (a) Surface scatter plot and (b) map showing a spatially varying covariation relationship between the length of, and number of teeth on, the apical part of the male titillator of E. ephippiger. Northeast populations show positive correlation between the average length and average number of teeth on the apical part of the titillator, whereas southeast populations show no correlation.

The PCA was calculated using the interpolated network node values of fourteen morphological traits, male calling song syllable number (song), and the commonest allele at four enzymic loci (phosphoglucomutase, tetrazoleum oxidase and two esterases). PC surfaces were subsequently generated from the nodes by creating a TIN surface as previously described (Figure 9).

PC1 is representative of a variety of traits including morphology, song and three allozymes; it explains 50% of the variation. The diversity and lack of functional similarity in the represented traits fits the expected pattern for primary drift or secondary contact scenarios described. The Pyrenees is an established barrier to advancing post-glacial populations (Rica and Recoder 1990, Hewitt 1996, Taberlet et al. 1998). Kidd and Ritchie (2000) also noted the decline marks the line of secondary contact between Iberian and eastern European forms of *E. ephippiger,* the Iberian from having advanced north of the Pyrenean watershed via along the Mediterranean coast as did *Genista scorpius* and *Psammodromus algirus* (Rica and Recoder 1990).

PC2 explains 25% of the variation and is representative of morphological traits highly correlated with overall body size, e.g. length of body, length of foretibia, body weight, etc. The similarity of represented traits combined with Kidd and Ritchie's (2000) observation that 80% of body length variation can explained by a multiple regression with environmental variables determining growth season quality suggests that PC2 pattern is of primary selective origin.

PC3 and PC4 explain 8.5% and 5% of the variation respectively. Similarly to PC1 they represent a variety of traits suggesting a primary drift or secondary contact origin. The very local distribution of high PC values is perhaps supportive of local drift and or selection in the coastal area. Their location either side of the putative PC1 secondary contact cline is intriguing. Also of great interest is the fact that some of these patches correspond closely with the "cruciger" form of *E. ephippiger*, which some authors have considered a distinct species or subspecies (Kidd and Ritchie 2000).

PCA has been demonstrated to have considerable potential for identifying multi-trait patterns. However, without additional data on modern and past environments, potential scenarios cannot be separated. PCA as used here is a whole map statistic which smoothes the local variation in covariation relationships (as already discussed). In addition, it does not provide significance levels for the patterns identified, however, significance levels may be determined through a combination of discriminant canonical variant analysis and bootstrapping (Kidd and Ritchie 2000).

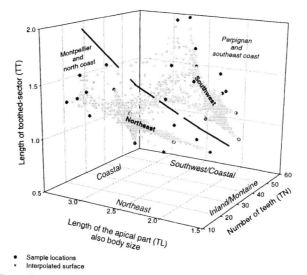

Figure 8. 3-D surface scatter plot of E. ephippiger titillator traits showing complex spatial covariance relationships.

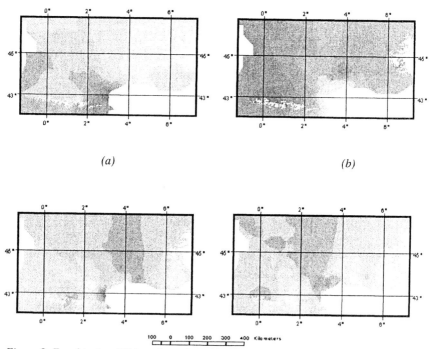

Figure 9. E. *ephippiger* PCA surfaces from NetSURF inverse distance interpolation, (a) PC1, (b) PC2, (c) PC3, (d) PC4. Maps have 7 equal interval classes of grey scale, dark = high, light = low. Barriers are unshaded.

THE FUTURE

The application of GISc techniques to regional trait patterns in *E. ephippiger* has revealed a complexity of patterns and covariation relationships not seen before, and has changed the evolutionary interpretation of geographic variation within this species. In addition it has allowed initial investigations into causal hypotheses. These successes encourage further developments to the system and avenues of research.

System Developments

A number of database developments that will assist in the examination of trait patterns are under investigation.

The addition of morphological and genetic data on individuals is a priority. This will allow more detailed comparisons between genetic and phenotypic patterns. Precise and accurate location of individuals, possibly by GPS survey, will allow a more hierarchical approach to pattern investigation. For example, variation can be related to scale using multilevel modelling tools (Jones 1991) or to habitat fragmentation through integration with remote sensed imagery and landscape connectivity statistics (Berry et al. 1998).

Data on other species will allow their patterns to be examined using the same techniques as *E. ephippiger* and will allow comparisons between species that should show congruence if they are the products of similar historical patterns of vicariance. For example, taxon distributions (Vogel et al. 1999), and palynological and ecological data (Kirkpatrick and Fowler 1998) have been used to hypothesise refuge location.

The collation of additional and better quality environmental data sets continues. In particular, climatic models of average monthly temperature and precipitation in the study area are under development. River attributes are also of interest as rivers are potential barriers to modern gene-flow. Investigations are under way into the availability and quality of palaeo-environmental data sets, which may help locate refugia and assess hypothesised migration routes.

Finally, explaining variation patterns in organisms requires a multidisciplinary approach that includes genetics, ecology, palaeontology, climatology and geography, to name a few. Its success depends on co-operation and the sharing of information. The GISc approach can facilitate data sharing by converting data in a wide variety of common digital formats that could then be published on the Internet using map-serving software.

Research Avenues

Increasingly, population geneticists are expressing concern with conventional methods for detecting patterns in intraspecific variation or population structuring (Slatkin 1985, Neigel 1997, Whitlock and MacCauley 1999). Almost all techniques in routine use are similar to (or direct developments of) Wright's F statistics, developed to detect whether populations deviate from the "Island Model", in which similar panmictic populations exchange migrants equally. Significant Fst values tell us little more than that there is variation in gene frequency among populations, i.e. all populations are not exchanging migrants at a sufficiently high level that gene frequencies are homogenised. The null hypotheses are so rarely met that the model has recently been rechristened the "Fantasy Island Model" (Whitlock and MacCauley 1999). One way of detecting the nature of population structure (rather than simply if it exists) is to test for isolation by distance, plotting pairwise estimates of migration or Fst values against geographic distance. This will detect if a cause of population structure is the (likely) fact that populations only exchange individuals with neighbouring populations, therefore gene frequencies are more likely to differ between distant neighbours. It is fair to say that there are no standard techniques for detecting more complex or historical patterns of intraspecific substructure, and that the assumptions of the island model are severe, with significant Fst possible for a whole range of alternative reasons (Whitlock and MacCauley 1999). If historical vicariance or contemporary barriers are suspected to be a major source of variation, there is little other than post-hoc partitioning of populations for further Fst analyses (or detection of clustering in isolation by distance plots) to find supporting evidence (Bossart and Prowell 1998). Variation among populations in the level of inbreeding (which increases any pairwise Fst values including such populations) is one major confounding factor in such analyses. If the evolutionary relatedness of marker loci used in analysis can be inferred, nested clade analysis (Templeton 1998) can be used to infer population histories.

Our GISc analyses of *E. ephippiger* presents an alternative methodology for the investigation of intraspecific variation. The analyses have clearly identified potential patterns to the population structure we now expect to find in the organism; the population groups detected by the analysis of interpolated trait surfaces produce clear predictions. This provides an external (rather than post-hoc) framework for the analysis of geographic variation of genetic markers. We now propose independent testing of the veracity of these subdivisions by two methods. Hierarchical Fst analysis with populations nested within the putative refugial groups will confirm if most of the variation in genetic variation reflects these patterns. Mantel tests comparing genetic variation with geographic distance, morphological or behavioural variation, or the presumed refugial origins of forms provides an alternative approach (Douglas and Endler 1982). The two analyses neatly reinforce one another here; GISci has given an external validity to the assemblages of populations tested for population substructure, and the patterning of genetic variation should verify if the clusters identified by GISci are indeed the best predictors of genetic variation.

ACKNOWLEDGEMENTS

The authors thank Matthijs Duijm, Leendert Oudman and their colleagues for the supply of *Ephippiger* data, Prof. Richard Healy for his support and critical reading of this chapter, and the two anonymous referees for their useful comments.

REFERENCES

Barton NH. The dynamics of hybrid zones. Heredity 1979 43:341-359.
Berry JK; Buckley DJ, McGarigal K. FRAGSTATS*ARC: Integrating ArcInfo with the FRAGSTATS Landscape Structure Analysis Program. Proceedings of the 1998 ESRI International User Conference;1998 Jul 27-Jul 31.
Bossart JL, Prowell DP. Genetic estimates of population structure and gene flow: limitations, lessons and new directions. Trends in Ecology and Evolution 1998; 13:202-206.
Brown RP, Thorpe RS. Within-island microgeographic variation in body dimensions and scalation of the skink *Chalcides sexlineatus* with testing of causal hypotheses. Biological Journal of the Linnean Society 1991a; 44:47-64.
Brown RP, Thorpe RS. Within-island microgeographic variation in the colour pattern of the skink, *Chalcides sexlineatus*: pattern and cause. Journal of Evolutionary Biology 1991b; 4:557-574.
Cavalli-Sforza LL, Menozzi P, Piazza A. *The history and geography of human genes.* Abridged paperback edition, Princeton, NJ: Princeton University Press, 1994.
Cesaroni D, Matarazzo P, Allegrucci G, Sbordoni V. Comparing patterns of geographic variation in cave crcikets combining geostatistic methods and Mantel tests. Journal of Biogeography 1997; 24:419-431.
Commission of the European Communities. *European solar radiation atlas, vol. 1 horizontal surfaces.* Brussels: Commission of the European Communities 1979.
Commission of the European Communities. Soil Map of the European Communities at 1:1,000,000. Luxemburg: The Office for Official Publications of the European Communities 1985. GNV153 – Soil map of the European (from CORINE). http://www.grid.unep.ch/datasets/gnv-data.html
Cooper SJB, Ibrahim KM, Hewitt GM. Postglacial expansion and genome subdivision in the European grasshopper *Chorthippus parallelus*. Molecular Ecology, 1995; 4:49-60.
Council of Europe and Commission of the European Communities. Map of the Natural Vegetation of the member countries of the European Community and the Council of Europe, scale 1:3,000,000. 2nd edition. The Office of Official Publications of the European Communities 1987. GNV154 - Natural Vegetation map of the European Communities (from CORINE). http://www.grid.unep.ch/datasets/gnv-data.html

Dixon LFJ, Barker R, Bray M, Farres P, Hooke J, Inkpen R, Merel A, Payne D, Shelford A. "Analytical photogrammetry for geomorphological research." In *Landform monitoring, modelling and analysis*. S Lane, K Richards, J Chandler eds. Chichester, UK: John Wiley & Sons Ltd., 1998.

Douglas, ME, Endler JA. Quantitative matrix comparisons in ecological and evolutionary investigations. Journal of Theoretical Biology 1982; 99:777-795.

Duijm M. On some song characteristics in *Ephippiger* (Orthoptera, Tettigonioidea) and their geographic variation. Netherlands Journal of Zoology 1990; 40:428-453.

Goodchild MF. Geographical information science. International Journal of Geographical Information Systems 1992; 6:31-45.

Grandcolas P. "Distribution and hybridization of species of the genus Ephippiger Bert. 1827 in the Briançon and Vallouise regions (Hautes-Alpes, France)". In *Evolutionary biology of Orthopteroid insects*, Baccetti B ed. Chichester, UK: Ellis Horwood Limited, 1987.

Efron B, Morris C. Stein's paradox in statistics. Scientific American 1997; 236:119-127.

Environmental Systems Research Institute. Digital Chart of the World 1993.

Environmental Systems Research Institute, Inc. ARC Version 7.1.2. Environmental Systems Research Institute Inc., 380 New York Street, Redlands, CA 92373-8100, USA 1997.

Hartley JC, Bugren MM. Colour polymorphism in Ephippiger ephippiger. Biological Journal of the Linnean Society 1986; 27:191-199.

Hartley JC, Warne AC. Taxonomy of the Ephippiger ephippiger complex (ephippiger, cruciger, cunii), with special reference to the mechanics of copulation. Eos 1984; 55:43-54.

Hewitt GM. Some genetic consequences of ice-ages, and their role in divergence and speciation. Biological Journal of the Linnean Society 1996; 58:247-276.

Hockham, L. Population genetics and reproductive biology of *Ephippiger ephippiger* (Orthoptera: Tettigoniidae). Ph.D thesis, University of St Andrews, UK, 1999.

Hutchinson MF. A locally adaptive approach to the interpolation of digital elevation models. Third International Conference/Workshop on Integrating GIS and Environmental Modeling; 1996a Jan 21-Jan 25; Santa Fe, New Mexico, USA. http://bbq.ncgia.ucsb.edu/conf/SANTA_FE_CD-ROM/program.html.

Hutchinson MF. "Thin plate spline interpolation of mean rainfall: getting the temporal statistics correct". In *GIS and environmental modeling: progress and research issues*. MF Goodchild, LT Steyaert, BO Parks, C Johnston, D Maidment, M Crane, S Glendinning eds. Fort Collins, CO: GIS World Books, 1996.

Jones K. *Muiit-level models for geographical research*. Concepts and techniques in modern geography 54. Norwich UK: University of East Anglia, 1991.

Kesteven J, Hutchinson M. Spatial modelling of climatic variables on a continental shelf. Third International Conference/Workshop on Integrating GIS and Environmental Modeling; 1996 Jan 21-Jan 25; National Center for Geographic Information and Analysis, Santa Fe, New Mexico, USA. 1996.
http://bbq.ncgia.ucsb.edu/conf/SANTA_FE_CD-ROM/program.html.

Kidd D. Using geographical information systems to investigate patterns of intraspecific variation in *Ephippiger ephippiger* (Orthoptera, Tettigonidea), MSc thesis, University of Huddersfield UK, 1998.

Kidd D, Ritchie M. Inferring the patterns and causes of geographic variation in Ephippiger ephippiger (Orthoptera, Tettigoniidae) using geographical information systems. Biological Journal of the Linnean Society 2000 accepted.

Kirkpatrick JB, Fowler M. Locating likely glacial forest refugia in Tasmania using palynological and ecological information to test alternate climatic models. Biological Conservation 1998; 85:171-182.

Lane S, Richards K, Chandler J eds. *Landform monitoring, modelling and analysis*. Chichester, UK: John Wiley & Sons Ltd., 1998.

Maurer BA. *Geographical population analysis: tools for the analysis of biodiversity.* Oxford, UK: Blackwell Scientific Publications, 1994.

Ministère des Transportes. *Atlas climatique de la France.* redite ed. Paris: Ministère des Transportes, 1980.

Neigel, JE. A comparison of alternative strategies for estimating gene flow from molecular markers. Annual Review of Ecology and Systematics 1997; 28:105-128.

Nichols RA. Genetical and ecological differentiation across a hybrid zone. Ecological. Entomology 1988; 13:39-49.

NOAA Baseline Climatological Dataset - Monthly Station Precipitation Data. National Oceanic and Atmospheric Administration. Washington DC: NOAA, 1995. http://www.ncdc.noaa.gov/onlinedata/climatedata/station.prcp.html.

NOAA Baseline Climatological Dataset - Monthly Station Temperature DataNational Oceanic and Atmospheric Administration. Washington DC: NOAA, 1995. http://www.ncdc.noaa.gov/onlinedata/climatedata/station.temp.html.

National Oceanic and Atmospheric Administration. ETOPO5 (Earth Topography - 5 Minute). NOAA Product Information Catalog. Washington DC: US Dept. of Commerce; 1988. http://www.ngdc.noaa.gov/mgg/global/seltopo.html.

Oudman L, Landman W, Duijm M. Genetic distance in the genus *Ephippiger* (Orthoptera, Tettigonioidea) – A reconnaissance. Tijdscrift voor Entomologie 1989; 132:177-181

Oudman L, Duijm M, Landman W. Morphological and allozyme variation in the *Ephippiger ephippiger* complex (Orthoptera, Tettigonioidea). Netherlands Journal of Zoology. 1990; 40:454-483.

Research Systems Inc. Interactive Data Language. Version 5.2. Research Systems Inc. 1999.

Rica JPM, Recoder PM. Biogeographic features of the Pyrenean range. Mountain Research and Development 1990; 10:235-240.

Ritchie MG. The shape of female mating preferences. Proceedings of the National Acadamy of Sciences USA 1996; 93:14628-14631.

Ritchie MG, Racey SN, Gleason JM, Wolff K. Variability of the bushcricket *Ephippiger ephippiger*: RAPD and song races. Heredity 1997; 79:286-294.

Slatkin M. Gene flow in natural populations. Annual Review of Ecology and Systematics 1985; 16:393-430.

Smouse, P. E., Long, J. C. Sokal, R. R. Multiple regression and correlation extensions of the Mantel test of matrix correspondence. Systematic Zoology 1986; 35:627-632.

Taberlet P, Fumagalli L, Wust-Saucy A-G, Cosson J-F. Comparative phylogeny and postglacial colonization routes in Europe. Molecular Ecology 1998; 7:453-464.

Templeton AR. Nested clade analyses of phylogeographic data: testing hypotheses about gene flow and population history. Molecular Ecology 1997; 7:381-397.

Thorpe RS, Baez M. Geographic variation within an island: univariate and multivariate contouring of scalation, size, and shape of the lizard *Gallotia gallotia*. Evolution 1987; 41:256-268.

USGS. GTOPO30, United States Geological Survey 1996. http://edcdaac.usgs.gov/gtopo30/gtopo30.html.

Vogel JC, Rumsey FJ, Schneller JJ, Barrett JA, Gibby M. Where are the glacial refugia in Europe? Evidence from Pteridophyes. Biological Journal of the Linnean Society 1999; 66:23-37.

Whitlock MC, McCauley DE. Indirect estimates of gene flow and migration: $Fst \neq 1/(4Nm+1)$. Heredity 1998; 82:117-125.

Young A. "Landscape structure and genetic variation in plants: empirical evidence." In *Mosaic Landscapes and Ecological Processes.* Hannsson L, Fahrig L, Merriam G. eds. London: Chapman & Hall, 1995.

Chapter 13

THE GIS REPRESENTATION OF WILDLIFE MOVEMENTS: A FRAMEWORK

Ling Bian
Department of Geography, State University of New York at Buffalo, Amherst, NY 14261-0023
lbian@geog.buffalo.edu

Keywords: wildlife movements, GIS data structure, individual-based modeling.

Abstract This chapter discusses a conceptual framework for the representation of wildlife movements in a GIS environment. Basic requirements for the GIS representation are identified in terms of movement mechanisms (forage quality, quantity, geometry, distribution, and visibility) and movement parameters (moving direction, distance, sinuosity, and velocity). Three GIS data structures, vector, raster, and object-oriented vector, are evaluated for their strengths and weaknesses in the representation of individual wildlife, the landscape, their characteristics, interactions between animals and with the landscape. Based on the evaluation, the chapter outlines an implementation design for the development of GIS functionality to support the representation of wildlife movements.

INTRODUCTION

Mobility is vital to the survival of wildlife populations. Wildlife movements are spatial phenomena and modeling the characteristics of movements involves the representation of changes in location with time. However, the representation of movement is a long-standing challenge in GIS because traditional GIS approaches do not effectively address dynamic phenomena (Peuquet 1994, Raper and Livingstone 1995, Tang et al. 1996). In addition, few recently developed object-oriented GIS software packages are equipped to represent movements and dynamic processes. While simulation systems pertaining to animal movements have been reported in ecological literature (Hyman et al. 1991, Johnson et al. 1992), most of these systems are not intended, or cannot be considered, to be a fully functional GIS.

The inadequacy of GIS in representing spatio-temporal dynamics has been attributed to both technical and conceptual difficulties (Peuquet 1994). Technically, the current computing paradigm of object-orientation has shown promising potential for the representation of dynamic processes. However, available techniques alone may not actually solve the representation problem without a comprehensive framework that reflects both ecological principles and spatial representation requirements.

The objectives of this paper are to (1) discuss a conceptual framework for the representation of wildlife movements in a spatial context, and (2) use this framework as the basis for a specification of GIS functionality for such representations. The conceptual framework addresses two basic issues -- the requirements for the representation of wildlife movements in a GIS environment, and the capabilities of different GIS data structures for meeting these requirements. The GIS functionality addresses the design and implementation concerns necessary to support the modeling of wildlife movements. Movements of large mammalian herbivores are of particular concern in the discussion because these animals share similar movement patterns. In addition, research findings on spatio-temporal characteristics of movement of large mammalian herbivores are available in the ecology literature and can be used as guidelines for the GIS representation.

BACKGROUND

Recent developments in ecological research have provided both a theoretical basis and management incentives for the development of GIS representations for wildlife movements. One such development is the rise of individual-based modeling (IBM) (DeAngelis and Gross 1992). IBM is a shift away from more traditional modeling approaches that rely on general equations, and towards approaches that emphasize individual behavior and localized interactions (Judson, 1994). This shift in scope has raised a series of research questions for the GIS community that concern the effective representation of individualized movements within a spatially-explicit framework.

Another ecological development germane to the GIS representation of wildlife movements is the on-going research into animal movement behavior in the last two decades. This research has produced a rich set of literature and proposes a family of random-walk models that are designed to simulate animal movements (Okubo 1980, Berg 1983, Kareiva and Shigesada 1983, Bovet and Benhamou 1988, Marsh and Jones 1988, McCullouch and Cain 1989). The models are based on simple rules or mathematical equations and assume a homogeneous landscape and random foraging behavior. These simplistic assumptions were developed to compensate for the paucity of field data in both landscapes and animal movements. Nevertheless, the models have had a profound influence on subsequent models that take into account the landscape heterogeneity and multiple spatial scales of movement (Hyman et al. 1991, Johnson et al. 1992, Turner et al.1993). The GIS representation discussed in this chapter is intended to support the implementation of these existing models.

A third development is the increasing accumulation of in-situ animal tracking data achieved primarily through the availability of telemetry technology and the global positioning systems (GPS) for defining location and geographic position with time. Associated are a series of research reports that contribute to the recent literature on animal movements (Loft et al. 1991, Ward and Saltz 1994, Gross et al. 1995, Focardi et al. 1996, Fritz and de Garine-Wichatitsky 1996, Moen et al. 1997, Etzenhouser et al. 1998). In these reports, the analysis of animal movements is based on heterogeneous landscapes and observed animal behavior. As a result, the research has shed new light on the fundamental understanding of animal movements, essential for the development of GIS representations of such movements in spatially-explicit modeling.

REPRESENTATION FRAMEWORK

Basic Representation Components

Ecology literature primarily focuses on the biological principles and processes of wildlife movement, whereas the GIS representation focuses on the spatial considerations for the representation of these principles and processes in a spatially-explicit and automated environment. A number of basic components emerge from the ecology literature that must be accounted for in a GIS representation: (1) the basic "participants" of wildlife movements; (2) characteristics of these participants; (3) interactions between the participants; and (4) movements of wildlife. The basic participants of wildlife movements are wildlife and the landscape, both of which must be included for a complete and realistic representation of movements. The characteristics associated with wildlife and the landscape constitute the driving force of the movements. The wildlife and landscape interact with each other and such interactions often trigger particular movements as a response. Several types of interactions exist, including intra-specific interactions between wildlife individuals, inter-specific interactions between the wildlife and other species, and interactions between wildlife and the landscape (Johnson et al. 1992, Westervelt and Hopkins 1999). The movement of wildlife depends on the combination of these participants, characteristics, and interactions. Figure 1 shows a conceptual relationship between wildlife and the landscape.

Wildlife, their characteristics, their interactions, and their movements need to be modeled at the level of the individual animal. Consequently, the GIS representation requires the identification of appropriate GIS data structures that can support the representation of individual animals and, on the other hand, the spatially extended landscape.

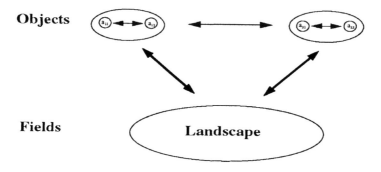

Figure 1. A schematic diagram of the representation of individual wildlife, the landscape, the intra- and inter-specific interactions, and the interaction between animals and the landscape. a11 and a12 represent individual animals 1 and 2 within species 1, respectively, while a21 and a22 represent two individual animals within species 2.

Recent developments in geographic information science have provided a theoretical framework that can encompass the GIS representation of wildlife individuals and the landscape. Two views of GIS representation, the object view and the field view, have emerged in recent literature (Goodchild 1992, Couclelis 1992, Frank 1996, Burrough 1996, Kemp 1997). Objects are perceived to be discrete phenomena with attributes that distinguish an object from other objects.

Fields, on the other hand, are continuous phenomena with attributes that vary continuously across space (Couclelis 1992, Goodchild 1992, Kemp 1997). A phenomenon can be perceived as either an object or a field depending on the purpose of the inquiry and the scale of observation (Couclelis 1992). It is conceptually straightforward to represent wildlife as objects and landscapes as fields. The GIS representation of wildlife movements thus requires both object and field representations that co-exist and interact. Interactions between species-specific wildlife and between wildlife and other species need to be represented as object-object operations, while the interactions between wildlife and landscapes need to be represented as object-field operations. The identification of wildlife and landscape representations has important implications for the selection of data structures and subsequent design, and implementation concerns for the GIS representation of animal movements.

GIS Data Structures

A number of GIS data structures can be useful for the representation of objects and fields, their characteristics, and their interactions. The data structures include traditional raster and vector structures and the recent object-oriented vector data structure. Each data structure has its strengths and weaknesses for the intended GIS representation of animal movements.

The use of the raster data structure and the associated cellular automata (CA) modeling approach has had a long tradition in ecological studies, including the modeling of animal movements. The simple data structure and simple geometry of data cells allow for easy management and computation of attributes. Regardless of whether they are perceived as objects or fields, geographic phenomena are represented through layers of attributes in the raster data structure. Such a structure favors the representation of spatially extended fields and their temporal dynamics.

The raster data structure, however, is more limited for the representation of objects than for the representation of fields. Objects, such as animals, may be represented as individual cells or groups of cells within a field, but, it is not possible to display the attributes of both the object and the field at the same time. It is also difficult to directly represent the multiple characteristics of objects because the raster data structure has limited capabilities to represent a comprehensive set of characteristics. Similarly, the raster data structure is equally, if not more, limited for the representation of object-object interactions that are based on the representation of individual objects and their characteristics.

Object-field interactions, on the other hand, can be reasonably supported by the CA approach. The regular geometry of raster data cells and the clear adjacency between cells make CA effective for supporting localized interactions and rule-based movements. The interaction is based on the state of a local cell and a number of adjacent cells. The movement of an animal, normally governed by a set of rules and dependent upon the state of local and adjacent cells, can be realized by a sequential change of state across the space over a series of time step (Clarke et al. 1997, Bian 2000).

However, the arbitrary geometry of the raster data structure also poses limitations to the representation of movements. This is because the limited number of possible exits from a cell are insufficient to represent realistic directions of animal movements. In addition, the fixed distance increments of the raster data structure make it difficult to represent the realistic moving distance. These limitations can affect the effectiveness of the raster data structure in the representation of animal movement (Tischendorf 1997).

Unlike the raster data structure, the traditional vector data structure represents geographic phenomena as layers of irregularly shaped geometric elements such as points, lines, or polygons. The traditional vector data structure is efficient for representing objects and fields as separate layers, each with multiple characteristics. Object-field interactions can be adequately supported by traditional vector GIS, through the use of spatial operations between data layers. Despite these capabilities, the traditional vector GIS approach is extremely limited in representing movements. This is because the data structure uses geographic location as the basis for organizing associated information such as attributes and topology. The representation of movements requires frequent updates of the location information, but attributes and topology must also be updated at the same time, although their values may remain the same. This is an extremely inefficient process within the vector data structure and renders it unsuitable for modeling movements.

In object-oriented vector GIS, the principles of object-orientation are applied to represent spatial information, while the geometry of the traditional vector data structure is retained. Thus, individual animals and elements of the landscape can be represented as objects, with each object maintaining its own set of properties and behaviors. As in the traditional vector data structure, objects and fields can be represented as separate entities, with their comprehensive characteristics stored as properties, and the interactions and movements represented as behavior of the animal objects.

In general, the object-oriented vector data structure is conceptually better suited for the representation of objects than it is for fields because object-orientation is designed for representing discrete phenomena rather than continuous phenomena (Wand 1989, Egenhofer and Frank 1992). As a result, the data structure is effective for the representation of object-object and object-field interactions, as long as objects are involved. The significant advantage of the data structure lies in its ability to support the representation of movements, where the identification, rather than the location, of an object is the basis for organizing information. Locations, therefore, can be treated as properties of an object and can be updated without affecting the identification, attributes, or the behavior of the object. In addition, the vector geometry allows the realistic representation of movement directions and distances without constraints of data resolution.

Understanding the basic components for the GIS representation and capabilities of GIS data structures facilitates the specific representations of wildlife movements, where wildlife movement is seen as the response of animals to various physiological and environmental mechanisms. A full understanding of the mechanisms and subsequent movement patterns is paramount for specific movement representations.

MOVEMENT MECHANISMS

Wildlife may move to acquire their resources needs, such as forage, water, and shelter. Of these driving mechanisms, foraging behavior has received the most attention because the foraging success of animals directly affects the survival, growth, and reproduction of individual animals, as well as population dynamics (Moen et al. 1997). The research on foraging-related movements has made noticeable progress, especially as instruments for field observation have become increasingly available. A rich set of literature has accumulated in ecology that analyzes the foraging behavior of animals, and rules and models have been developed to describe this behavior.

Foraging behavior is usually interpreted as the outcome of the interaction between two general categories of factors, namely the forage preference of animals and the availability of the forage in the landscape. In response to the spatial distribution of resource availability, wildlife may develop particular strategies and patterns to maximize energy intake and minimize the energy costs associated with searching for and processing food. Animals also tend to minimize intra-specific competition and the risk of predation (Ward and Saltz 1994, Gross et al. 1995, Focardi et al. 1996, Fritz and de Garine-Wichatisky 1996, Moen et al. 1997, Etzenhouser et al. 1998).

The physical process of foraging includes searching for forage and feeding. The search activity includes choosing and moving towards a feeding station (i.e. a location at which animals stop for forage), while the feeding activity includes cropping and chewing (Spalinger and Hobbs 1992, Focardi et al. 1996, Moen et al. 1997). These activities differ in space and time. Temporally, the activities occur in sequence, although chewing and searching may overlap in time. Spatially, searching for forage is the activity that mostly involves movement, while the feeding process is considered to be stationary.

For the purpose of analysis, the choice of a feeding station and movement towards that station are treated separately, and each is believed to be governed by a set of rules (Ward and Saltz 1994, Focardi et al. 1996, Moen et al. 1997). Search rules determine where to forage, while movement rules define the movement parameters, such moving direction, distance, and so on. Numerous rules and models have been developed in an attempt to simulate the search behavior. The factors most frequently considered in these rules and models include distribution of forage availability in the landscape, intra-specific competition, inter-specific interactions, and their temporal changes.

Interactions with Landscapes

In recent years, the effect of landscape heterogeneity on foraging behavior has been extensively researched (Hyman et al. 1991, Johnson et al. 1992, Spalinger and Hobbs 1992, Gross et al. 1995, Moen et al. 1997, Etzenhouser et al. 1998). This research effort has yielded not only an understanding of the effect of landscape heterogeneity but also a number of forage variables that are particularly relevant to the search strategies and movement patterns of animals. These variables are discussed below.

(1) Forage Quality. Many wildlife species demonstrate strong dietary preference of particular plant species, while others show a pattern of feeding on dominant forage species that may not be the most preferable. Both strategies are believed to maximize the forage intake. The presence of these plant species may cause animals to follow a directed movement pattern rather than an absolute random search (Ward and Saltz 1994, Fritz and de Garine-Wichatitsky 1996, Moen et al. 1997, Etzenhouser et al. 1998). Plant species and their dominance in the landscape, as well as other variables that indicate forage potential are thus important in the representation of landscapes.

(2) Forage Quantity. Forage quantity, which is often represented as forage density or biomass, has been repeatedly found to be related to a high concentration of foraging, short move distances, and other search behavior that favors forage efficiency (Hyman et al. 1991, Spalinger and Hobbs 1992, Ward and Saltz 1994, Gross et al. 1995, Focardi et al. 1996, Fritz and de Garine-Wichatitsky 1996, Moen et al. 1997, Etzenhouser et al. 1998). Basic statistical indices, such as the mean, standard deviation, and minimum threshold of forage density/biomass, can be effective for the representation of forage quantity (Stenphen and Krebs 1986, Focardi et al. 1996).

(3) Forage Geometry. The size, shape, and complexity of individual plants or clusters of plants have been suggested to be related to wildlife search and movement behavior, although response to the forage geometry differs between wildlife species (Etzenhouser et al. 1998). A number of indices, such as fractals, can represent plant geometry and spatial organization of forage in the landscape.

(4) Forage Distribution. Spatial patterns of forage, whether they be spatially aggregated or dispersed, discrete or continuous, are strongly related to the choice of feeding stations and the movement pattern of animals (Hyman et al. 1991, Spalinger and Hobbs 1992, Focardi et al. 1996, Moen et al. 1997, Etzenhouser et al. 1998). The spatial distribution of forage can be

an indication of forage abundance and the complexity of landscape structure. Complex landscape structures may imply physical barriers and force increased sinuosity of movement (Etzenhouser et al. 1998). The spatial distribution of forage can be represented by a number of readily calculable indices.

(5) Forage Visibility. Forage that are visible to animals or distinguishable against unwanted plants in the landscape, can lead to localized and short-distance foraging movements (Hyman et al. 1991, Gross et al. 1995, Gross et al. 1995, Etzenhouser et al. 1998). Indices such as plant species contrast and viewing area such as a viewshed can be effective for the representation of forage visibility (Gross et al. 1995, Etzenhouser et al. 1998).

The variables that indicate forage characteristics can be readily represented as landscape attributes. Existing GIS approaches based on all three data structures (i.e. raster, vector, and the object-oriented vector data structure) support the representation in a set of more or less effective ways.

Intra-Specific and Inter-Specific Interactions

Intra-specific interactions include two types of social behavior, namely grouping and repulsion (Hyman et al. 1991). Certain species of wildlife forage in groups in order to minimize the risk of predation (Fritz and de Garine-Wichatitsky 1996), but do so at the expense of the foraging efficiency of individuals in the group (Crawley 1983, Fritz and de Garine-Wichatitsky 1996). The group size has been found to have a major influence on foraging decision and, subsequently, movement patterns. This is because intra-specific competition forces animals to be highly selective for plants with high forage potential. Another consequence of group foraging is that the movement patterns of individual animals within the group exhibit increased sinuosity compared to animals that forage individually, although the movement pattern of the group can remain directed (Etzenhouser et al. 1998). Repulsion refers to aggression, dominance, and avoidance behaviors within or between social groups. Repulsion may lead to a foraging pattern that is more dispersed than that of independent foraging (Hyman et al. 1991).

Inter-specific interactions includes inter-specific competition, repulsion, and predation. When in competition with cattle grazing, certain wildlife species have been found to alter their diet choice and reduce the use of favorable forage areas that are also preferred by cattle. They have also been found to be less selective with regard to forage areas that are unfavorable, but are avoided by cattle (Loft et al. 1991). Few field observations are available for movements under inter-specific impacts. Thus, modeling of such interactions has primarily relied upon assumptions of behavior.

The representation of interactions must take into account several considerations. For group foraging and movement, the representation should include the identification of both groups and individuals within the group, different movement patterns for individuals and for groups, and spatial and temporal associations between individuals within a group. Both CA and object-oriented vector GIS can accommodate the considerations, but object-oriented vector GIS approach is more flexible and robust for the representation.

MOVEMENT PARAMETERS

Movement Parameters

Movements occur at all levels of spatial scale, such as long-range seasonal migration, daily movement at the home-range scale, and foraging-related localized movements. Most field studies have focused on the foraging-related movements that occur between consecutive feeding stations. These field studies have provided observations and analyses of movement patterns in relation to wildlife-landscape, and intra- and inter-specific interactions. Movement patterns are usually described by several parameters, such as moving direction, distance, speed, and sinuosity. These parameters are discussed below.

(1) Moving Direction. For foraging-related localized movements, the moving direction usually concerns the direction of the present movement path with reference to the direction of the preceding path. Traditional movement models have treated the movement of individuals as either a simple random-walk, if the direction of the present path is independent of the preceding path (Okubo 1980, Berg 1983), or as a correlated random walk, if directions of two successive paths are related (Kareiva and Shigesada 1983, Bovet and Benhamou 1988, Marsh and Jones 1988, McCullouch and Cain 1989). Recent field observations and analyses have suggested that animal movement paths deviate considerably from random walks (Ward and Saltz 1994, Gross et al. 1995, Focardi et al. 1996, Moen et al. 1997, Etzenhouser et al. 1998). The observed moving directions are mostly directed so that animals tend to maintain a previous moving direction, while the turning direction (left or right turns) can be independent between consecutive paths (Ward and Saltz 1994, Focardi et al. 1996). Along a directed general path, animals have exhibited nearest-neighbor or best-neighbor foraging behavior. That is, animals may either move to a nearest feeding station that provides forage above a threshold or they move to the station with the most available forage within a detection distance. These movement strategies are believed to increase feeding efficiency, and

differences in efficiency can become apparent when the forage density declines (Gross et al. 1995, Moen et al. 1997).

(2) Moving Distance. The moving distance between feeding stations tends to increase with dispersed forage distribution (Ward and Saltz 1994, Focardi et al. 1996). In the landscape with aggregated forage distribution, the moving distance tends to be short and localized within a patch where forage is abundant. Longer distance movements occur between patches when animals leave a present patch and move to a different patch for foraging (Ward and Saltz 1994, Gross et al. 1995, Focardi et al. 1996).

(3) Moving Sinuosity. When moving between two consecutive feeding stations, animals tend to walk in straight lines (Ward and Saltz 1994, Gross et al. 1995, Focardi et al. 1996). The moving paths over a series of feeding stations can be sinuous or linear (Focardi et al. 1996, Etzenhouser et al. 1998). The path sinuosity may increase as animals respond to a high forage density, to aggregated forage distribution, or to complex landscape structure (Focardi et al. 1996, Etzenhouser et al. 1998). Regardless of the sinuosity, animals seldom back track into areas that have already been visited (Focardi et al. 1996, Ward and Saltz 1994).

(4) Moving Velocity. Few field studies have recorded the actual moving speed of animals. However, analyses based on field observations have reported that forage density, forage quality, and landscape structure may affect the ratio between the time animals spend on feeding and the time spent on searching (Focardi et al. 1996, Etzenhouser et al. 1998). The ratio is often used to measure feeding efficiency.

Defining the movement parameters is one of the three steps of a complete movement that requires approximately three steps. In the first step, knowledge of landscape conditions and the presence of other animal individuals is acquired. In the second step, movement decisions are made according to certain rules and based on the evaluation of these conditions. In the third step, the actual movement is defined through the specification of movement parameters. Existing GIS approaches provide query functions to support the first step, while both CA and object-oriented vector GIS can support the implementation of the second and the third steps. As discussed earlier, the object-oriented vector data structure is less constrained than the raster structure for the realistic representation of movements.

Temporal Scales

In the representation of movements, temporal scales are as important as spatial scales. The point in time when movements begin and the duration of the movements are often used to describe the temporal aspects of movements. Long-term movements may be attributed to various factors. Seasonal changes in forage conditions, for example, can cause the seasonal migration of animals (Schoen and Kirchhoff 1990). Inter-specific competition such as that between wildlife and cattle can cause seasonal shifts in forage area for wildlife (Loft et al. 1991). In addition, seasonal intra-specific grouping can also contribute to seasonal changes in foraging area. An example of this last type of seasonal change is the grouping of female deer in the Fall to focus on different foraging areas from male deer (Marchinton and Hirth 1984, Hyman et al. 1991).

Home-range scale movements are always related to daily activities such as feeding, ruminating, and bedding. These activities occur at different times of a day and with different durations, each involving specific locations in the landscape. For example, feeding may occur on the open grass at dawn (Moen et al. 1997). The most extensively studied foraging-related movements, including searching and feeding, occur at the minute-time scale. Since detailed field data for seasonal and daily movements are rare, efforts to model these movements have typically relied on assumed rules.

The representation of the timing and duration of movements requires the incorporation of these variables into movement rules. Both CA and object-oriented vector GIS can accommodate these. For the representation of landscape dynamics at various temporal scales, the raster data structure and the associated CA approach have the most advantages.

IMPLEMENTATION DESIGN

The implementation design needs to consider the representation of animal objects and landscape fields, the attributes associated with animals and the landscape, the interactions between animals and between animals and the landscape, and the movements of animals. The design considerations for each of these representation components are addressed in subsequent sections.

(1) Animal Objects and the Landscape Field. Each individual animal should be treated as an object. A group of animals that forage together should be treated as an aggregation object with individual animals as members of the aggregation. The object-oriented vector data structure is the most appropriate data structure for these representations. In contrast, all three data structures (i.e. raster, vector, and the object-oriented vector data structure) can support the representation of continuous landscape field.

(2) Properties of the Landscape. Landscape properties should include attributes that represent forage quality, quantity, geometry, distribution, and visibility. The storage and query of these attributes can be supported by existing GIS approaches. Vector GIS (both the traditional and the object-oriented) approaches can support more complex attribute manipulations

through database management tools but are limited for the representation of dynamic landscapes.

(3) Properties of Animals. A minimum of six sets of properties are required for the representation of animal movements. These include (a) demographic states, such as species, age, and sex; (b) physiological states of individual animals, such as body mass; (c) movement states that include all movement parameters; (d) spatio-temporal states, such as location and time; (e) forage states that include all landscape attributes; and (f) social states, such as the presence of other animals, that indicate competition or threat. Group objects may have their unique properties such as group size. The demographic and physiological states are important for modeling the forage requirement and consumption. The movement and spatio-temporal states are directly relevant to the modeling of movements. The forage and social states are external states (not possessed directly by animals) and animals must obtain these states by querying the condition of both the landscape and other animals. The object-oriented vector data structure is ideal for accommodating these representations.

(4) Interactions. The interactions between animals and the landscape require that the landscape attributes be queried and the forage states of animals be updated. The inter- and intra-specific interactions require similar querying and updating of the states of other animals. Object-oriented vector GIS can support the representation of interactions by treating them as behavior of objects. With some limitations, CA has also been used for the representation of animal-landscape interactions.

(5) Movements. Particular movements depend on the query of the landscape and other animals conditions, rule-based choice of feeding stations, and application of movement parameters. These functions can be treated as behaviors in the object-oriented vector GIS. In addition, CA has been traditionally used to implement these functions.

Most implementations of movement models in ecology have employed the CA approach. However, Westervelt and Hopkins (1999) and Bian (2000) have proposed a hybrid approach. It employs the object-oriented vector data structure to represent objects, object-object interactions, and movements, while the raster data structure is used to represent the landscape and to support object-field interactions. The hybrid approach takes full advantage of each data structure, and is particularly well suited to representing landscapes that are spatially variable and temporally dynamic. The object-oriented vector GIS can also support the representation of object-field interactions. It is similar to the hybrid approach, except that a vector field is used to represent landscapes, especially those that are stable. As a new development, the object-oriented GIS approach has seen the least implementation of movement functions, thus demanding considerable development to realize the advantages associated with the data structure.

All necessary GIS functionalities should be implemented for the purpose of general movement modeling. The functionalities, therefore, should be able to adopt customized variables and rules and, in the mean time, be compatible with existing GIS functions. A new technical development, namely component-ware, is emerging as the new direction for the development of compatible software. With this technique, spatial data and functions are developed as components that can match each other and can be accessed through the internet (Buehler and McKee 1996).

SUMMARY

The object-oriented vector data structure emerges as the most appropriate data structure for the representation of animals, their characteristics, interactions, and movement. In contrast, several choices exist for the representation of landscapes. The hybrid approach, that combines the raster and object-oriented vector data structures, takes most advantages of the relative strengths of each data structure for addressing behavioral characteristics of animals and their interactions with the landscape.

The conceptual framework and functionality design discussed in this chapter can benefit a range of natural resources management studies such as wildlife population dynamics, rare and endangered species conservation, and wildlife reintroduction. The discussion can also be extended to broad issues in biogeography and ecology. Plant ecology, for example, involves the representation of individual plants or clusters of plants, their characteristics, and interactions between plants and with the surrounding environment (Baveco and Lingeman 1992, Liu and Ashton 1998). An understanding of GIS data structures is directly relevant to such studies.

The aforementioned GIS representation facilitates a new approach to integrating GIS with environmental modeling because the approach addresses the integration from a representational perspective rather than a mere technical perspective. It can have a broad implication in the effort to improve ecological modeling, foster further development in geographic information science, and enrich the collective knowledge of the dynamic environment.

ACKNOWLEDGMENTS

This research was partially funded by the National Science Foundation under Award No. SBR88-10917, and by Intergraph Corporation under Education Agreement ML0016. The author would like to thank ZhiXiao Xie for his assistance in the research. The reviewers provided valuable comments and suggestions that greatly improved the manuscript.

REFERENCES

Baveco JM, Lingeman R. An object-oriented tool for individual-oriented simulation: host-parasitoid system application. Ecological Modeling 1992; 61: 267-286.

Berg HC. *Random walks in biology*. Princeton, NJ: Princeton University Press, 1983.

Bian L. Object-oriented representation for modeling mobile objects in an aquatic environment. International Journal of Geographical Information Science 2000; 14(7): 603-623.

Bovet P, Benhamou S. Spatial analysis of animal movements using a correlated random walk model. Journal of Theoretical Biology 1988; 131: 419-433.

Buehler K, McKee L. The Open GIS Guide. http://www.opengis.org The Open GIS Consortium, Inc., 1996.

Burrough PA. "Natural objects with indeterminate boundaries." In *Geographic Objects with Indeterminate Boundaries*, PA Burrough, AU Frank eds. London: Taylor and Francis, 1996.

Clarke KC, Hoppen S, Gaydos L. A self-modifying cellular automaton model of historical urbanization in the San Francisco Bay area. Environment and Planning B 1997; 24: 247-261.

Couclelis H. "People manipulate objects (but cultivate Fields): beyond the raster-vector debate in GIS." In *Theories and Methods of Spatio-Temporal Reasoning in Geographic Space*, AU Frank, I Campari, U Formentini eds. Berlin: Springer-Verlag, 1992.

Crawley MJ. *Herbivore: The Dynamics of Animal-Plant Interactions*. Berkeley and Los Angeles: University of California Press, 1983.

DeAngelis DL, Gross LJ. *Individual-based Models and Approaches in Ecology*. New York: Chapman and Hall, 1992.

Egenhofer MJ, Frank AU. Object-Oriented Modeling for GIS. URISA Journal 1992; 4: 3-19.

Etzenhouser MJ, Owens MK, Spalinger DE, Munden SB. Foraging behavior of browsing ruminants in a heterogeneous landscape. Landscape Ecology 1998; 13: 55-64.

Focardi S, Marcellini P, Montanaro P. 1996. Do ungulates exhibit a food density threshold? a field study of optimal foraging and movement patterns. Journal of Animal Ecology 1996; 65: 606-620.

Frank AU. "The prevalence of objects with sharp boundaries in GIS." In *Geographic Objects with Indeterminate Boundaries*, PA Burrough, AU Frank eds. London: Taylor and Francis, 1996.

Fritz H, de Garine-Wichatitsky M. Foraging in a social antelope: effects of group size on foraging choices and resource perception in impala. Journal of Animal Ecology 1996; 65: 736-742.

Goodchild MF. Geographical Data Modeling. Computers and Geosciences 1992; 18: 401-408.

Gross JE, Zank C, Hobbs NT, Spalinger DE. Movement rules for herbivores in spatially heterogeneous environments: responses to small scale pattern. Landscape Ecology 1995; 10: 209-217.

Hyman JB, McAninch JB, DeAngelis DL. 1991. "An individual-based model of herbivory in a heterogeneous landscape." In *Quantitative Methods in Landscape Ecology*, MG Turner, RH Gardner eds. Heidelberg: Springer-Verlag, 1991.

Johnson AR, Wiens JA, Milne BT, Crist TO. Animal movements and population dynamics in heterogeneous landscapes. Landscape Ecology 1992; 7: 63-75.

Judson OP. The rise of the individual-based model in ecology. TREE 1994; 9: 9-14.

Kareiva P, Shigesada N. Analyzing insect movement as a correlated random walk. Oecologia, 1983; 56: 234-238.

Kemp KK. Fields as a framework for integrating GIS and environmental process models. Part 1: representing spatial continuity. Transactions in GIS 1997; 1: 219-234.

Liu J, Ashton PS. FORMOSAIC: an individual-based spatially explicit model for simulating forest dynamics in landscape mosaics. Ecological Modeling 1998; 106: 177-200.

Loft R, Menke JW, Kie JG. Habitat shifts by mule deer: the influence of cattle grazing. Journal of Wildlife Management 1991; 55: 16-26.

Marchinton RL, Hirth DH. "Behavior." In *While-Tailed Deer: Ecology and Management*, LK Hall ed. Harrisburg, PA: Stackpole Books, 1984.

Marsh LM, Jones RE. The form and consequence of random walk movement models. Journal of Theoretical Biology 1988; 133: 113-131.

McCullouch CE, Cain ML. Analyzing discrete movement data as a correlated random walk. Ecology 1989; 70: 383-388.

Moen R, Pastor J, Cohen Y. A spatially explicit model of moose foraging and energetics. Ecology 1997; 78: 505-521.

Okubo A. *Diffusion and Ecological Problems: Mathematical Models*. Berlin: Springer-Verlag, 1980.

Peuquet D. It's about time: a conceptual framework for the representation of temporal dynamics in geographic information systems. Annals of the Association of American Geographers 1994; 84: 441-461.

Raper J, Livingstone D. Development of a geomorphological spatial model using object-oriented design. International Journal of Geographical Information Science 1995; 9: 359-383.

Schoen JW, Kirchhoff M. Seasonal habitat use by Sitka black-tail deer on Admiralty Island, Alaska. Journal of Wildlife Management 1990; 54: 371-378.

Spalinger DE, Hobbs NT. Mechanisms of foraging in mammalian herbivores: new models of functional response. The American Naturalist 1992; 140: 325-348.

Stenphen DW, Krebs JR. *Foraging Theory*. Princeton, NJ: Princeton University Press, 1986.

Tang AY, Adams TM, Usery EL. A spatial data model design for feature-based geographical information systems. International Journal of Geographical Information Systems 1996; 10: 643-659.

Tischendorf L. Modeling individual movements in heterogeneous landscapes: potential of a new approach. Ecological Modeling 1997; 103: 33-42.

Turner MG, Wu Y, Romme WH, Wallace LL. A landscape simulation model of winter foraging by large ungulates. Ecological Modeling 1993; 69: 163-184.

Wand Y. "A Proposal for a Formal Model of Objects." In *Object-Oriented Concepts, Databases, and Applications* W Kim, FH Lochovsky eds. New York: ACM Press, 1989.

Ward D, Saltz D. Foraging at different spatial scales: dorcas gazelles foraging for lilies in the Negev desert. Ecology 1994; 75: 48-58.

Westervelt JD, Hopkins LD. Modeling mobile individuals in dynamic landscapes. International Journal of Geographic Information Science 1999; 13: 191-208.

Chapter 14

STRATIFIED SAMPLING FOR FIELD SURVEY OF ENVIRONMENTAL GRADIENTS IN THE MOJAVE DESERT ECOREGION

Janet Franklin[1], Todd Keeler-Wolf[2], Kathryn A. Thomas[3], David A. Shaari[1], Peter A. Stine[4], Joel Michaelsen[5] and Jennifer Miller[1]
[1]*Department of Geography, San Diego State University, San Diego, CA 92182-4493, USA*
janet@typhoon.sdsu.edu
[2]*California Department of Fish and Game, 1416 9th Street, Sacramento, CA 95814, USA*
[3]*USGS Forest and Rangeland Ecosystem Science Center, Colorado Plateau Field Station, Northern Arizona University, PO Box 5614, Bldg 24, Flagstaff, AZ 86011-5614, USA*
[4]*USGS, Western Ecological Research Center, Sacramento, CA 95819, USA*
[5]*Department of Geography, University of California, Santa Barbara, Ellison Hall 3611, Santa Barbara, CA 93106-4060, USA*

Keywords:	environment gradients, sampling, Mojave Desert Ecoregion.
Abstract	Environmental gradients, represented by mapped physical environmental variables within a GIS, were classified and used to allocate a two-stage random stratified sample for field survey of vegetation in the Mojave Desert Ecoregion, California. The first-stage sample was allocated randomly and with unequal proportions among 129 environmental classes defined by the intersection of climate and geology digital maps at 1 km resolution. The second-stage sample was selected for each 1 km cell in the primary sample by defining six terrain classes related to desert vegetation patterns and randomly locating one plot location per class per cell. The total number of observations (1133) was determined by the resources available for the survey. This approach allowed the vegetation survey to be planned efficiently, alternate samples to be located, and vegetation types to be defined quantitatively. The sample allocated surveyed broad scale environmental gradients effectively, and the objective of oversampling rare environmental classes and undersampling common classes was achieved in most cases. It did not succeed, however, in capturing replicates of rarer plant alliances. We suggest sampling efforts should be weighted even more heavily toward rare environments and plant communities for this objective.

INTRODUCTION

Scientists and resource agencies are increasingly emphasizing biological inventory and monitoring of large regions. Information on the distribution and abundance of plants and animals is required for biodiversity assessment and ecosystem management at the landscape scale, and earth system science research (Scott et al. 1993, Bojorquez-Tapia et a.l 1995, Bourgeron et al. 1995, Davis et al. 1990, 1995, Franklin and Woodcock 1997, Smith et al. 1997, Scott and Jennings 1998, Austin 1998). Resource managers within the Mojave Desert have collaborated to develop regional geographic data to support better land use decisions. With guidance from the Desert Managers Group we are developing a digital map of actual vegetation for this purpose. This would be a difficult and expensive task if it were carried out using conventional vegetation mapping methods over such a large area. We have therefore sought to employ techniques that would improve the efficiency and effectiveness of the effort.

In a series of papers, researchers in Australia presented a procedure for allocating a sample of locations for field data collection intended to estimate the spatial distributions of species (Gillison and Brewer 1985, Austin and Heyligers 1989, 1991, Margules and Austin 1994, Austin 1998, reviewed by Bourgeron et al. 1994). The approach, termed gradient directed sampling, consists of stratified random sampling, where stratification is based on those mapped environmental variables believed to represent broad-scale environmental gradients that influence species distributions (Whittaker 1973). We used this method to allocate a two-stage random stratified sample of locations for collecting data on vegetation composition in a portion of the Mojave Desert Ecoregion within California. The sampling is intended to quantitatively define the composition of vegetation communities at the alliance level within the National Vegetation Classification System (Grossman et al. 1998) based on classification and ordination, and to provide data for mapping existing vegetation using predictive modeling (Franklin 1995). In this paper we describe and evaluate the procedure used to define environmental gradients and allocate the sample.

BACKGROUND

It has long been noted that the study of distributions in geographic space, based on sampling, must be concerned with both the accuracy (bias) and precision (efficiency) of sample estimates. For example, Berry and Baker (1968) showed that for estimating proportions of different land covers, or their spatial distribution depicted in a map, greatest efficiency is achieved using systematic sampling when spatial autocorrelation is present and strongest at near distances. However, when spatial patterns were complex and varied, adding stratification and randomization to systematic sampling improved efficiency and accuracy. Sampling has been discussed by Cochran (1977), Congalton (1991), and Thompson (1992) discusses sampling for map accuracy.

The conceptual framework for gradient directed ecological survey rests on the link between gradients in the physical environment and biotic distributions (Whittaker 1960, 1973, Nix 1982). The method of gradient directed sampling was developed to estimate the spatial pattern of plant species distributions and diversity (richness) within a large area more accurately than "ad hoc" (opportunistic) collection of biological distribution data (Margules and Austin 1994). Gillison and Brewer (1985) found that, when the objective of a survey is to encounter as much biological diversity as possible (rather than to statistically estimate the proportions of the survey area covered by different species), stratified sampling along environmental gradients is more efficient than random sampling – the same amount of diversity is encountered in a smaller sampling area. Nelder et al. (1995) compared several methods for evaluating the representativeness of vegetation surveys.

Previous work on gradient directed sampling emphasized sampling within representative belt transects oriented along environmental gradients ("gradsects"). Sampling along transects can reduce time and travel costs during field work, especially in very large and roadless areas. We did not use a transect survey design, and instead employed a geographic information system (GIS) to select samples unrestricted to geographic belts, but instead constrained by environmental criteria and accessibility.

MATERIALS AND METHODS

We adopted the protocol described by Austin and Heyligers (1989) with some minor modifications:
1. Identify regional environmental variables influencing plant distributions in the study area;
2. Choose best available data (digital maps) for environmental stratification;
3. Stratify the area for sampling by classifying the environmental variables into discrete categories or ranges and combining them;
4. Identify local variables to be used, and data depicting them, for a second level of stratification at the local scale;
5. Decide on effort to be used for sampling rarer environmental strata versus adding more replicates to common strata.

Following these steps, a two-stage random stratified sample with variable proportions was allocated. This set the goals for the field survey, identifying the locations to be sampled, and alternate locations. This approach was intended to sample the floristic diversity of the region efficiently – we expected a wide range of vegetation alliances and plant species to be recorded in the sample. We evaluated this by comparing a) the stratified sample that was allocated to b) those locations actually surveyed, to c) a supplemental (purposively located) sample, and to d) a collection of vegetation plots from prior surveys. We compare the distribution of these samples among environmental classes (representativeness), and the number of species, alliances and replicates per alliance (efficiency) in each sample.

Study Area

A portion of the Mojave Desert Ecoregion within California spanning approximately 5 million ha (Figure 1 – on the CD-ROM) was selected for the first phase of the Mojave Vegetation Mapping Program (Central Section) of the Mojave Desert Ecosystem Program (www.mojavedata.gov).

This constituted our study area. The Mojave Desert forms an ecological transition zone between the Great Basin and Sonoran deserts in North America. It is dominated by basin and range physiography, with extremes of elevation (-86 to 3368 m) and climate. Low to mid-elevation basins experience cool to mild winters and hot summers, with cold winters and mild summer temperatures at the highest elevations. There are large diurnal and seasonal fluctuations in temperature. Precipitation in the Mojave results mainly from winter frontal storms; however, summer monsoon precipitation may provide a significant contribution to the yearly total in the eastern Mojave. Precipitation is low because the Mojave lies in the rainshadow of the Sierra Nevada and Transverse ranges.

Landforms in the California Mojave include alluvial fans, bajadas and alluvial plains (42%), rocky highlands (45%), washes (5%), playas (2.5%) and sand sheets and dunes (3.5%) (www.mojavedata.gov). Vegetation is dominated by xeromorphic scrub formations, with more limited areas of wash and wetland vegetation, conifer woodland, desert grassland, and other communities (Vasek and Barbour 1977, Schoenherr 1992). Previous regional vegetation mapping of the Mojave Desert in California (Thomas 1996) described forty-one vegetation types using the Holland (1986) classification. Three of these types occur over 73% of the Mojave Desert: Creosote Bush Scrub, Mojave Mixed Woody Scrub, and Desert Saltbush Scrub. Many vegetation types occur with limited extent (e.g. the smallest twenty-three cover 4.5% of the land area).

Environmental Variables

Gradsect surveys have used precipitation, temperature and rock type as the environmental variables influencing plant distributions at broad spatial scales, and topography or terrain position at finer scales (Gillison and Brewer 1985, Austin and Heyligers 1989, Bourgeron et al 1995). Other variables such as landform, soil type, elevation, and general vegetation types could also be used (Nelder et al. 1995). A GIS is used to assemble the environmental variables and allocate the sample; therefore, digital maps of those variables must be available or developed. Also, because samples are allocated to environmental classes defined by the intersection of maps of classified environmental variables, it is desirable to capture environmental gradients in a small number of variables, and categories of those variables. Environments can be then sampled with replication (assuming that resources for the survey are limited). Vegetation and plant species distributions and richness have been linked to precipitation, temperature, elevation, geologic substrate, topography, landform, soil texture, depth to caliche layer, salinity gradients and ground water depth in the Mojave and nearby deserts (Hunt 1966, Beatley 1974, Vasek and Thorne 1977, Thorne 1982, Yeaton et al. 1985, Pavlik 1989, Montana 1990, McAuliffe 1994, Thomas 1996).

Nix (1982) and others (Mackey et al. 1988, 1989, Mackey 1994) have described five primary environmental regimes that influence biotic distributions: solar radiation, temperature, moisture, mineral nutrients and biotic interactions. We selected mean summer maximum and winter minimum temperature, mean summer and winter precipitation (Table 1), and lithology or rock type (Table 2) to represent the first four at the broad scale. Other bioclimatic variables could have been selected such as annual or seasonal radiation, potential evapotranspiration, water deficit, or growing degree days (see for example Mackey et al. 1988, Leathwick 1998). However, we felt that seasonal temperature and precipitation adequately captured the bioclimatic regimes in this temperate desert. We then developed a simple classification of terrain to represent topographic influences on vegetation patterns at a finer scale (Table 3).

Climate Maps

The four climate variables just described (Table 1) were interpolated to a 30 arc-sec (roughly 1 km) grid using 30-year averages from 104-136 climate stations. Interpolations were based on a two component statistical model similar to universal kriging (Bailey and Gatrell 1998, Venables and Ripley 1999). The first component consisted of multiple regressions between the climate variable of interest and latitude, longitude and elevation. This component was designed to capture the large-scale variation, or trend, in the climatic variable. The residuals from the linear model predictions at the station locations were autocorrelated, and standard geostatistical models were fit to the variograms of the residuals (Bailey and Gatrell 1998). Since the presence of autocorrelated residuals violates the assumptions for ordinary least squares, the linear regression models were refit using generalized least squares with residual covariance matrices based on the spatial autocorrelation models (see Michaelsen in prep. for details). Cross validation was used to diagnose any problems with misfitting of either component of the models, as well as to flag potentially erroneous station data.

In effect, model predictions were based on the large-scale relationships captured by the linear regression models with adjustments made for deviations of nearby stations from the overall linear regression parameters. Over much of the study area, stations were separated by distances greater than the characteristic autocorrelation distances of 50-260 km, so most estimates were based solely on the linear regression models (Michaelsen in prep.). The standard errors of the models vary spatially depending on the distance from the predicted location to the nearest stations (ranges given in Table 1).

Table 1. Classification of mapped climate variables, including standard error of spatial regression model used to interpolate these variables from climate station data (see text).

Variable	Mean and Range	Classes	Standard Error
Mean monthly minimum winter (January) temperature	-0.0 °C (-11.3 to 4.8 °C)	< –7 °C -7 to -4.5 °C -4.5 to –2 °C -2 to 0.5 °C 0.5 to 2 °C ≥ 2 °C	+/- 2.2 °C
Mean monthly maximum summer (July) temperature	35.5 °C (16.6 to 44.4 °C)	< 35 °C ≥ 35 °C	+/- 1.1 °C
Mean total winter (Nov-Apr) precipitation	124 mm (45-579 mm)	< 100 mm 100-175 mm >175 mm	+/- 30%
Mean total summer (May-Oct) precipitation	30 mm (11-146 mm)	< 40 mm ≥ 40 mm	+/- 30%

Histograms and maps of these gridded climate surfaces were inspected in order to select the class intervals for each variable that best captured climatic gradients in the Mojave in the smallest number of classes. January minimum temperature (Figure 2a) and elevation (Figure 1) are correlated in the study area (r=-0.78), and so winter temperature was finely divided into six class intervals in order to stratify this broad orographic gradient.

Summer and winter temperature are correlated (r=0.59). Summer maximum temperature was divided into two classes using a threshold value of 35 °C. Winter precipitation is correlated with elevation (r=0.67) but also shows a west-east gradient (Figure 3a) that was captured in three class intervals (Table 1). Summer precipitation shows relatively higher values in the eastern Mojave (Figure 3b), and we captured this distinction using a threshold value of 40 mm to divide this variable into two classes (Table 1). Austin and Heyligers (1989) used a similar *ad hoc* procedure to select class intervals for continuous climate variables.

Geology Map

The 1:750,000 scale geologic map of California (Jennings 1985) was the best available digital data on substrate or lithology for the study area when the study began. This digitized map, gridded to 1 km resolution, depicted about 22 categories in the study area that were aggregated to eight classes (Table 2) thought to best represent gradients (water availability, nutrients) affecting plant species distributions. When geologic map classes are aggregated in this way, alluvium (unconsolidated bajadas and alluvial fans) is the most extensive substrate (Figure 4 – on the CD-ROM, Table 4).

Terrain Classes

Because terrain (hillslope position, slope angle, slope aspect) exerts a strong influence on plant distributions at a finer spatial scale than bioclimatic gradients (Moore et al. 1991, Franklin 1995), terrain classes can be used for the second stage of stratification (Austin and Heyligers 1989). This is important in the Mojave where vegetation composition can change dramatically over short distances as a function of terrain position. Again, our goal was to define a simple set of terrain classes to be used for allocating the second stage sample.

Digital terrain models (30 m resolution, corresponding to 1:24000 scale topographic quadrangles) were available from the USGS (http://rmmcweb.cr.usgs.gov/elevation/). The level 2 (r.m.s.e. < ½ contour interval) quadrangles for the Mojave were acquired by the Bureau of Land Management (BLM). They were mosaicked and seams and other obvious errors were edited (T. Zmudka, pers. comm.). This dataset formed the basis of our terrain stratification.

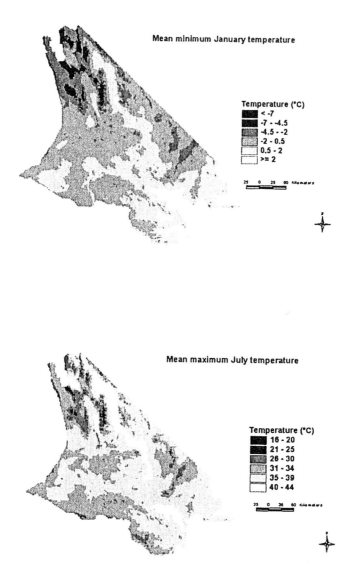

Figure 2. Mean (a) minimum winter (January) and (b) maximum summer (July) temperature, interpolated to a 1 km grid, for the California Mojave.

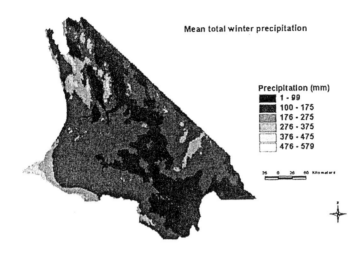

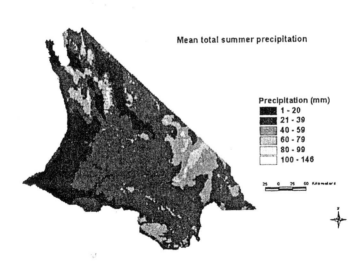

Figure 3. Mean total (a) winter (Nov-Apr) and (b) summer (May-Oct) precipitation, interpolated to a 1 km grid, for the California Mojave.

Table 2. Classification of geologic substrate, based on aggregation of classes in the geologic map of California, 1:750,000 scale (Jennings 1985).

Class	Description (source map categories combined to form this class)
1. Alluvium	Unconsolidated bajadas and alluvial fans; Quaternary, mainly Holocene marine and non-marine origin.
2. Older Alluvium	Old bajada and fans surfaces, consolidated into fanglomerates.
3. Eolian Sand	Extensive sand deposits, sand sheets.
4. Playa	Quaternary playas.
5. Weakly lithified sedimentary rock	"Badlands" -- sedimentary rock of various ages and marine/non-marine origin.
6. Silicic-intermediate rock	Includes igneous plutonic, volcanic/metavolcanic, sedimentary/metasedimentary rock of various ages of silicic-intermediate composition (granite, diorite, rhyolite, andesite, gneiss, etc.).
7. Mafic-ultramafic rock	Includes igneous plutonic, volcanic/metavolcanic, sedimentary/metasedimentary rock of various ages of mafic-ultramafic composition (schist, basalt, gabbro, etc.).
8. Carbonate rock	Sedimentary/metasedimentary rock of various ages with carbonate composition (limestone, dolomite, marble, etc.).

Table 3. Terrain classification: rules applied to topographic variables derived from DEM to assign classes for second stage stratification.

Class	Description (hierarchical decision rules)
Drainage	Flow Accumulation greater than threshold value of 100 cells (9 ha); else
Flat (<1% slope)	Slope less than 1%; else
Gentle Slope (1-10% slope)	Slope less than 10%, else
Northeast Aspect	Slope $\geq$ 10% and aspect 0-90°; else
Southwest Aspect	Slope $\geq$ 10% and aspect 180-270°; else
Neutral Aspect	Slope $\geq$ 10% and aspect 90-180° or 270-360°

Slope aspect, angle and flow accumulation (upslope catchment or contributing area) were calculated using ARC/INFO GIS software. Aspect was derived based on the direction of the maximum rate of change, and slope angle on the maximum difference in elevation, for a 3x3 window. The terrain was classified into six categories (Table 3): drainages (areas of high flow accumulation, corresponding to washes and streams); flat terrain (relatively finer soils and higher soil moisture); gentle slopes (corresponding to most bajada surfaces); and steeper slopes of upland formations, divided into three aspect classes corresponding to higher (southwest), lower (northeast) and intermediate values of insolation (southeast, northwest). Hillslope position, slope curvature, and other terrain variables also are correlated with plant distributions, but this simple scheme likely captured first order effects of terrain on vegetation, nested within the climate-geology stratification, in this desert landscape. Those effects are the influence of slope angle and drainage basin position on soil texture and moisture and the influence of slope aspect on solar insolation and evapotranspiration. Each of the 1100 (33x33) 30-m terrain cells falling within the 1 km^2 cells selected for sampling was classified into one of these classes using the decision rules outlined in Table 3.

Allocation of Sample

Overlaying the five categorized macroscale environmental variables (Tables 1 and 2) could potentially have produced $2 \times 6 \times 2 \times 3 \times 7 = 504$ unique combinations or "environmental classes." Only 167 combinations occurred in the study area. While we did not restrict sampling to belt transects it was necessary to restrict it to public lands. Therefore, we needed to determine if the distribution of environmental classes on public lands was representative of the entire study area. The proportion of land area in each of the environmental classes on public land differed from the proportions in the whole area by at most only a few percent (Table 4). One hundred and sixty of the environmental classes occurred on public lands (seven very small classes occurred only on private lands).

➢ We estimated that resources were available to survey 1000-2000 vegetation plots, and, because there are potentially 1-6 terrain classes per environmental cell, we allocated ~500 cells to the first-stage sample.
➢ Thirty-one of these classes constituted seven cells or less (7 km^2) and were not allocated any cells.
➢ Nine comprised 8-10 km^2 and were allocated one cell that the field crew would attempt to survey, time permitting.
➢ 120 classes contained at least 11 km^2 and were allocated at least one cell.

- Classes with an area of 11-100 km^2 (0.02-0.20% of the study area) were allocated 1-2 cells each.
- Classes comprising 100-1000 km^2 were allocated 2-9 cells each (roughly one per 100 km^2).
- Classes ranging from 1000-5000 km^2 in area were allocated 10-15 cells each.

A computer program was written to select cells randomly from the environmental grid until at least 15 cells per environmental class were selected (establishing both the sample and alternate cells). A total of 501 cells were allocated in the first stage of sampling (Figure 5 – on the CD-ROM).

The same program was used to allocate the second-stage sample – the actual plot locations. At least two plots were selected per terrain class from the terrain grid within each 1-km^2 cell in the sample (again providing an alternate location). Terrain classes were only allocated a plot if they comprised at least 5% of the 1 km^2 cell. Owing to constraints of time, personnel, and accessibility, 1133 vegetation plots in 348 environmental cells were actually surveyed during the 1997-98 field season.

Collection of Vegetation Data in the Field

The UTM coordinates of locations to be sampled were provided to the crew. Global positioning systems (GPS) with 5-10 m precision were used to navigate to the sample location. Field crews were able to adjust their location by up to 90 m so they did not locate a plot on a boundary between distinctive vegetation stands. All perennial plant species were identified and cover estimated in a 1000 m^2 circular plot (vegetation survey methods are described more fully in Thomas et al. in prep., see also www.csps.org). These data were analyzed using classification and ordination methods to define vegetation alliances for the Mojave ecoregion using the standards set by the National Vegetation Classification System (Grossman et al 1998). Thus, each plot could be assigned an alliance label to evaluate the effectiveness of the gradient directed stratification. The survey was conducted primarily in 1998 so this dataset will be referred to as the 1998 survey.

Table 4. Proportions (percent) of the total land area in the study area, the public lands within the study area, and the allocated sample, in each environmental variable class.

Environmental Variable	% Study area	% Public Land	% Sample
Geology			
Alluvium	48.7	45.9	31.9
Older Alluvium	3.3	3.6	5.2
Eolian Sand	2.0	2.0	2.6
Playa	3.7	3.7	4.9
Weakly lithified sedimentary rock	0.5	0.4	1.0
Silicic-intermediate rock	25.4	26.7	29.5
Mafic-ultramafic rock	9.6	10.4	13.7
Carbonate rock	6.7	7.4	11.1
Jan Minimum Temperature			
< –7 °C	1.0	1.1	1.8
–7 to –4.5 °C	5.6	6.3	9.2
–4.5 to –2 °C	13.5	14.9	20
–2 to 0.5 °C	35.6	37.4	36.1
0.5 to 2 °C	27.6	23.7	23.2
≥ 2 °C	16.7	16.6	9.7
Jul Maximum Temperature			
< 35 °C	38.3	39.0	49.1
≥ 35 °C	61.7	61.0	50.9
Winter Precipitation			
<100 mm	39.0	39.6	30.6
100-175 mm	51.4	50.9	56
≥ 175 mm	9.6	9.5	13.4
Summer Precipitation			
<40 mm	77.2	76.3	69.2
≥ 49 mm	22.8	23.7	30.8
Total Area (km^2)	55738	49159	501

Evaluation of the Sample

The distribution of the sampled plots can be compared to the class proportions of the five broad scale environmental variables and the fine-scale terrain classes in the study area. This provides an evaluation of how well we achieved our goal of acquiring replicate samples of both common and rare environmental classes (the environmental "representativeness" of the sample). Also the "efficiency" of sampling can be evaluated by the number of replicates per vegetation alliance, and the number of additional species sampled by each additional plot. A sample that has "efficiently" identified all vegetation alliances in an area would have a relatively even number of replicates per alliance – neither over-sampling common assemblages nor under-sampling rare ones. An efficient sample would also show the number of new species recorded with each additional plot to approach an asymptote (see Nelder et al. 1995).

The representativeness and efficiency of the gradient directed sample were compared to an existing dataset for the same region, referred to as the "retrospective" data. These data consisted of 676 vegetation plots from five separate surveys collected between the 1970s and 1990s. Surveys occurred in selected portions of the study area, but together ranged over a large area (Figure 5 - on the CD-ROM). They used various sampling schemes, plot sizes and data collection methods, but all surveys provided species abundance data and could be assigned to an alliance. We expected the gradient directed sample to capture more vegetation diversity more efficiently than these surveys.

RESULTS

The first stage sample allocated to the broad scale environmental gradients was, on the whole, representative of the geology and climate variable classes that were used as a basis for stratification (Figures 6-8). Allocation of samples to the geology, precipitation and temperature classes was roughly proportional to class area but incorporated the explicit bias to undersample large classes and oversample smaller classes. Only for winter precipitation were there "too many" cells allocated to the largest class (100-175 mm) and too few to the 0-99 mm class (Figure 8). The cells actually sampled (348 of the 501 allocated) even more effectively oversampled rare environmental classes (Figures 6-8).

This contrasts with the distribution of the retrospective data among the environmental variable classes. Although it comprised a reasonably large sample, the retrospective dataset included several surveys allocated using different schemes, and tended to sample proportional to class area (Figures 6-8). In some cases the retrospective data even over-represent widespread environments (see January minimum temperature, Figure 7, and winter precipitation, Figure 8), and under-represent rare classes (Figure 6; eolian sand and carbonate rock).

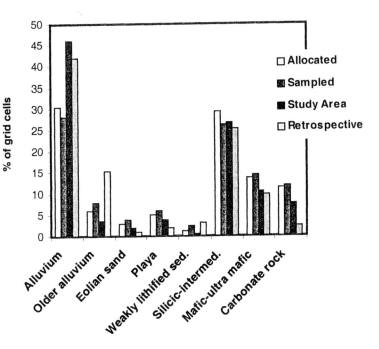

Figure 6. Sampling results for the geology classes –proportion (% grid cells) of the study area and proportion of the sample that was allocated, actually sampled (1998), and of the retrospective data.

The second stage sample, the plots actually surveyed, was representative of the terrain classes that occurred within the first-stage sample and incorporated the explicit bias, except that the "flat" class (<1% slope) was undersampled (Figure 9). This was due to an error in the terrain processing prior to allocating the sample. Figure 10 shows that the most extensive and the rarest environmental classes from the first stage sample are sampled independent of area, while the classes of intermediate size are sampled proportional to area.

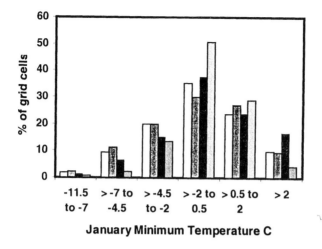

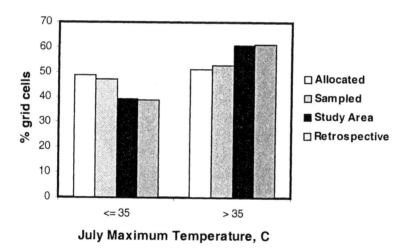

Figure 7. Sampling results for the temperature classes (see caption Figure 6).

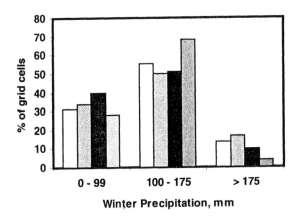

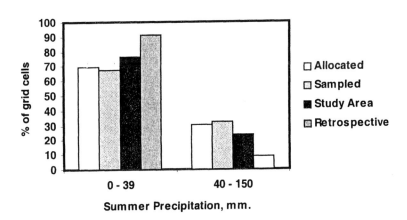

Figure 8. Sampling results for the precipitation classes (see caption Figure 6).

However, although the plots represented the environmental gradients, 49% of the plots in the 1998 survey (n=1133) belonged to the two most common alliances in the Mojave – *Larrea tridentata-Ambrosia dumosa* (411) and *Larrea tridentata* (145). A total of 68 alliances were defined ultimately, and 43 of those were designated for mapping. Of those 43, only 34 alliances had five or more replicates in the 1998 survey (Figure 11). Thus, these data were inadequate for describing rarer alliances.

An additional 76 plots were surveyed in 1999. They were chosen from among the remaining allocated points, or located purposively, to obtain replicates of under-represented alliances, and visit unsampled geographical areas. In addition, another study focusing on desert wash vegetation yielded 128 plots in 1999 (J. Evens, unpublished data). When these 204 plots were combined with the 1998 survey (increasing the sample by 18%) four additional alliances were identified, and only five of the (now) 47 alliances had less that five replicates (Figure 11). The 676 plots in the retrospective dataset were classified using a dichotomous key developed for the alliances (Thomas et al in prep.). Sixty-three percent of the retrospective plots represented the most common alliance (*Larrea tridentata-Ambrosiadumosa*), only 34 alliances had at least one plot, and only 17 alliances had at least 5 plots (Figure 11).

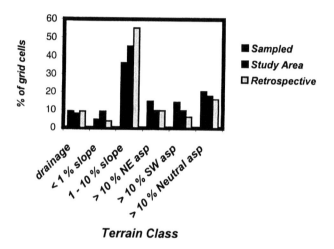

Figure 9. Second stage sampling results for terrain classes -- proportion of the study area and proportion of the sample that was actually sampled (1998).

There were 624 species analyzed from the 1998 survey plots (183 species occur in only one plot and another 76 in only two). The cumulative distribution of species richness in plots appears to be reaching an asymptote (Figure 12). When the 204 additional plots sampled in 1999 were included, the number of perennial plant species sampled increased by 8% to 672 (192 species with only one occurrence and another 72 with only two), and the shape of the species accumulation curve is the same. It has been estimated by Thomas (1996) that the total number of vascular plant species in the California Mojave is 1100-1200 (based on the digital floristic database for California, http://galaxy.cs.berkeley.edu/calflora/). Note that while the number of perennial species listed in the 1998-99 surveys has since been adjusted to 462 (correcting for duplicate listings of subspecies and varieties, and annuals that were erroneously included), we do not expect that the general shape of the species accumulation curve would change.

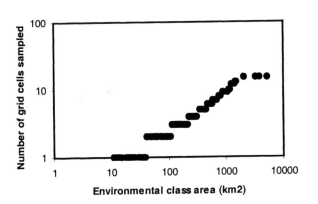

Figure 10. Comparison of the number of cells sampled in each of the coarse-scale environmental (geology-climate) classes with class area, shown on semi-log scale.

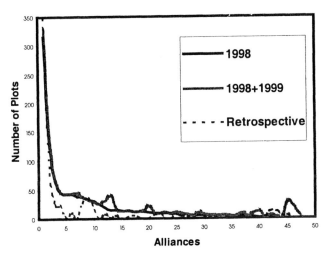

Figure 11. Number of plots representing each of 47 vegetation alliances identified in the sample for the a) 1998 survey, b) 1998+1999 surveys, and c) retrospective dataset.

The distribution of species dominance within the sample, as measured by the rank order of the log of average species cover (mean percent ground cover across all plots), shows a distribution that falls somewhat in between geometric (with a few species strongly dominating many sites) and lognormal (with a large number of moderately abundant species, relative to very abundant or rare ones; Figure 13). The two most abundant species, *Larrea tridentata* and *Ambrosia dumosa*, have one to four orders of magnitude greater average cover than all other species, suggesting a geometric distribution. Whittaker (1965) observed that plant communities in harsh physical environments tend to exhibit a geometric dominance-diversity curve. Our results, however, are based on a sample of the flora and should be interpreted with caution.

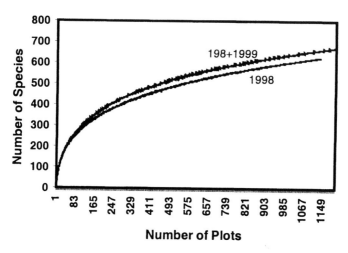

Figure 12. Cumulative number of perennial vascular plant species in plots from the 1998 and 1998+1999 surveys.

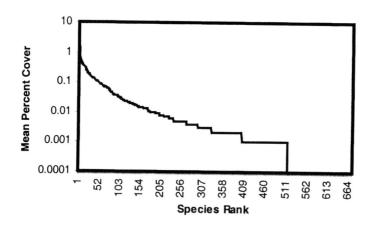

Figure 13. Rank order of plant species average cover for all plots in the sample (1998 and 1999 surveys combined), showing species dominance-diversity.

DISCUSSION

A biological survey of a large region is often expected to serve a number of goals. This study is no exception. Data on vegetation composition at sampled locations were intended to:

A) Describe quantitatively all vegetation alliances found in the region – therefore, survey floristic diversity as completely as possible, and sample alliances with replication;

B) Examine the relationship between floristic composition (at the alliance level and below) and environmental variables using ordination techniques – thus, sample environmental gradients as widely as possible (more observations in widespread alliances would presumably reveal greater variability at the association level);

C) Calibrate predictive models so that survey data can be "interpolated," based on relationships between vegetation and environmental variables, in order to develop a vegetation map – again, for this purpose, sampling proportional to the area of environmental classes would probably allow common alliances with wider tolerances (characterized by species with broader niches) to be modeled more accurately, but replication (including geographical) is required for all alliances to be modeled.

Achieving these goals using the gradient directed sampling strategy depends on the relationship between species distributions and the gradients used to stratify the sample. It also depends upon the size of the area and the characteristics of the flora – characteristics that could be expressed in the dominance-diversity relationship, for example (Westman 1985, p. 448). The Mojave desert flora could probably be characterized by a geometric distribution – a large concentration of resources (relative importance – biomass, density or cover) in a few species, with comparable numbers of moderately abundant and rare species (Figure 13). Also, because mountain ranges within the basin and range topography of the Mojave constitute habitat islands and Pleistocene refugia (Brown 1971, Mead 1982, Grayson and Livingston 1993), these environmental classes of more limited extent (Figures 1-3) probably require greater survey effort. We expect them to have higher plant diversity because of relict diversity, endemism (on carbonate substrates), higher available moisture, and local habitat diversity due to the complexity of terrain and substrate.

Further, our ability to develop predictive models (goal C above) is related to the detail of the vegetation classification (goal A) and the resolution of the map. There are alliances described by the vegetation survey that typically occur in stands smaller than the minimum mapping unit (5 ha). Our strategy was to allocate a stratified sample along environmental gradients, with unequal proportions (adjusted to allocate relatively more observations to rare classes and fewer to common ones). This appears to have been appropriate for meeting goals B and C. However, given the geometric distribution of plant species dominance, sampling equally among environmental classes would probably have surveyed more species with greater efficiently (goal A).

Even sampling with equal numbers of observations would not necessarily capture more species efficiently if the mapped environmental' gradients are not strongly correlated with plant distributions. This is the case when 1) key variables or their proxies are not available in map form (for example landform or soils), or 2) mapped variables do not adequately depict important gradients due to inappropriate map scale (e.g., terrain features derived from 30-m DEMs do not model microtopographic features that may be critical predictors of desert plant distributions). Further, biotic factors affecting plant distributions are not taken into account in gradient directed sampling. Finally, a greater overall sampling effort may be required to i) collect a larger sample for common types (required to define sub-alliance associations and predict their broad distribution); ii) encounter more of the rare alliances present in the study area; and, iii) provide replicate observations of those rare types required both for quantitative definition of the alliance and for predictive modeling.

ACKNOWLEDGEMENTS

Funding and in-kind support were provided by the Department of Defense Legacy Program, Environmental Protection Agency, Bureau of Land Management Barstow and California Districts, National Park Service, California State University Desert Studies Center and University of California Granite Mountain Reserve. Many people have contributed to this study, especially R. Dokka, C. Everly, E. Ezcurra, R. Fullerton, S. Hathaway, T. Pavlis, J. Thorne and T. Zmudka, and we thank them.

REFERENCES

Austin MP. An ecological perspective on biodiversity investigations: examples from Australian eucalypt forests. Annals of the Missouri Botanical Garden. 1998; 85:2-17.

Austin MP, Heyligers PC. Vegetation survey design for conservation: gradsect sampling of forests in North-eastern New South Wales. Biological Conservation. 1989; 50:13-32.

Austin MP, Heyligers PC. "New approach to vegetation survey design: gradsect sampling." In *Nature conservation: cost effective biological surveys and data analysis*, CR Margules, MP Austin eds. East Melbourne, Australia: CSIRO, 1991.

Bailey TC, Gatrell AC. *Interactive spatial data analysis*. Harlow, UK: Longmans, 1998.

Beatley JC. Effects of rainfall and temperature on the distribution and behavior of *Larrea tridentata* (creosote bush) in the Mojave Desert of Nevada. Ecology. 1974; 55:245-261.

Berry BJL, Baker AM. "Geographic sampling". In *Spatial analysis: a reader in statistical geography*, BJL Berry, DF Marble eds. New Jersey: Prentice-Hall, YEAR.

Bojorquez-Tapia LA, Azuara I, Ezcurra E, Flores-Villa O. Identifying conservation priorities in Mexico through Geographic Information Systems and modeling. Ecological Applications. 1995; 5:215-231.

Bourgeron PS, Engelking LD, Humphries HC, Muldavin E, Moir WH. Assessing the conservation value of the Gray Ranch: rarity, diversity and representativeness. Desert Plants. 1995; 10/11:5-51.

Bourgeron PS, Humphries HC, Jensen ME. "General sampling design considerations for landscape evaluation." In *Eastside forest ecosystem health assessment – Volume II: Ecosystem management: principles and applications*, ME Jensen, PS Bourgeron eds. General Technical Report PNW-GTR-318. Portland OR: US Department of Agriculture, Forest Service, Pacific Northwest Research Station, 1994.

Brown JH. Mammals on mountaintops: non-equilibrium insular biogeography. American Naturalist. 1971; 105:467-478.

Cochran WG. *Sampling techniques*. New York: John Wiley and Sons., 1977.

Congalton RG. A review of assessing the accuracy of classification of remotely sensed data. Remote Sensing of Environment. 1991; 37:35-46.

Davis FW, Stine PA, Stoms DM, Borchert MI, Hollander A. Gap analysis of the actual vegetation of California: 1. The southwestern region. Madrono. 1995; 42:40-78.

Davis FW, Stoms DW, Estes JE, Scepan J, Scott JM. 1990. An information systems approach to the preservation of biological diversity. International Journal of Geographic Information Systems. 1990; 4:55-78.

Franklin J. Predictive vegetation mapping: geographic modeling of biospatial patterns in relation to environmental gradients. Progress in Physical Geography. 1995: 19:474-499.

Franklin J, Woodcock CE. "Multiscale vegetation data for the mountains of Southern California: spatial and categorical resolution." In *Scale in remote sensing and GIS*, DA Quattrochi, MF Goodchild eds. Boca Raton, FL: CRC?Lewis Publishers, 1997.

Gillison AN, Brewer KRW. The use of gradient directed transects or gradsect in natural resources survey. Journal of Environmental Management. 1985; 20:103-127.

Grayson DK, Livingston SD. Missing mammals on Great Basin mountains: Holocene extinction and inadequate knowledge. Conservation Biology. 1993; 7:527-532.

Grossman DH, Faber-Langendeon D, Weakley AS, Anderson M, Bourgeron P, Crawford R, Goodin K, Landaal S, Metzler K, Patterson K, Pyne M, Reid M, Sneddon L. *International Classification of Ecological Communities: Terrestrial vegetation of the United States Volume 1. The National Vegetation Classification System: Development, Status and Applications*. http://consci.tnc.org/library/pubs/class/index.html. Washington DC: The Nature Conservancy, 1998.

Holland RF. *Preliminary descriptions of the terrestrial natural communities of California*. Sacramento CA: The Resources Agency, Nongame Heritage Program, California Department of Fish and Game, 1986.

Hunt CB. *Plant ecology of Death Valley, California*. U.S. Geologic Survey Professional Paper No. 509. Washington DC: USGS, 1966.

Jennings CW. An explanatory text to accompany the 1:750,000 scale fault and geologic maps of California. Sacramento CA: California Department of Mines and Geology

Bulletin 201. California Department of Conservation, Division of Mines and Geology, 1985.

Leathwick JR. Are New Zealand's *Nothofagus* species in equilibrium with their environment? Journal of Vegetation Science. 1998; 9:719-732.

Mackey BG. A spatial analysis of the environmental relations of rainforest structural types. Journal of Biogeography. 1994; 20:303-336.

Mackey BG, Nix HA, Hutchinson MF, Macmahon JP, Fleming PM. Assessing the representativeness of places for conservation reservation and heritage listing. Environmental Management. 1988; 12:501-514.

Mackey BG, Nix HA, Stein JA, Cork SE, F. T. Bullen FT. Assessing the representativeness of the wet tropics of Queensland World Heritage Property. Biological Conservation. 1989; 50:279-303.

Margules CR, Austin MP. Biological models for monitoring species decline: the construction and use of databases. Philosophical Transactions of the Royal Society of London B. 1994; 344:69-75.

McAuliffe JR. Landscape evolution, soil formation, and ecological patterns and processes in Sonoran desert bajadas. Ecological Monographs. 1994; 64:111-148.

Mead JI. Late Quaternary environments and biogeography in the Great Basin. Quaternary Research. 1982; 17:39-55.

Michaelsen, J. in prep. Optimal interpolation of climate station data with the aid of a digital elevation model.

Montana C. A floristic-structural gradient related to land forms in the southern Chihuahuan Desert. Journal of Vegetation Science. 1990; 1:669-674.

Moore ID, Grayson RB, Ladson AR. Digital terrain modeling: a review of hydrological, geomorphologic and biological applications. Hydrological Processes. 1991; 5:3-30.

Nelder VJ, Crossley DC, Cofinas M. Using geographic information systems (GIS) to determine the adequacy of sampling in vegetation surveys. Biological Conservation. 1995; 73:1-17.

Nix HA. "Environmental determinants of biogeography and evolution in Terra Australis." In *Evolution of the flora and fauna or arid Australia,* WR Barker, PJM Greenslade eds. Adelaide: Peacock Press, 1982.

Pavlik BM. Phytogeography of sand dunes in the Great Basin and Mojave Deserts. Journal of Biogeography. 1989; 16:227-238.

Schoenherr AA. *A natural history of California.* Berkeley: University of California Press, 1992.

Scott JM, Davis F, Csuti B, Noss R, Butterfield B, Groves C, Anderson H, Caicco S, D'Erchia F, Edwards Jr TC, Ulliman J. Wright RG. Gap analysis: a geographical approach to protection of biological diversity. Wildlife Monographs. 1993: 123(January):1-41.

Scott JM, Jennings MD. Large-area mapping of biodiversity. Annals of the Missouri Botanical Garden. 1998; 85:34-47.

Smith TM, Shugart HH, Woodward FI. *Plant functional types: their relevance to ecosystem properties and global change.* Cambridge: Cambrigde University Press, 1997.

Thomas KA. Vegetation and floristic diversity in the Mojave Desert of California: a regional conservation evaluation. Ph.D. Dissertation. University of California, Santa Barbara, 1996.

Thomas KA, Stine PA, Keeler-Wolf T, Franklin J. The Mojave vegetation database: Central section. Technical Report No. # to Department of Defense Legacy Program Mojave Desert Ecosystem Program. Flagstaff AZ: USGS, in prep.

Thompson SK. *Sampling.* New York: John Wiley & Sons, 1992.

Thorne RF. The desert and transmontane plant communities of southern California. Aliso. 1982; 10:219-257.

Vasek FC, Barbour MG. "Mojave desert scrub vegetation." In *Terrestrial Vegetation of California,* M G Barbour, J Major eds. New York: John Wiley & Sons, 1977.

Vasek FC, Thorne RF. "Transmontane coniferous vegetation." In *Terrestrial Vegetation of California,* M G Barbour, J Major eds. New York: John Wiley & Sons, 1977.

Venables WM, Ripley BD. *Modern applied statistics with S-Plus.* New York: Springer-Verlag, 1999.

Westman WE. *Ecology, impact assessment and environmental planning.* New York: John Wiley & Sons, 1985.

Whittaker RH. Vegetation of the Siskyou Mountains, Oregon and California. Ecological Monographs. 1960: 30:279-338.

Whittaker, RH. Dominance and diversity in land plant communities. Science. 1965; 147:250-260.

Whittaker, RH. Direct gradient analysis. In *Handbook of Vegetation Science 5: Ordination and Classification of Communities,* RH Whittaker ed.. The Hague: Dr W. Junk, 1973.

Yeaton RI, Yeaton III RW. Waggoner JP, Horenstein JE. The ecology of *Yucca* (Agavaceae) over an environmental gradient in the Mojave desert: distribution and interspecific interactions. Journal of Arid Environments. 1985; 8:33-44.

Chapter 15

DEVELOPMENT OF VEGETATION PATTERN IN PRIMARY SUCCESSIONS ON GLACIER FORELANDS IN SOUTHERN NORWAY

Günter A. Grimm
School of Geography, University of Oxford, Mansfield Road, Oxford OX1 3TB, U.K.
gunter.grimm@geography.ox.ac.uk

Keywords: Pattern analysis, autocorrelation, fractal dimension, fragmentation index, primary succession.

Abstract: This study evaluates spatial patterns exhibited by dominant woody plant species on terrain of increasing age in front of retreating glaciers. It analyses distribution patterns in terms of their correlation to geomorphological features on the glacier forelands, dispersal mechanisms and the time component. It is envisaged that a mix of these factors plays a major role in the development and change of the distribution patterns of the selected species and that this role should be incorporated in the interpretation of the more traditional analyses of primary successions in those areas (e.g. gradient analysis, correlation of environmental factors with vegetation). Vegetation pattern development on five glacier forelands in southern Norway was analysed using a fragmentation index, an autocorrelation measure and fractal dimension. Results show that the dispersal mode of the investigated plants (wind or animal dispersed) influences the way distribution patterns develop over a time span of about 250 years. It is also evident that there are pronounced differences between high and low altitude glacier forelands in terms of the vegetation patterns developing over time. Finally, it is argued that a suite of analytical tools and methods is more appropriate for interpreting the ecological processes on glacier forelands than the use of any single method.

INTRODUCTION

Chronosequence studies have been successful in elucidating some of the processes and pathways involved in primary successions on glacier forelands (Matthews 1979, Matthews and Whittaker 1987, Walker and Chapin III 1987). There is no doubt about the usefulness of this approach and the wealth of information it produced on succession processes in those areas. However, in addition to the conventional correlation of vegetation with environmental factors and the idea of a sequential development of plant communities, the spatial arrangement of the plant species on glacier forelands holds another key to the understanding of primary succession (Friedel 1938, Lüdi 1958, Jochimsen 1962, Böhmer 1999, Matthews 1999). Matthews' (1992) geoecological approach to succession research on glacier forelands stresses the importance of both autogenic (biological) and allogenic (abiotic) processes (see also White 1979). A combination of both leads to the distribution patterns exhibited by colonising plant species. These patterns are expected to reflect some identifiable life history traits of the plants, which can be used to explain some of the phenomena that conventional methods like gradient analysis cannot address.

A variety of spatially explicit analytical techniques, some of which have only become available over the last 15 years to plant ecologists, were employed in this study to elucidate the potential role of spatial patterns in research into primary succession on glacier forelands.

This chapter examines the spatial patterns exhibited by dominant woody plant species along a time gradient on glacier forelands in southern Norway. Specifically, it addresses the hypothesis that animal dispersed species enter the forelands with their biotic vectors, advance large distances across the "virgin" terrain and establish small, isolated nuclei, which grow and over time coalesce into larger more continuous distribution areas. Wind dispersed species on the other hand are expected to advance in a more homogenous front across the foreland and as time passes become more fragmented by competition with other species. It has been pointed out that dispersal mechanisms are not necessarily reflected in colonisation ability (e.g. Ryvarden 1971, 1975) but these mechanisms are expected to establish themselves in different distribution patterns.

STUDY AREA

Five glacier forelands in southern Norway were selected (Figure 1). Three low altitude and two high altitude forelands were included in the study. This was done to make sure that different successional pathways, which have been shown to exist on glaciers at varying altitudes, were covered. The low altitude glacier forelands are all in front of outlet glaciers of the Jostedalsbreen icefield in southern Norway. The area is underlain by Precambrian gneiss with a few outcrops of ultrabasic rocks (Sigmond et al. 1984). Nigardsbreen is on the south-eastern side of the Jostedalsbreen icefield (long. 7°15' E, lat. 61°40' N). The foreland covers an area of 3.2 km2 over an altitudinal range from 250 m to 355 m asl. Dating of the moraines is from Bickerton and Matthews (1992) and sets the framework for the estimated dates of the surveyed blocks. Austerdalsbreen (long. 7° E, lat. 61°36' N) is on the southeastern side of the icefield. With an 310 m to 400 m elevational range, the foreland lies slightly higher than Nigardsbreen and at 3 km2 it covers a slightly smaller area. The dating for Austerdalsbreen is taken from Petch and Whittaker (1997). The last low altitude glacier is Bødalsbreen (long. 7°7' E, 61°46' N) on the north-western side of the icefield. Bødalsbreen covers an area of 0.45 km2 at about 600 m asl. Moraine dates for Bødalsbreen stem from Bickerton and Matthews (1992).

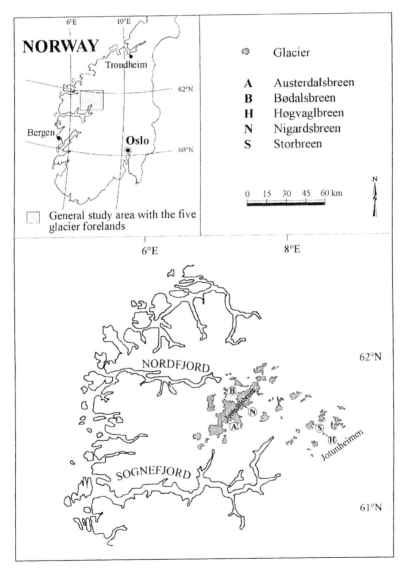

Figure 1. Location of the study sites.

All three low altitude glacier foreland lie below the treeline and are covered by subalpine Betula forest interspersed with pre-alpine Alnus glutinosa forest fragments and additions of sub-alpine heath species. Both of these mature forest types are commonly found at this altitude in western Norway (Odland 1981, Moen 1987). The high altitude glaciers are both situated in the Jotunheimen area, which is part of the Smørstabbtinden massif to the east of the Jostedalsbreen. The bedrock underlying the Jotunheimen area has been described by Battey and McRitchie (1973, 1975) as pyroxene-granulite gneiss. The Storbreen foreland (long. 8°10' E, lat. 61°35' N) lies above the treeline in the alpine vegetation zone and is dominated by heath communities with e.g. Betula nana, Arctostaphylos alpina, Vaccinium myrtillus, and several species of Salix (Moen 1987). It extends from 1150 m to 1550 m and covers an area of about 2 km2. Dates for the moraine sequence at Storbreen are taken from Matthews and Whittaker (1987). The Høgvaglbreen foreland (long. 8°22' E, lat. 61°31' N) spans an altitude from 1400 to 1600 m asl. and covers an area of about 0.8 km2. It also lies entirely above the tree line in the alpine vegetation zone and is covered by a very species poor heath community dominated by Salix herbacea and Loiseleuria procumbens. Its moraine sequence has been dated by Erikstad and Sollid (1986). These dates provide the basis for the estimation of time since deglaciation for the surveyed blocks.

METHODS

Field Sampling

Sampling was conducted from 26 June to 2 August in the Summer of 1996. Dominant woody species were recorded in 30 blocks, ranging in size from 10 x 15 m to 21 x 30 m. The location of the sites was selected subjectively, so as to cover a variety of terrain ages on the different forelands. The presence/absence of the species within the blocks was recorded in 0.25 m2 cells (see Figure 2) and this information was used to compile distribution maps of the species.

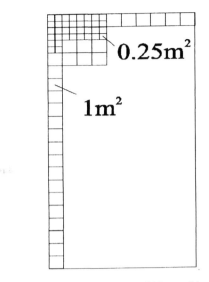

Figure 2. Schematic outline of sampling quadrats within one block.

ANALYTICAL METHODS

Spatial properties of species distributions can be described in their most basic way by the area/perimeter index (area being the number of 0.25 m² cells and perimeter the number of 0.5 m border segments of a given distribution). The index can be used for different species at the same location to assess dominance. Furthermore it allows for the direct comparison of the fragmentation of the distribution of a species in a given area. When the values are compared for the same species in blocks of increasing age, any change of these indices during successions can be used to estimate underlying ecological and geophysical processes.

A well known feature of spatial data is autocorrelation. It is brought about by various factors, such as restriction to suitable patchy environments, dispersal mechanisms, clonal growth, etc. This means that the identification of autocorrelation for a certain data set does not necessarily mean that the underlying processes are known, although good guesses usually can be made. The autocorrelation values of the species in this study were calculated with Moran's I using IDRISI software (Eastman 1987-1995). The distribution maps were analysed using the King's case (all eight surrounding cells contribute to the calculation of Moran's I). Diagonally neighboring cells are given a weight of 0.7071, whereas vertically and horizontally adjacent cells are given a weight of 1. Moran's I is the measure to describe the propensity of cells to be surrounded by similar or dissimilar cells. It can vary from -1 to 1, with values close to -1 describing very rough surfaces where cells are surrounded by very dissimilar cells and values close to 1 for smooth surfaces on which cells are surrounded by similar cells.

Fractal dimension is another tool available for ecologists to evaluate spatial properties of patterns. There have been many different approaches with the use of fractals mainly in geology and in the geophysical sciences. In the original form fractals are a measure of self-similarity (Mandelbrot 1983). Most of the studies using fractals are concerned with characterising a stationary state of a system (e.g. Burrough 1981, Phillips 1985, Palmer 1988, De Cola 1989, Fitter and Stickland 1992, and Virkkala 1993). In this study, fractals are being used to investigate the temporal change of distribution patterns, thereby helping to test hypotheses about the importance of ecological processes on primary successions. The values for the fractal dimension were calculated using De Cola's (1989) approach ($D_{De\ Cola}$; Equation 1). It provides a measure of the perimeter-complicatedness of an area with values for $D_{De\ Cola} \sim 1$ for smooth perimeters and $D_{De\ Cola} > 1$ for complicated perimeters. In this study the fractal dimension for the distribution of a species within a block was calculated by using the average of the fractal dimensions of all the individual patches within the block.

$$D_{DeCola} = \frac{\ln(p) - \ln(4)}{\ln(\sqrt{s})} - \varepsilon \qquad \text{Equation (1)}$$

where   p = perimeter
        s = area
        ε = error term

Another way to calculate the fractal dimension of a distribution area has been suggested by Virkkala (1993). In his approach the slope of the regression of the logarithm of the number of area units with the logarithm of the scale is used as D. This represents the quantity of squares of a given area occupied by a species and is restricted to between 1 and 2 ($D = 2$ means a continuous cover of squares at all scales). Gautestad (1994) suggested improvements to the calculation of D since he shows that Virkkala's calculation of D rests on a misinterpretation of the theoretical relationship between spatial resolution, number of gridcells covered by the species, and the fractal dimension. It is this modification that is used to calculate D in this study. Using this transformation, the new range of D_{Gau} is $0 \leq D_{Gau} \leq 2$ ($D_{Gau} = 0$ for widely spaced distributions and $D_{Gau} = 2$ for total cover).

RESULTS

The mean of the perimeter/area index for animal and wind dispersed species shows no distinct trend over successional time for all forelands (see Table 2). If these mean values are investigated for the individual forelands

Table 2. Mean perimeter/area index values for animal and wind dispersed species ordered chronologically. Aus refers to Austerdalsbreen, Bod to Bødalsbreen, Nig to Nigardsbreen, Hog to Høgvaglbreen and Sto to Storbreen forelands. Aus, Bod and Nig are low altitude (l) and Hog and Sto are high altitude (h) forelands.

Foreland block name	Estimated age at time of investigation	Mean of animal dispersed species	Mean of wind dispersed species
Sto8 (h)	45	2.09	2.50
Bod3 (l)	65	1.70	2.07
Aus3 (l)	91	2.31	1.91
Hog5 (h)	119	-	0.98
Aus2 (l)	126	1.50	2.07
Sto6 (h)	144	2.88	2.58
Nig4 (l)	149	1.59	2.15
Nig7 (l)	149	2.15	2.26
Nig5 (l)	155	1.80	1.89
Hog4 (h)	156	2.00	1.59
Hog6 (h)	160	-	0.66
Aus1 (l)	161	2.11	2.01
Hog3 (h)	165	2.80	1.65
Nig2 (l)	166	1.73	2.25
Nig8 (l)	166	1.92	1.68
Nig1 (l)	172	1.69	2.19
Bod2 (l)	176	1.34	2.09
Nig9 (l)	182	1.45	1.83
Sto3 (h)	186	2.08	2.15
Sto4 (h)	186	1.96	2.02
Hog2 (h)	186	4.00	1.89
Sto1 (h)	216	2.35	1.96
Sto2 (h)	216	2.68	1.65
Sto5 (h)	216	2.59	1.91
Bod1 (l)	229	1.42	1.79
Nig3 (l)	231	1.88	2.13
Nig6 (l)	231	1.69	2.20
Hog1 (h)	9996	4.00	2.27
Nig10 (l)	9996	1.84	2.23
Sto7 (h)	9996	1.53	2.34

still no conclusive pattern of change over successional time can be shown. It becomes clear, however, that 81% of blocks on low altitude forelands show higher values for the wind than for animal dispersed species. In contrast 67% of blocks on high altitude forelands show higher values for the animal than for wind dispersed species.

Since the mean is strongly influenced by extreme outliers, investigation of the behaviour of individual species was expected to reveal more about the pattern change over successional time. The four dominant animal dispersed species (*Empetrum hermaphroditum, Vaccinium uliginosum, V. myrtillus* and *V. vitis-idaea*) show a consistent pattern of indices from higher values on young ground within the low altitude forelands to lower numbers on older ground (Table 3).

Table 3. Perimeter/area indices for four animal dispersed (a) species. Aus refers to Austerdalsbreen, Bod to Bødalsbreen and Nig to Nigardsbreen forelands. The age is estimated maximal time since deglaciation.

Species	Aus3 (91 yrs)	Aus2 (126 yrs)	Aus1 (161 yrs)	Bod3 (65 yrs)	Bod2 (176 yrs)	Bod1 (229 yrs)	Nig7 (149 yrs)	Nig2 (166 yrs)	Nig6 (231 yrs)
Vaccinium uliginosum (a)	2.34	1.13	0.61	1.67	1.56	1.20	2.48	2.26	1.97
V. myrtillus (a)	1.32	0.60	0.96	1.91	1.25	0.98	3.18	2.37	2.19
V. vitis-idaea (a)	3.56	0.64	-	1.81	1.57	0.69	1.55	1.21	1.67
Empetrum hermaphroditum (a)	0.41	0.08	0.40	0.99	0.17	0.12	1.38	1.08	0.94

For the dominant wind dispersed species there is neither a common trend across the forelands, nor a general pattern for the change of the index across the species. The only consistent feature is that the values for *Salix phyllicifolia* are consistently higher (i.e. it is present in many small, highly fragmented patches) than for the other species. Visual comparison of the distribution maps confirms that this species is far less abundant and only present in small patches in the surveyed blocks.

As for the perimeter/area index, mean Moran's I values for the two dispersal groups for all the forelands show no distinguishable trend through successional time (Table 4). Temporal patterns discernible on the individual forelands include an increase of the values for the wind dispersed species on the Austerdalsbreen foreland with age. On the Bødalsbreen foreland, the values for the animal dispersed species decrease with age. When the values for the animal and wind dispersed species on low altitude forelands are compared, 75% of the blocks show consistently higher values for animal than wind dispersed species in the same blocks. On the high altitude forelands, however, no such trend can be distinguished.

For three of the dominant animal dispersed species (*Vaccinium uliginosum, V. myrtillus* and *V. vitis-idaea*) Moran's I increases consistently with time since deglaciation at Austerdalsbreen and Bødalsbreen (see Table 4). The development of autocorrelation of *Empetrum hermaphroditum*, another dominant animal dispersed species is, however, not as consistent. The fractal dimension values (D_{Gau} or $D_{De\ Cola}$) exhibit no distinct trend over time when data for all forelands are combined.

Table 4. Mean autocorrelation values (Moran's I; King's case) for animal and wind dispersed species ordered chronologically.

Foreland block name	Estimated age at time of investigation	Mean of animal dispersed species	Mean of wind dispersed species
Sto8	45	0.31	0.21
Bod3	65	0.49	0.41
Aus3	91	0.28	0.33
Hog5	119	-	0.35
Aus2	126	0.43	0.39
Sto6	144	0.17	0.21
Nig4	149	0.52	0.37
Nig7	149	0.38	0.28
Nig5	155	0.41	0.42
Hog4	156	0.44	0.18
Hog6	160	-	0.33
Aus1	161	0.37	0.48
Hog3	165	0.22	0.25
Nig2	166	0.47	0.38
Nig8	166	0.30	0.31
Nig1	172	0.49	0.36
Bod2	176	0.47	0.31
Nig9	182	0.58	0.45
Sto3	186	0.00	0.27
Sto4	186	0.41	0.32
Hog2	186	0.37	0.33
Sto1	216	0.20	0.30
Sto2	216	0.24	0.41
Sto5	216	0.30	0.35
Bod1	229	0.45	0.36
Nig3	231	0.46	0.38
Nig6	231	0.47	0.33
Hog1	9996	0.00	0.16
Nig10	9996	0.38	0.30
Sto7	9996	0.35	0.06

Table 5. Autocorrelation values (King's case) for four animal dispersed species. Aus refers to Austerdalsbreen and Bod to Bødalsbreen. The age is estimated maximal time since deglaciation.

Species	Aus3 (91 yrs)	Aus2 (126 yrs)	Aus1 (161 yrs)	Bod3 (65 yrs)	Bod2 (176 yrs)	Bod1 (229 yrs)
Vaccinium uliginosum	0.28	0.63	0.83	0.51	0.57	0.65
V. myrtillus	0.33	0.47	0.62	0.44	0.56	0.59
V. vitis-idaea	0.09	0.82	-	0.47	0.57	0.81
Empetrum hermaphroditum	0.59	0.34	0.57	0.54	0.23	0.09

DISCUSSION

One consistent pattern of the D_{Gau} values is that in 69% of the blocks on low altitude forelands the values for animal dispersed species are higher than for wind dispersed species. On the high altitude forelands, 91% of blocks show higher D_{Gau} values for wind than for animal dispersed species.

Primary succession on glacier forelands is the result of a complex of interwoven processes. Simple chronological ordering of the blocks from all forelands would not be expected to produce clear trends for the four spatial pattern descriptors (perimeter/area index, Moran's I, and two fractal dimension measures). The results presented in Tables 1, 3, and 5 demonstrate that indeed none of the descriptors shows a clear trend when analysed in this general way. There are a number of problems associated with chronosequence approaches that could explain a lack of temporal correlation. The chronological ordering of the blocks in itself can be affected by difficulties in the accurate dating of recently deglaciated terrain (Matthews 1992: 11). Other factors like disturbances, which have been known to reset successional seres, varying starting conditions and diverging or parallel pathways of succession also play important roles (e.g. Fastie 1995, Vetaas 1997). It also has to be noted that the earliest stages of the succession on the forelands were not sampled, which reduces the overall length of the time gradient, making the detection of temporal developments more difficult. Since regional differences are known to exist in primary successions on glacier forelands (for summary see Matthews 1992: 220), the mixture of data from low and high altitude forelands might well obscure any trend.

The next step in the analysis of the data was therefore to investigate the pattern development for the three low and two high altitude forelands separately. For the mean perimeter/area index of all animal and all wind dispersed species no time trend for the individual forelands could be found. Some of the other descriptors however show distinct trends. On the

Bødalsbreen foreland the pattern descriptor that directly supports the hypothesis of differential colonisation for wind and animal dispersed species, is the increase of the values of D_{Gau} for animal dispersed species. This increase can be interpreted as an increase of overall cover of the species with time. D_{Gau} values for wind dispersed species however also increase over time. This result may be an indication that the wind dispersed species on this foreland have not yet reached the cover levels which produce the necessary competitive pressures to cause the distribution to break up into more fragmented areas. The temporal decrease of Moran's I values for animal dispersed species on the Bødalsbreen foreland also does not directly support the hypothesis of differential colonisation. It is, however, conceivable that by being dominated in its earlier successional stages by two species (i.e. *Calluna vulgaris* and *Empetrum hermaphroditum*) both wind and animal dispersed species have to compete for space to establish themselves in the dominant groundcover. This would subsequently lead to the observed trends of the formation of more continuous, less fragmented areas for the investigated species. Confirmation of this hypothesis can be found in the decreasing Moran's I values for the two dominant species. They are also the only species that show an increase in the number of unconnected patches over time indicating a fragmentation of their distribution with a parallel decrease in their average patch size from > 550 m² to < 10 m².

On the Austerdalsbreen foreland there is an increase of the Moran's I values for wind dispersed species as well as an increase in $D_{De\ Cola}$ values for animal dispersed species. Both trends are opposing the expected trends of the hypothesis of differential colonisation. They suggest that wind dispersed species develop from more patchy to more continuous distributions whereas the perimeter of distribution areas of animal dispersed species becomes more complicated (convoluted) with time. On this foreland, as on the Bødalsbreen foreland, one species (*Empetrum hermaphroditum*) dominates the three blocks (average area for each block is 235, 581.25, and 159.9 m²). With one species so dominant, all other species will be restricted in their potential to develop their distributions. It is only through time that they can enlarge their distributions competitively against the dominant species, which would explain the trend in the pattern descriptors.

On the Høgvaglbreen foreland, D_{Gau} values for animal dispersed species decrease over time. These values are, in this specific case, just the values of one animal dispersed species, namely *Empetrum hermaphroditum*. Since this species is only present in a few cells in the surveyed blocks, scarcity of distribution values may distort the fractal dimension. Caution is therefore advised for the use of these descriptors for very scarce distributions.

Table 6. Fractal dimension values ($D_{De\ Cola}$, after De Cola, 1989 and D_{Gau}, after Gautestad, 1994) animal dispersed and wind dispersed species. Aus refers to Austerdalsbreen, Bod to Bødalsbreen and Nig to Nigardsbreen forelands.

Foreland block name	Estimated age at time of investigation	$D_{De\ Cola}$ animal dispersed	$D_{De\ Cola}$ wind dispersed	D_{Gau} animal dispersed	D_{Gau} wind dispersed
Sto8	45	1.15	1.16	1.05	1.55
Bod3	65	1.13	1.19	0.93	0.64
Aus3	91	1.15	1.21	0.97	1.03
Hog5	119	-	1.19	-	1.80
Aus2	126	1.16	1.19	1.21	0.81
Sto6	144	1.10	1.14	0.49	1.17
Nig4	149	1.17	1.12	1.06	0.63
Nig7	149	1.18	1.05	0.88	0.26
Nig5	155	1.20	1.09	1.46	0.58
Hog4	156	1.11	1.19	0.86	1.25
Hog6	160	-	1.42	-	1.92
Aus1	161	1.17	1.20	1.12	1.01
Hog3	165	1.22	1.26	0.68	1.37
Nig2	166	1.24	1.24	1.14	0.68
Nig8	166	1.19	1.25	1.18	1.15
Nig1	172	1.17	1.12	0.95	0.85
Bod2	176	1.18	1.12	1.16	0.77
Nig9	182	1.19	1.11	1.09	0.76
Sto3	186	1.27	1.20	1.12	1.39
Sto4	186	1.19	1.25	1.09	1.33
Hog2	186	1.00	1.14	-	1.16
Sto1	216	1.17	1.15	1.18	1.21
Sto2	216	1.16	1.18	0.70	1.39
Sto5	216	1.15	1.19	1.02	1.33
Bod1	229	1.15	1.18	1.34	1.04
Nig3	231	1.15	1.13	1.19	0.61
Nig6	231	1.15	1.18	1.04	0.41
Hog1	9996	1.00	1.02	0.29	2.00
Nig10	9996	1.23	1.10	1.09	0.59
Sto7	9996	1.21	1.10	1.39	1.87

None of the other forelands shows any signs of discernible trends for the average spatial descriptors of animal and wind dispersed species over time. Blocks on the individual forelands contain a variety of different environments, the ratio of which alter for the series of blocks surveyed. This contributes to a situation in which true replicate samples for a sound statistical analysis could not be collected. The data is therefore bound to be influenced by unusual effects making a detection of temporal trends hard.

To investigate the hypothesis of differential colonisation further, the pattern development of four animal dispersed species (*Vaccinium myrtillus*, *V. uliginosum*, *V. vitis-idaea,* and *Empetrum hermaphroditum*) was examined individually. The four selected species exhibit a general decrease of the perimeter/area index on the low altitude forelands (Table 3). This is in keeping with the expected development from many smaller patches to fewer, larger patches. It seems that the mean value was unduly influenced by some very rare species like *Sorbus aucuparia* and *Juniperus communis*. The Moran's I values for the three *Vaccinium* species on the low altitude forelands also show the expected increase, with *Empetrum hermaphroditum* showing no trend. As with the D_{Gau} values for this species on the Høgvaglbreen foreland discussed earlier, the Moran's I values for very dense distributions might be influenced by the prevalence of the species in these areas and should be used with the appropriate care.

When the average values of the pattern descriptors for animal and wind dispersed species within each block are compared, a consistent pattern emerges with either wind or animal dispersed species showing higher values on low and high altitude forelands (Table 6).

Table 7. Relation of mean descriptor values for animal (Animal) and wind (Wind) dispersed species on low and high altitude glacier forelands.

	Perimeter/area index	Autocorrelation (Moran's I, Kings case)	Fractal dimension D_{Gau}	Fractal dimension $D_{De\ Cola}$
Low altitude forelands	Wind > Animal	Animal > Wind	Animal > Wind	No trend
High altitude forelands	Animal > Wind	No trend	Wind > Animal	Wind > Animal

The descriptors strongly suggest that in the same blocks on low altitude glacier forelands, wind dispersed species are present in smaller, more patchy and more widely spaced areas than animal dispersed species. In contrast to this wind dispersed species in blocks on high altitude glacier forelands tend to exist in larger, more continuous patches with rather more convoluted perimeters. This observation lends credibility to the previously mentioned supposition that primary succession on low and high altitude glacier forelands are fundamentally different (e.g. Matthews 1992: 221). Lying below the treeline, general ground cover on low altitude forelands is relatively high. Most of the cover is provided by the three *Vaccinium* species, *Empetrum hermaphroditum,* and *Calluna vulgaris.* Assuming that the wind dispersed species (*Betula* and *Salix spec.*) can potentially reach all "suitable" sites for colonisation in an area, they rely on a gap in the dense ground cover and are probably constrained by moisture for germination (Vetaas 1994). This might explain why the majority of the wind dispersed species show the observed patchy spatial pattern on low altitude forelands.

In contrast, the environment on high altitude forelands is generally more severe and other than in sheltered areas no full ground cover develops. This situation provides more empty "suitable" sites for colonisation for wind dispersed species and, combined with strong winds that provide an effective way of distributing the seeds, this can lead to the more continuous, less patchy distribution pattern of wind dispersed species on high altitude forelands. It is in these hostile environments that the biotic vectors of animal dispersed species instigate the nuclei of distributions for animal dispersed species, which form small, highly fragmented and more widely spaced patterns. Unfortunately no prediction can be made about the processes in the pioneer zone on these glaciers, since the youngest terrains have not been sampled.

CONCLUSION

Dynamic patchiness of biological systems has been shown to exist at many spatial scales (e.g. Pickett and White 1985, White and Pickett 1985). Physical environmental factors, such as disturbance, in combination with biological controls such as dispersal mechanisms and species interactions, are the driving forces behind such patchiness. It has been shown in this study that by using a suite of spatial analytical methods, an effective approach to the analysis of the development of distribution patterns amongst the woody species on glacier forelands in southern Norway is possible. Using different indices and spatial statistics does not merely duplicate information, but adds to the elucidation of different processes important for the development of the patterns under investigation. Careful interpretation of the results allows us to gain more information about primary succession on glacier forelands.

However, some of the pattern descriptors used in this study seem to have problems with very dense or very sparse distribution patterns. This urges caution in the use of those indices in those cases. Furthermore a better understanding of the theoretical background for the use of these indices is needed.

For some of the indices, a relation between area and perimeter enters the equation. In the case of this study the surveyed blocks were quite small in comparison to the sort of satellite image scenes that have previously been analysed with these methodologies. The problem of boundary effect is therefore expected to be quite prominent within the analysis of these data.

This will influence all the calculations, which depend on an accurate estimation or measurement of the length of the real distribution boundary.

Despite these problems, this study shows that pattern development can be explained by the dispersal mechanism of the plants involved. On low altitude forelands, with a lack of empty, suitable sites for colonisation, wind dispersed species tend to show a pattern of smaller, more patchy distributions whereas, on high altitude forelands, the more homogeneously advancing wind dispersed species show more continuous, less fragmented distribution patterns than animal dispersed species.

To add further to the understanding of primary succession in these areas, expansion of the investigation of the pattern development into the pioneer zones of the relevant glacier forelands could provide us with even more insight into the processes involved.

ACKNOWLEDGEMENTS

I thank three anonymous referees for their comments on earlier versions of this chapter. The fieldwork was carried out on the Jotunheimen Research Expedition 1996. This paper is Jotunheimen Research Expeditions Contribution No. 139.

REFERENCES

Battey MH, McRitchie WD. A geological traverse across the pyroxene granulites of Jotunheimen in the Norwegian Caledonides. Norsk Geologisk Tidsskrift. 1973; 53:237-265.
Battey MH, McRitchie WD. The petrology of the pyroxene-granulite facies rock of Jotunheimen. Norsk Geologisk Tidsskrift. 1975; 55:1-49.
Bickerton RW, Matthews JA. On the accuracy of lichenometric dates: an assessment based on the Little Ice Age moraine sequence of Nigardsbreen, southern Norway. The Holocene. 1992; 2:227-237.
Böhmer HJ. *Vegetationsdynamik im Hochgebirge unter dem Einfluß natürlicher Störungen.* Berlin: J. Cramer, 1999.
Burrough PA. Fractal dimensions of landscapes and other environmental data. Nature. 1981; 294:240-242.
De Cola L. Fractal analysis of a classified Landsat scene. Photogrammetric Engineering and Remote Sensing. 1989; 55:601-610.
Eastman JR. (1987-1995) IDRISI for Windows Ver. 1.0.003. Computer Program, Worcester MA: Clark University, 1987-95.
Erikstad L, Sollid JL. Neoglaciation in South Norway using lichenometric methods. Norsk Geografisk Tidsskrift. 1986; 40:85-105.
Fastie CL. Causes and ecosystem consequences of multiple pathways of primary succession at Glacier Bay, Alaska. Ecology. 1995; 76:1899-1916.
Fitter AH, Stickland TR. Fractal characterization of root system architecture. Functional Ecology. 1992; 6:632-635.

Friedel H. Die Pflanzenbesiedlung im Vorfeld des Hintereisferners. Zeitschrift für Gletscherkunde, für Eiszeitforschung und Geschichte des Klimas. 1938; 26:215-239.
Gautestad A, Mysterud I. Fractal analysis of population ranges: methodological problems and challenges. Oikos. 1964; 69:154-157.
Jochimsen M. Das Gletschervorfeld - Keine Wüste. Jahrbuch des Österreichischen Alpenvereins. 1962; 87:135-142.
Lüdi W. Beobachtungen über die Besiedlung von Gletschervorfeldern in den Schweizeralpen. Flora. 1958; 146:386-407.
Mandelbrot BB. *The fractal geometry of nature*. San Francisco: Freeman, 1983
Matthews JA. The vegetation of the Storbreen gletschervorfeld, Jotunheimen, Norway. I. Introduction and approaches involving classification. Journal of Biogeography. 1979; 6:17-47.
Matthews JA. *The ecology of recently-deglaciated terrain; A geoecological approach to glacier forelands and primary succession*. Cambridge: Cambridge University Press, 1922.
Matthews JA. "Disturbance Regimes and Ecosystem Response on Recently-Deglaciated Substrates." In *Ecosystems of Disturbed Ground*, LR Walker ed. Amsterdam: Elsevier Science Publishers B.V., 1999.
Matthews JA, Whittaker RJ. Vegetation succession on the Storbreen glacier foreland, Jotunheimen, Norway: A review. Arctic and Alpine Research. 1987; 19:385-395.
Moen A. The regional vegetation of Norway; that of Central Norway in particular. Norsk Geografisk Tidsskrift. 1987; 41:179-225.
Odland A. Pre- and subalpine tall herb and fern vegetation in Rødal, Western Norway. Nordic Journal of Botany. 1981; 1:671-690.
Palmer MW. Fractal geometry: a tool for describing spatial patterns of plant communities. Vegetatio. 1998; 75:91-102.
Petch JR, Whittaker RJ. Chronology of the Austerdalen glacier foreland, Norway. Zeitschrift für Geomorphologie. 1997; 41:309-317.
Phillips JD. Measuring complexity of environmental gradients. Vegetatio. 1995; 64:95-102.
Pickett STA, White PS. "Patch Dynamics: A Synthesis." In *The ecology of natural disturbance and patch dynamics* STA Pickett, PS White eds. New York: Academic Press, 1985.
Ryvarden L. Studies in seed dispersal I. Trapping of diaspores in the alpine zone at Finse, Norway. Norwegian Journal of Botany. 1971; 18:215-226.
Ryvarden L. Studies in seed dispersal II. Winter dispersal at Finse, Norway. Norwegian Journal of Botany. 1975; 22:21-24.
Sigmond EMO, Gustavson M, Roberts D. Berggrunnskart over Norge. (Map 1:1 000 000). Trondheim: Norges Geologiske Undersøkelse, 1984.
Vetaas OR. Primary succession of plant assemblages on a glacier foreland - Bødalsbreen, southern Norway. Journal of Biogeography. 1994; 21:297-308.
Vetaas OR. Relationships between floristic gradients in a primary succession. Journal of Vegetation Science. 1997; 8:665-676.
Virkkala R. Ranges of northern forest passerines: a fractal analysis. Oikos. 1993; 67:218-226.
Walker LR, Chapin III FS. Interactions among processes controlling successional change. Oikos. 1987; 50:131-135.
White PS. Pattern, process and natural disturbance in vegetation. The Botanical Review. 1979; 45:229-299.
White PS, Pickett STA "Natural Disturbance and Patch Dynamics: An Introduction." In *The ecology of natural disturbance and patch dynamics* STA Pickett, PS White eds. New York: Academic Press, 1985.

Chapter 16

MULTI-SCALE ANALYSIS OF LANDCOVER COMPOSITION AND LANDSCAPE MANAGEMENT OF PUBLIC AND PRIVATE LANDS IN INDIANA

Tom P. Evans[1], Glen M. Green[2] and Laura A. Carlson[2]
[1]Department of Geography, Indiana University, Student Bldg 120, Bloomington, IN 47405
evans@indiana.edu
[2]Center for the Study of Institutions, Population and Environmental Change, Indiana University,
408 N Indiana

Keywords: land cover, forest cover dynamics, land management, multi-scale analysis.

Abstract A multi-scale analysis was conducted in Indiana (Midwest United States) to examine social and biophysical factors contributing to landcover pattern and composition at different spatial extents. Historical data shows that Indiana has experienced dramatic landcover change over the past two centuries. A state-level analysis found that while forest covered more than 85% of Indiana prior to European-American settlement, almost all of these forests were cleared for agriculture following settlement. Forests have since regrown in part such that they now cover nearly 20% of the state. In recent years, analysis at state and regional levels has found a strong relationship between topography and landcover, especially for agricultural and forested land. A parcel level analysis was conducted for 250 private landholdings in Monroe County, linking household level survey data to Landsat Thematic Mapper imagery. This local level analysis allows parcel land management practices and associated land holder characteristics to be linked to specific landscape outcomes. This research demonstrates the utility of a multi-scale analysis linking social and biophysical factors in landcover change studies.

INTRODUCTION

Indiana, located in the Midwest United States, has experienced dramatic landcover change over the past 200 years. These landcover changes have largely been the result of European-American colonization in the first half of the 19th century and the subsequent social and economic changes that accompanied settlement, all within the context of the biophysical environment. This chapter focuses on deforestation and afforestation in Indiana over this time period and examines how social, institutional and biophysical factors have interacted to affect the present landcover pattern and composition.

Researchers from a variety of disciplines have studied how forest cover change affects global carbon cycling and the availability of species habitat in various ecosystems (Ojima et al. 1994, Dale et al. 1998, Vitousek et al. 1997). While social scientists are reasonably well represented in this area of research, much of the research in the area of forest cover change has emphasized biophysical factors as predictors of change. While biophysical factors are clearly a critical component to the process of landcover change, social and institutional elements may be of equal or greater importance in many areas. It is important to consider the interaction between social and biophysical factors to assess the major forces contributing to landcover change, particularly in heavily managed or disturbed environments.

There has been a recent call for a greater emphasis on the social and institutional factors contributing to landcover change, and population-environment relationships in general, to complement the research already present on the biophysical side landcover change processes (National Research Council 1998, Pebbley 1998, Rindfuss and Stern 1998). One method to integrate data from these seemingly disparate data sources is through the integration of satellite imagery with social survey data in a spatially explicit environment or GIS (Entwisle et al. 1998). Researchers in a variety of ecoregions have examined population-environment interactions using remotely sensed imagery (Green and Sussman 1990; Entwisle et al. 1998, Wood and Skole 1998).

The scale of analysis is of particular importance when combining social and biophysical data. Regional scale analyses are highly suitable to observing processes occurring over large spatial extents (O'Neill et al. 1997, Wood and Skole 1998). However, such studies are limited in their ability to link land management decisions and specific land use practices to landscape outcomes without associated finer scale data (Lambin and Ehrlich 1997). Remotely sensed image data can be readily obtained from a variety of sources and can now provide a relatively rapid method of generating a continuous coverage of landcover and landcover change for large areas. However, collecting social data that corresponds to this landcover data usually requires a tremendous investment in time and labor. Thus, most regional scale studies use government census data that is highly aggregated and limited to the questions that are part of the survey instrument (questionnaire). For example, United States census data contains a variety of demographic and economic data, but provides relatively little information related to specific land management practices.

Local scale analyses are typically limited in their spatial extents due to the costs of data collection. However, such studies are able to construct social survey instruments that specifically address processes affecting landcover change (See for example Entwisle et al. 1998, Entwisle et al. 1999, McCracken et al. 1999). Local-scale analyses used in conjunction with regional scale analyses may allow the finer scale findings to be "scaled up" to determine whether processes observed at the local-level are representative of processes occurring over larger spatial extents. In addition, Walsh (1999) found that population-environment relationships are often scale dependent, which suggests that a multi-scale approach may be needed to observe the range of processes at work in heavily disturbed environments.

Considerable attention has been focused on areas exhibiting rapid deforestation, such as the Brazilian Amazon (Dale et al. 1993, Moran and Brondizio 1998, Wood and Skole 1998). Researchers have suggested that the deforestation is so severe that the forests may be unable to regenerate to their previous state (Shukla and Sellers 1990). These studies focus on deforestation "hotspots," where massive forest cover loss has occurred over a short period of time. In the past, Indiana experienced dramatic deforestation at rates approaching those now present in the Brazilian Amazon. While the social and biophysical differences between Indiana and the Amazon are vast, examining landcover changes over a long period of time lends insight into possible future landcover scenarios for areas now undergoing rapid change.

This chapter presents analysis which links remotely sensed data with spatially referenced social and biophysical data to examine the composition of landcover at three spatial scales. Figure 1 shows the three study areas used in this research. A state level analysis is used to examine the composition of forest cover over the range of topographic slopes found in the state of Indiana. An intermediate scale analysis is used to examine similar relationships for a nine county portion of the state where topography is relatively homogenous compared to the topographic differences observed across the entire state. Finally, a parcel level analysis for Monroe County, Indiana is used to link specific landcover compositions to household level survey data including land management practices and household socio-economic characteristics.

DEFORESTATION AND AFFORESTATION IN INDIANA 1800-2000

Indiana's topography varies greatly from north to south because of the differential affects of various geologic processes. The limestone, shale and sandstone bedrock that underlies the forests of Indiana were deposited in shallow Paleozoic seas as distinct layers between 500 and 280 million years ago (Howe 1997). During the creation of the Appalachian Mountains these formations were tilted toward the Illinois basin to the west. Subsequent erosion has exposed these various lithologies as outcrops, with the more resistant lithologies weathering to parallel series of rugged hilly terrain running primarily north south.

During glacial periods of both the Illinoian (140,000 years BP) and the Wisconsinan (21,000 years BP), continental ice sheets advanced from the north, responding to changing climatic regimes and covered large portions of Indiana (Melhorn 1997). The first glacial advance covered four fifths of the state scouring the landscape across all but the south central portion that includes Monroe County. In contrast, during the second advance glaciers covered only the northern two thirds of the state, terminating along a broad east west oriented front. These glaciers deposited a vast array of surficial material that now covers much of the underlying bedrock of the northern half the state to a depth averaging 12-15 m. Thus, the glaciation of Indiana during the Pleistocene, which involved both the scouring away of bedrock and the deposition of glacial till, resulted in a loss of topographic relief across much of the state, and a preferential smoothing of relief across much of northern Indiana. These glacial deposits provided the flat, nutrient rich parent material for the development of soils that now boast extensive agricultural production. In contrast, the highly dissected hilly terrain in the south-central part of the state is now characterized by relatively thin and poor soils covering steeper slopes.

Indiana Forests Prior to Settlement

Much of Indiana was surveyed prior to settlement by the Government Land Office (GLO) that was formerly a branch of the federal government. The GLO survey facilitated the transfer of the newly acquired lands of the Northwest Territory from the federal government to private individuals. These surveys also provide essential information on pre-settlement vegetation. It is estimated from survey data that when European settlement of Indiana began in the early 1800s, more than 87% of the state was covered with forest of some type across a wide range of topographic differences (Lindsey et al. 1965; Lindsey 1997). Wetland and dry prairie, located primarily in the northwestern quarter of the state, made up the remaining 10% and 3%, respectively (see Figure 2 – on the CD-ROM).

Though often referred to as "virgin," these forests were most likely somewhat altered by populations of native peoples indigenous to the area (or by those who were recently displaced by earlier waves of European settlement). Estimates of the numbers and geographic extent of native peoples and associated disturbances vary and their impact is not yet clearly understood as it relates to the nature of land cover at the time of pioneer settlement (Parker 1997). What modification of land cover prior to European settlement was accomplished by Native-Americans and was likely limited to mostly floodplain areas in the form of clearings for swidden agriculture.

Forest Change with Settlement

Indiana settlers in the early 19th century, generally advancing from south to north, felled or girdled trees with simple iron hand tools, and set fires to kill understory species. Initial plantings by colonists were done between the stumps and standing snags. In many cases, the fields lost productivity after a few years use and the farmers moved on to clear new land, thus increasing the spatial impact of settlement on the forest (Parker 1997). The practice of allowing hogs and cattle to run free to graze on acorns and roots in the woods and degraded fields further impacted the ecosystem.

By 1860, the state's population had increased to 1,350,428 (Parker 1997). As colonists spread throughout the state, altering the physical environment in their path, social and economic conditions began to change with the beginnings of industrialization. Mechanized farming was beginning to take over from more traditional hand and animal powered methods, and the expansion of the canal and then rail systems allowed more extensive settlement of the northern part of the state as well as increased opportunities for farmers to market their goods further from home (Gray 1982). The flat to gently rolling, rich glacial till soils in the northern part of the state were well suited for this type of agriculture. In southern Indiana much of the terrain was too steep for large farm machinery and railroad building, thus farming was limited to relatively small-scale areas (Sieber and Munson 1992).

In addition to agriculture, the establishment of sawmills contributed to the rapid clearing of Indiana's forests. Advancements in transportation and other technologies, in particular the steam sawmill, aided these businesses as well, and by the late 1800's, timber production had become a major industry in the state. In 1899, Indiana led the nation in timber extraction with 1,036,999,000 board feet (Parker 1997).

This combination of agricultural clearing and timber extraction reduced Indiana's forested land to approximately 560,000 ha (~1,390,000 acres), or about 6% of the state by the early 1920's (Nelson 1998). Since that time, the extent of forest cover has increased to over 1.6 million ha (4 million acres) (Nelson 1998). Still, nearly all forest that existed prior to European colonization was cleared within the last 200 years. Today, Indiana retains only an estimated 0.06% of its old growth forest from its estimated original forest cover at time of European settlement (Davis 1993, Lindsey 1997). If selectively logged dry upland forest and glade areas are not included, Indiana may only retain approximately 0.01% of its old growth forest cover (Davis 1993).

Land Cover and Topographic Relief

The present distribution of forest cover differs dramatically from that of 1820. Figure 2 (on the CD-ROM) shows a land cover classification image for the state in 1992 (30m resolution). These vegetation classes were compared to topographic relief values (83m resolution) calculated from 1:250,000 scale digital elevation models (DEM) available from the U.S.G.S. Currently, forests are preferentially distributed on land of higher topographic relief, while agricultural land occupies flatter terrain. Table 1 shows the topographic distribution within the state of Indiana. The percentage of land in agricultural uses (which accounts for nearly 60% of Indiana's land cover today) decreases from more than 70% of flat land (slopes < 3%) to about 10% of land at slopes greater than 10%. In contrast, the percentage of land covered by forest increases from 10% on flat to gently rolling topography to more than 65% of land with slopes exceeding 10%.

While the areal extent of all land in Indiana declines rapidly with increasing slope, the proportion of land under state and federal management increases sharply with increasing relief. The amount of land under different management types is also distributed differently with respect to relief, with some agencies having holdings concentrated on flatter land and some on topographically steep terrain. For example, Indiana's state and federal forests are located primarily in areas of steep topography, while U.S. Fish and Wildlife landholdings associated with wetlands or near wetlands in Indiana are mostly contained in areas with little relief. This is due to a combination of environmental and social factors, and the mandate of these organizations.

Table 1. Distribution of Slopes – Indiana State

Slope Category (%)	Area (%)
S < 1	66.2
1 ≤ S < 5	28.1
5 ≤ S < 10	4.4
10 ≤ S < 15	1.08
S > 15	0.3

Abandonment and Forest Regrowth

After settlement, farms in the southern part of the state began to suffer environmental devastation from long-term use of the poor soils of that region. This is probably not surprising given what are now viewed as agricultural practices which cause severe degradation. This hardship was exacerbated by drought and the economic disruption associated with the Great Depression, "By the 1920's and 1930's the economic disadvantage of the terrain was made far more devastating by the effects of massive land erosion in the hilly regions. Many landowners could not pay taxes and others were forced to abandon their holdings because of the condition of the land" (Sieber and Munson 1992).

Federal legislation provided the statutory basis for the location and purchase by the federal government of "forested, cut-over, or denuded lands within the watersheds of navigable streams" (16 U.S.C. 516) to create and expand National Forests in the Eastern states. Spurred by the alarming economic conditions in the southern part of the state, the Indiana General Assembly in 1935 passed legislation (IC 4-20.5-16-1) authorizing the federal government to purchase land from willing sellers for inclusion in what was to eventually become the Hoosier National Forest, much of which is contained in our nine county area. The state forests in Indiana were created largely from lands similar to those outlined in the federal legislation. Like the land acquisitions for the Hoosier National Forest, studies of erosion and economic conditions guided the acquisition of property, keeping the focus on the southern hill country (Sieber and Munson 1992). These state and federal policies, which were aimed primarily at easing the economic devastation of the Depression, resulted in public forest on areas of steep topography near streams and watersheds in southern Indiana.

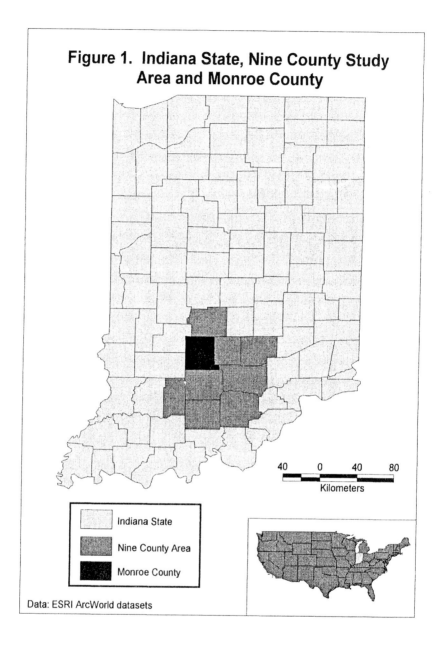

Figure 1. Indiana State, Nine County Study Area and Monroe County

These public land holdings however, explain only a portion of the forest that exists in Indiana today. In total, state and federal forest agency lands in Indiana make up less than 8% of the total forested lands in the state. Currently Indiana, a state of 23,367,240 acres, has a little over 1.7 million ha (4.2 million acres) of forested land: approximately 19% of the land base

(Nelson 1998). Today, 87% of Indiana's forest cover is on private land, most of which is owned by farmers (Peterson 1998).

It seems natural to conclude that agricultural land will be located in areas with little topographic relief and forests will occupy areas not suitable for cultivation, or areas of steeper slopes. Yet, at one time early in the 20th century 95% of Indiana may have been under agricultural land-uses. The dramatic landcover changes evident in Indiana's history demonstrate how both socio-economic and biophysical factors are critical components to landcover analysis for both public and private lands at the state level. However, at this scale of analysis, it is difficult to link specific land management practices to specific landscape outcomes. The following sections present results from regional and parcel level analyses of landcover in Southern Indiana.

REGIONAL AND PARCEL-LEVEL ANALYSIS

Data Processing

A series of 1:24,000 scale USGS DEM quadrangles were acquired in USGS SDTS format and converted into Arc/INFO Grid raster surfaces. Individual DEM quadrangles were mosaiced into a single surface with a cell resolution of 30m. Gaps between quadrangles ranged from 1-6 cells and were filled using a low pass filter. Slope angle and slope aspect surfaces were generated from the elevation grid using the average maximum technique within a 3x3 window (Burrough 1986).

A September 6, 1997 Landsat 5 Thematic Mapper image was acquired and processed to construct a landcover classification covering the nine county study area. The image was geometrically registered, radiometrically calibrated, atmospherically corrected and topographically normalized (cosine corrected). Radiometric calibration converted image DNs to at-sensor reflectance values using procedures outlined in (Teillet and Fedosejevs 1995) which relies on time dependent sensor gains and offsets derived by Thome (1995) by observing White Sands, New Mexico throughout Landsat 5 operations. At-sensor reflectance values were converted to surface reflectance values by performing a simple subtraction of the additive atmospheric component by assuming dark lakes have the reflectance similar to that measured by Bartolucci (1997).

Landcover classes were derived from the topographically normalized TM image using the maximum likelihood supervised classification technique (ERDAS Imagine 8.3). Topographic maps (1:24,000 scale) were used to identify training sample locations for residential, commercial and water classes. Training sample fields for advanced secondary successional forest were derived from the Ecological Classification System of Van Kley (1993). Initial secondary successional forest training fields were obtained from forest plot data spatially referenced using GPS receivers. These initial successional forest sites were formerly clear cuts, young tree plantations and both upland and lowland abandoned agricultural fields. Agricultural areas (row crops and pasture) were identified in Monroe County in coordination with an on-site parcel-level survey of private landholders in summer 1998 (see below).

A total of 233 training sample locations were established, half of which were randomly selected and used as input to the supervised classification of the Landsat TM image. Table 2 reports the classification accuracy statistics for each class. The users accuracy measure reports the predictive power of each class, while the producers accuracy measure reports what percent of the actual landscape was correctly classified. Overall accuracy was 86%, with the initial secondary succession being the most mis-classified landcover class. There was significant class confusion between the initial secondary successional forest class and the advanced secondary successional forest class. This class confusion is in part due to the discrete categories used to represent this continuous gradient of forest growth.

Table 2. Landcover classification accuracy for 1997 Landsat 5 TM image.

Landcover Class	Producers Accuracy (%)	Users Accuracy (%)
Agriculture-Pasture	75.00	100.00
Commercial	100	86.36
Residential	100	95.45
Initial Secondary Forest	45.45	50.00
Advanced Secondary Forest	93.33	80.00
Water	76.92	100.00

Regional Level Analysis for a Nine County Area

Southern Indiana is characterized by hilly terrain in contrast to the northern part of the state that is predominantly flat and topographically homogenous. Landcover composition and pattern varies dramatically between the northern and southern sections of the state, as does landuse. A regional level analysis was conducted in a nine county area in the south-central part of the state (see Figure 1) to examine the relationship between topography and landcover at a finer spatial scale for an area with a range of topographic gradients. This area is characterized by a mosaic of agricultural and forest land-uses, including the state's largest federal and state managed forests as well as large private holdings. Slopes in this nine county area range from 0-56°, although most slopes are relatively moderate.

Table 3 shows the landcover proportions within the entire nine county area, on private landholdings in the nine county area, and by three slope categories within private landholdings. The proportion of forested land within the entire nine county area (36.8%) is greater than that on private landholdings (30.6%), indicating that private landholdings in aggregate have a lower proportion of advanced successional forest cover than publicly managed lands. The amount of forest cover in the nine county area is uncharacteristically high for southern Indiana because of state and federal forests located in the nine county area. These forests are actively managed and used for timber production as are private parcels in the study area. Thus, there is a large proportion of land in some stage of forest succession both in the entire nine county study area and on private landholdings.

Table 3. Landcover Proportions in Nine County Area by Slope Category (all values are %)

Landcover Class	Nine County Area		Low Slopes	Moderate Slopes	Steep Slopes
	Total	Private	$S < 4°$	$4° \leq S \geq 10°$	$10° < S$
Agriculture/Pasture	25.2	29.0	38.3	18.3	6.5
Advanced Sec. Forest	36.8	30.6	16.3	46.2	69.3
Initial Sec. Forest	22.2	23.8	25.5	23.1	15.3
Residential	9.6	10.4	11.7	9.0	7.0
Commercial	5.4	6.0	8.0	3.4	1.8
Water	0.8	0.2	0.3	0.1	0.1

While advanced successional forest is the predominant landcover class in the nine county area, this class only accounts for 16.3% of the area on slopes less than 4°. Likewise, agriculture and pasture accounted for 38.3% of the area on slopes less than 4° and only 6.53% of the area greater than 10° in slope.

Parcel Level Analysis - Monroe County

Land management decisions in Monroe County are made at the household level on private parcels. Thus, a parcel level analysis can be used to link household socio-economic characteristics and specific land management decisions to landscape outcomes. A parcel level analysis was conducted to link information from a household level social survey to parcel site characteristics including landcover and topography. Spatial metrics were computed for each individual parcel from the 1997 Landsat TM derived classification using FRAGSTATS 2.0 (McGarigal and Marks 1994). Landscape metrics were computed for each parcel area and class level metrics were computed for the mature forest class and the aggregated agriculture/pasture class.

The state and regional level analyses indicate that topography (specifically slope) is a critical factor affecting landcover pattern. In Indiana, topographic characteristics determine likely landuses and associated landcover patterns. A parcel level analysis will show to what extent individual land managers are representative of the landcover-topography relationship evident at the regional and state levels.

Household Level Land Management Survey

A household level survey was conducted of private landholders on parcels greater than 5 acres in the summer of 1998. A random sample of 250 households was generated and stratified by parcel size to insure that landholders across a range of parcel sizes would be included in the sample. The household survey focused on land management questions, but also contained sections reporting the social and demographic structure of the household. A more complete analysis of this parcel level study is forthcoming elsewhere (Koontz 1999). The following section reports representative examples of linking this household survey data to landcover characteristics at the parcel level.

Landcover Pattern and Composition on Private Landholdings

Respondents were asked to rank different benefits of forests on their parcels. Those reporting aesthetics as one of the top three reasons had an average of 59ha of forest on their parcels. Respondents reporting timbering as one of the top three reasons had an average of 46ha of forest on their parcels. T-test statistic indicates this difference to be statistically significant at the 0.05 confidence interval (t=-2.499, p= 0.0069).

More respondents reported aesthetics (84) among their top three reasons for having forest cover on their parcel than timbering (51). Only seven of the 250 respondents reported both aesthetics and timbering to be among their top three reasons for having forested land on their parcel. There was no correlation between parcel size and proportion of a parcel under forest cover, nor was there a relationship between income and amount of forest cover on parcels.

In Monroe County, forest cover is more dominant than agriculture on both private and public landholdings. Figure 3 shows the ratio of forested land to agriculture/pasture land within individual private parcels.

Examination of this figure shows a higher concentration of private parcels dominated by forest cover than those of other mixes of forest and agriculture/pasture. The mean difference between the percent forest and percent agriculture/pasture is 26.5%. In contrast to the northern part of the state, private land holdings that have agriculture as the major land management activity (73 of the 250 parcels) also have forest patches of considerable size (mean 1.3ha, max 11ha). Approximately 27% of the parcels that were mostly agricultural had no forest cover present on the parcel (8% of the total survey). This may be due to the heterogeneity of topography within much of Monroe County at the scale of the individual property parcels.

Monroe County is currently experiencing two major landcover transitions. First, suburban development is expanding into old farmland surrounding the major urban areas in the county as land prices increase. Secondly, rural areas of the county are undergoing a transition from parcels where agriculture is the main activity to parcels used primary for residential purposes. Both agricultural and residential type parcels have forest cover, but residential parcels tend to have more forest cover. If agricultural production continues to become less economically viable, this trend towards rural residential land will likely continue.

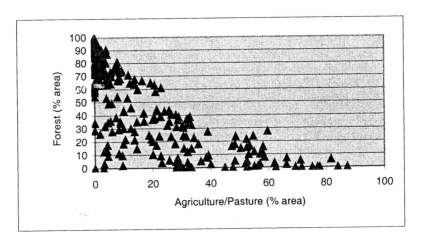

Figure 3. Landcover composition by parcel, Forest and Agriculture/Pasture % area.

DISCUSSION

The availability of the parcel boundary dataset was critical to this analysis because it allowed us to make a reliable link between the social survey data and landcover in a particular location. In many areas a spatial representation of land ownership is very difficult to obtain, if such data exists at all (Entwisle et al. 1998). When only hardcopy parcel boundary data are available, a substantial labor and time investment is required to develop a spatially referenced digital dataset (McCracken et al. 1999). However, in contrast to regional level analyses, these types of local-scale products allow analyses to be performed at the level of the specific decision making processes that affect landcover change.

A series of processes have produced an environment that has experienced afforestation in the 9 county study area: 1) establishment of state and federally managed forests in the mid-1900's and 2) farm failure in the depression era and subsequent forest regrowth on slopes marginal for agriculture, and 3) economic changes, transition from economy dominated by agrarian landuses to residential land ownership. At the same time, there is some loss of forest cover due to expansion of urban areas. However, over the last 80 years forest cover regrowth has outpaced the rate of urban expansion in Indiana.

Management of landcover on private lands presents different institutional challenges than on federal/state lands. Private lands account for 87% percentage of all forest cover in Indiana, making monitoring of these lands critical. Mean forest patch size on private lands is often smaller than on publicly managed lands (Turner et al. 1996) because private parcels are typically much smaller than publicly managed parcels. Thus, individual forest patches often are contained within multiple private landholdings. This does not imply that forest management of public lands is less complex than that on groups of private land holdings. There are a variety of agents affecting land management decisions on public lands (e.g., district foresters, NGO's, purchasers of timber from these managed forests and residents on private lands surrounding these forests). Private properties are affected by different agents, and further analysis of our 250 private parcel respondents will more thoroughly define these agents and their behavior.

CONCLUSION

Indiana has undergone dramatic land cover changes over the last 200 years. While Indiana's forests were vast and minimally disturbed prior to European-American settlement, Indiana has since lost almost all of these old-growth forests. The amount of forested area in Indiana has increased in the 20th century, yet forests still cover less than a quarter of their former extent and are now probably composed of a significantly higher proportion of young regrowth.

What forests exist in Indiana today can be considered to be in some successional stage of forest regrowth rather than like the mature forests of

200 years ago. The pattern of landcover today is a function of both social and biophysical factors. Topography is clearly a critical factor controlling land-use and landcover. However, topography alone does not explain landcover pattern and composition. Forest cover in the state has gone from nearly 95% prior to 1820 to 6% in 1910 to 20% in 1992, suggesting that social, economic and institutional changes over this period are also driving land-use and landcover. In these heavily disturbed environments, the ecology of the landscape cannot be separated from social and institutional processes.

This research demonstrates how remotely sensed and GIS data can be used to link land management decisions and landholder characteristics to landcover at the local-scale. This research also shows the importance of a multi-scale analysis in examining the social and biophysical factors contributing to landcover change, including historical data and institutional data. Local-scale relationships can be examined in the context of relationships observed at broader levels of analysis, which allows case study areas to be made applicable to larger geographic extents.

ACKNOWLEDGEMENTS

This research is supported by NSF funding (SBR 9521918) and the Center for the Study of Institutions, Population and Environmental Change (CIPEC) at Indiana University. Charlie Schweik, Lin Ostrom, Derek Kauneckis, Tom Koontz and J. C. Randolph are acknowledged for their substantial contributions to the larger set of projects associated with this research. State and federal forestry officials associated with Morgan Monroe and Yellowwood Forests are acknowledged for provision of data and field assistance. The Office of the Tax Assessor for Monroe County, IN is gratefully acknowledged for providing parcel boundary data for Monroe County.

REFERENCES

Bartolucci LA, Robinson BF, Silva LF. Field Measurement of the Spectral Response of Natural Water. Photogrammetric Engineering and Remote Sensing. 1997; 43:595-598.

Burrough PA. *Principles of Geographical Information Systems for Land Resource Assessment*, Oxford: Clarenden Press, 1986.

Dale V, O'Neill RV, Pedlowski M, Southworth F. Causes and Effects of Land-Use Change in Central Rondonia, Brazil. Photogrammetric Engineering and Remote Sensing. 1993; 59: 997-1005.

Dale VH, King AW, Mann LK, Washington-Allen RA, McCord RA. Assessing Land-Use Impacts on Natural Resources. Environmental Management. 1998; 22:203-211.

Davis MB. *Old Growth in the East*. Richmond VT: The Cenozoic Society, 1993.

Entwisle B, Rindfuss RR, Walsh SJ, Evans TP. Satellite Data and Social Demographic Research, Paper presented at 1998 Population Association of America conference; 1998 April 2-4; Chicago.

Entwisle B, Walsh SJ, Rindfuss RR, Chamratrithirong A. "Land-Use/Land-Cover and Population Dynamics, Nang Rong, Thailand." In *People and Pixels*, D Liverman, EF Moran, RR Rindfuss, PC Stern eds. Washington DC: National Academy Press, 1998.

Gray R. The Canal Era in Indiana. Transportation and the Early Nation: Paper presented at the 4th Indiana American Revolution Bicentenial Symposium, Indiana Historical Society, Indianapolis IN, 113-134, 1982.

Green GM, Sussman R. Deforestation of the Eastern Rain Forest of Madagascar from Satellite Images. Science. 1990; 248: 212-215.

Howe RC. "The Terrain and Its Origin." In *The Natural Heritage of Indiana*, MT Jackson, ed. Bloomington IN: Indiana University Press, 1997.

Koontz TM. Explaining Private Land Use Decisions and Outcomes in a Midwest County: A Micro-Level Approach, in review.

Lambin E, Ehrlich D. The Identification of Tropical Deforestation Fronts at Broad Spatial Scales. International Journal of Remote Sensing. 1997; 18: 3551-3568.

Lindsey A. "Walking in Wilderness." In *The Natural Heritage of Indiana*, MT Jackson, ed. Bloomington IN: Indiana University Press, 1997.

Lindsey A, Crankshaw A, William B, Qadir AS. Soil Relations and Distribution Map of the Vegetation of Presettlement Indiana. Botanical Gazette. 1965; 126: 155-163.

McCracken S, Brondizio E, Nelson D, Moran E, Siqueira A, Rodriguez-Pedraza C. Remote Sensing and GIS at Farm Property Level: Demography and Deforestation in the Brazilian Amazon. Photogrammetric Engineering and Remote Sensing. 1999; 65: 1311-1320.

McGarigal K, Marks B. *FRAGSTATS: Spatial pattern analysis program for quantifying landscape structure, v. 2.0*, Forest Science Lab, Oregon State University, Corvallis, Oregon, 1994.

Melhorn WN. "Indiana on Ice: The Late Tertiary and Ice Age History of Indiana Landscapes." In *The Natural Heritage of Indiana*, MT Jackson, ed. Bloomington IN: Indiana University Press, 1997.

Moran EF, Brondizio E. "Land-Use change after deforestation in Amazonia." In *People and Pixels*, D Liverman, EF Moran, RR Rindfuss, PC Stern eds. Washington DC: National Academy Press, 1998.

National Research Council. *Global environmental change: research pathways for the next decade.* Washington DC: National Academy Press, 1998.

Nelson J. Indiana's Forests: Past, Present and Future. Woodland Steward. 1998; 7.

Ojima DS, Galvin KA, Turner II BL. The global impact of land-use change. Bioscience. 1994; 44: 300-304.

O'Neill R, Hunsaker C, Jones KB, Riitters K, Wickham J, Schwartz P, Goodman I, Jackson B, Baillargeon W. Monitoring Environmental Quality at the Landscape Scale. Bioscience. 1997; 47:513-519.

Parker GR. "The Wave of Settlement." In *The Natural Heritage of Indiana*, MT Jackson, ed. Bloomington IN: Indiana University Press, 1997.

Pebbley AR. Demography and the Environment. Demography. 1998; 35: 377-389.

Peterson J. Forests and Forestry in Indiana: Answers to Questions of Public Interest and Concern. Evergreen. 1998; 9.

Rindfuss RR, Stern PC. "Linking Remote Sensing and Social Science: The Need and the Challenges." In *People and Pixels*, D Liverman, EF Moran, RR Rindfuss, PC Stern eds. Washington DC: National Academy Press, 1998.

Shukla JCN, Sellers P. Amazon deforestation and climate change. Science. 1990; 247: 1322-1325.

Sieber E, Munson CA. *Looking at History: Indiana's Hoosier National Forest Region, 1600 to 1950*, Bloomington IN: Indiana University Press, 1992.

Teillet PM, Fedosejevs G. On the Dark Target Approach to Atmospheric Correction of Remotely Sensed Data. Canadian Journal of Remote Sensing. 1995; 21: 374-387.

Thome KJ. Absolute calibration of Landsat-5 Thematic Mapper. Proceedings of the International Geoscience and Remote Sensing Symposium, Pasadena CA, 2295-2297, 1994.

Turner MG, Wear DN, Flamm RO. Land Ownership and Land-Cover Change in the Southern Appalachian Highlands and the Olympic Peninsula. Ecological Applications. 1996; 6: 1150-1172.

Van Kley JE. An Ecological Classification System for the Central Hardwoods Region. Ph. D. thesis, Purdue University, 1993.

Vitousek P, Mooney H, Lubchenco J, Melillo J. Human Domination of Earth's Ecosystems. Science 1997; 277: 494-499.

Walsh SJ, Evans TP, Welsh W, Entwisle B, Rindfuss R. Scale-Dependent Relationships between Population and Environment in Northeast Thailand. Photogrammetric Engineering and Remote Sensing. 1999; 65: 97-105.

Wood CH, Skole D. "Linking Satellite, Census, and Survey Data to Study Deforestation in the Brazilian Amazon." In *People and Pixels*, D Liverman, EF Moran, RR Rindfuss, PC Stern eds. Washington DC: National Academy Press, 1998.

Chapter 17

SHIFTING CULTIVATION WITHOUT DEFORESTATION: A CASE STUDY IN THE MOUNTAINS OF NORTHWESTERN VIETNAM

Jefferson Fox[1], Stephen Leisz[2], Dao Minh Truong[3], A. Terry Rambo[1], Nghiem Phuong Tuyen[3], and Le Trong Cuc[3]
[1] East-West Center, 1601 East-West Road, Honolulu, Hawaii 96848, USA - FoxJ@EastWestCenter.org
[2] CARE International in Vietnam, 63 To Ngoc Van, Hanoi, Vietnam
[3] Center for Resource & Environmental Studies (CRES), Vietnam National University, Hanoi, 167 Bui Thi Xuan, Hanoi, Vietnam

Keywords: Shifting cultivation, Vietnam, land cover change, land cover fragmentation.

Abstract To assess the role of shifting cultivation as a driving force of land cover change we examined the social, cultural, economic, and spatial dynamics of land use in a Vietnamese village. Instead of the denuded landscapes associated with shifting cultivation, the landscape of Tat hamlet is composed of a heterogeneous mosaic of fields, pastures, and forest patches in various stages of secondary succession. Failure to see secondary forests, let al.one the benefits of secondary forests, has led to government policies encouraging permanent agriculture - most of which have failed. Failure to account for the effects of landscape heterogeneity also means that significant effects of land cover change are not being recognized.

INTRODUCTION

The driving force behind environmental policies in many developing countries around the world is a set of powerful, widely perceived images of environmental change. These images include extensive deforestation, plummeting biodiversity, the collapse of sustainable agricultural systems, widespread soil loss, and the mining of natural resources caused by rapidly growing populations. So self-evident do these phenomena appear that their prevalence is generally regarded as common knowledge among development professional, international donor agencies, and non-governmental organizations. These images have acquired the status of conventional wisdom. Yet as Leach and Mearns (1996) and Fairhead and Leach (1996) have shown so convincingly in Africa, these images may be deeply misleading.

Shifting cultivation, or swidden farming, is often held to be the principle driving force for deforestation in tropical Asia (Myers 1993).[i] While this idea can be traced back to early colonial times (Poffenberger 1990), national governments in Southeast Asia, notably in Indonesia, the Philippines, Thailand, and Vietnam, have been particularly prone to blame shifting cultivators, usually members of ethnic minorities, for the rapid loss of forests suffered by these countries in the past 20 years (Do Van Sam 1994, Dove 1984, Le Trong Cuc 1996, Rambo 1996). In Vietnam, the official view of shifting cultivation has been particularly negative, reflecting a combination of ethnocentric assumptions about the cultural superiority of wet rice farming held by the numerically dominant *Kinh* (lowland Vietnamese) with the Marxist view that swiddening represents a primitive stage in the cultural evolutionary sequence (Jamieson 1991, Rambo 1995). In official discourse, shifting cultivation is invariably identified as the primary cause of deforestation.[ii] Putting an end to shifting cultivation has for the past 30 years been one of the major objectives of national policies for upland development (Chu Huu Quy 1995, Le Duy Hung 1995).

Such views of shifting cultivation in Southeast Asia are not restricted to professional circles. They are also popularized in the news and current affairs media and help to build support among the general public for field operations designed to halt this form of land use. These orthodoxies assign to swidden farmers a particular role as agents, as well as victims, of environmental change. If current trends are to be reversed, it is implied, local land use practices have to be transformed and made less destructive. Yet the development polices and programs that result commonly prove to be at best neutral and at worst deleterious in their consequences for rural people and for the natural resource based on which their livelihoods depend.

Complicating the story is the fact that swidden agriculture and its associated stages of secondary vegetation are not captured well in traditional land cover classes of agriculture, plantation, forest, etc. Potter et al. (1994) suggest that as much as 26% of all land in Southeast Asia falls into the "other" category where other is defined as shrub, brush, pasture, waste and other land use categories, many of which are actually secondary vegetation. Likewise Kummer and Turner (1994) suggest that approximately 33% of the land cover in the Philippines falls into the "other" category and that this category grew by more than 20,000 hectares (ha) between 1948 and 1987. In Thailand, the field plan of the Tropics Program of the GEWEX Asian Monsoon Experiment (GAME-Tropics) is based on a land use breakdown in which fully 49% of the non-forested land in northern Thailand is "unclassified" (GAME-Tropics 1996).

This paper seeks to describe and classify the agroecosystem of a hamlet in northern Vietnam in which long-term practice of shifting cultivation has not resulted in extensive deforestation—but it has altered the character of the vegetative cover. Through the analysis of this agroecosystem we seek to develop a better understanding of the swidden agricultural system found in this hamlet and the effects of this system on land cover change in the area over the last 40 years. Through re-examining how we classify land cover in a swidden agriculture system, we seek to question ideas of ecological equilibrium or "climax vegetation community." We argue that the failure to understand the swidden agricultural system and its effects on forest regeneration has led scientist to overestimate the amount of "deforestation" that has occurred in Southeast Asia. Failure to understand secondary forests has also led to government policies, mostly failures, for settling swidden agriculturists

This research is part of a joint project on sustainable rural resource management and conservation of biodiversity in Vietnam being carried out by the East-West Center Program on Environment (ENV) and the Center for Natural Resources and Environmental Studies (CRES) of the Vietnam National University, Hanoi. The goals of this project are to find more sustainable ways to manage upland ecosystems and conserve biological diversity in Vietnam's extensive uplands. Since January 1992, CRES-EWC research teams have made repeated field trips to the hamlet. The present paper draws heavily on data collected during these visits.

METHODOLOGY

The project required accurate and large-scale land cover classification of a remote area of approximately 6,500 ha in northern Vietnam where little ancillary data were available, where it is difficult to obtain cloud-free satellite imagery, and where land cover is extremely heterogeneous and fragmented into small plots (e.g. < 60 m x 60 m). To meet this requirement we designed a methodology that integrated the development of a spatial database—based on topographic maps, aerial photographs, satellite images and a digital elevation model with information on elevation, slope, and aspect—with information on historical land cover and land use practices collected through interviews with farmers and other key informants. This database served as a framework for analyzing changes in land cover and forest patterns through time, as well as a tool for analyzing the information and insights collected in semi-structured informal interviews.

Spatial Database

The spatial database was developed using 1952 (nominally 1:40,000) aerial photographs[iii] as well as the most recent cloud-free Landsat TM image of the study site (June 24, 1995) available from the Royal Thai Remote Sensing Center. Fieldwork was conducted in 1997 and 1998. We conducted a normalized difference vegetation index (NDVI) analysis and an unsupervised classification of the TM image in order to estimate the number of land cover classes we could expect to find in the field and to estimate the optimal number of ground truth points needed for conducting a supervised classification and accuracy assessment. Land cover could be broken into 6 to 8 classes running a continuum from agriculture (e.g. rice paddy areas and active swidden) to closed-canopy forest with the exact make up of the vegetation along this continuum not known. A minimum of 55 to 60 randomly selected ground-truth points needed to be collected for each land cover class (Lillesand and Kiefer 1994, Congalton and Green 1999) or approximately 330 and 480 points. The field site was an isolated area with dense vegetation and steep terrain, lacked electricity, researchers could not spend extended time in the field because of local political considerations, and field work was being conducted 2 to 3 years after the date of the imagery. Consequently we developed a field methodology for collecting as many ground-truth points in as random a manner as possible and with as much ancillary information as we could about land cover at the time the image was acquired.

We began by conducting a walking survey of the area in conjunction with local farmers and resource managers. Farmers were interviewed regarding their farming systems, the history of land cover dynamics in the area, and specific information about individual plots of land. This information was crucial for relating the present land cover to the land cover found on the historical images. After setting up the GPS base station[iv] we walked transects using GPS receivers to collect ground-reference points. We used a rangefinder, compass and a clinometer to survey points up to 600 m from the GPS surveyed point. Field boundaries, paths, streams, and roads were surveyed, and, where possible, community borders. Figure 1 shows the ground-truth points collected in the Tan Minh village area. A total of 320 ground truth points were collected.

Socio-Economic Database

We interviewed farmers and resource managers at each ground-truth point about land cover on that piece of land at the time the imagery was collected. If the interviewee could not remember, we noted that information and discussed the general land cover rotation of the area. Photographs of each ground-truth point and points mapped with the rangefinder were taken and later tied to the reference point in the spatial database. Table 1 is a portion of the database table created from the ground reference points.

We also conducted interviews in 42 households (60% of total households) as well as with provincial, district, village, and hamlet government officials. Researchers documented changes in national and regional policies influencing land use (e.g. tenure, taxation, credit, import and export regulations) as well as changes in infrastructure (roads and markets). We interviewed residents of the village to learn more about the socio-economic factors contributing to their decision to create or maintain forest fragments in their area. Key informants were used to assess, among other factors, local peoples' perception of the forest.

After analyzing the field interviews, we determined the number of land cover classes identified by the local informants. A description of the vegetation within each class was noted and, if available, the local name of the class was also recorded. Each land cover class was given a code number and this was then added as an attribute in the GIS database. These classes were used in classifying the image.

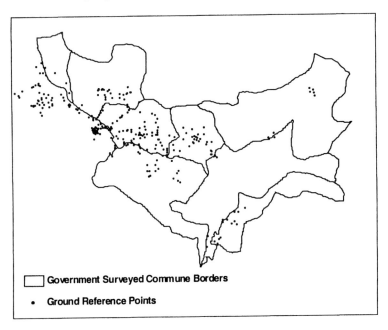

Figure 1. Ground-truth points in Tan Minh Commune.

Crosschecking Data Accuracy and Building an Attribute Database

All information was crosschecked to verify its accuracy. GPS surveyed points were checked by taking repeated measurements of a number of the points on different days and at different times during the day and then comparing the results to obtain a measure of repeatability and, correspondingly, a best estimate of the accuracy of the GPS surveying. The distance from the GPS surveyed point to each extended point, as well as the bearing and slope of the extended points, were measured three times with the rangefinder. We did a further check by measuring from one GPS surveyed point to another using the compass, clinometer and rangefinder. The two results were compared after differential processing was done and the coordinate geometry calculated. Field interviews and farming system information were also crosschecked by independently interviewing different individuals about each location, the farming system rotation used, and the types of land cover found in the fields at each point in the rotation.

We differentially corrected the GPS data by averaging all the base station data to a single point. We assumed that this point was accurate to +/- 2 m. We then differentially corrected the data from the rover units. Given that the equipment used has a manufacturer specified accuracy of +/- 1 to 5 m absolute error, and adding error from the initial reference position, we estimated the error for the rover unit was approximately +/- 3 to 7 m. We calculated coordinates for the extended points by using coordinate geometry and recorded these points on a spreadsheet with the GPS points. This information was then converted into a database (see Table 1) and points, lines and polygons were generated using Arc/Info GIS software. Attribute information such as current land cover, land cover at the time of image acquisition, information about the farming system, and any other information available regarding the history of land cover at each field location was added to the data base. All photographs were scanned and included in the database. Other information, such as topographic maps of the area, were digitized and added to the GIS database.

Image Analysis

The satellite image was registered to the map base that has been created from the collected field survey points. We aimed for an RMS error of +/- 1 pixel. After registering the image to the map base, the field information was lain over the satellite image. Training sets were chosen using between 5 to 10 ground truth points per class per image to extract spectral training set

Table 1. Land cover database sample*.

POINT	LOC. Descrip.	PHOTO #	LAND-COVER 1995	LAND-COVER 1998	LONG.	LAT.	ALT.
Enh							
11_27_1	border of Enh and Tam Luong hamlet, Doan Ket, cut through hill		2		507244.347	2316003.969	594.879
11_27_2	on road		2		507424.585	2315894.945	603.182
11_27_3	cassava, then at 8 year point, old trees	1_12	4	4 or 5	507578.260	2316078.088	
11_27_4	ferns and bamboo		3	4	507445.106	2315838.563	
11_27_5	on road at curve		3		508078.845	2315500.266	505.463
11_27_6	trees and bamboo, fallow for 7 years	1_15	3	4	508113.331	2315464.555	
11_27_7	Bamboo		3	3	508056.896	2315428.474	
11_27_8	cassava in foreground then bamboo	1_14	2	1&4	508036.050	2315466.831	
11_27_9	big trees	1_13	6	5 (6)	507961.609	2315496.172	
11_27_10	on road		3		508216.712	2315583.459	489.833

signatures for each class (a total of 62 points were used to derive the training sets). Finally, we conducted a supervised classification of the subset images using Erdas Imagine's maximum likelihood classifier algorithm. We then ran a supervised classification of the image using the training sets with a maximum likelihood algorithm.

We conducted an accuracy assessment of the classification results. Two types of assessment were done. The first was a strict assessment that classified a checkpoint as correct only if it fell exactly within a correctly classified pixel. The second was a more liberal assessment that accounted for the error propagated through the system. When all possible errors were summed, each reference point could be off by 1.5 pixels—3 to 7 m from the ground survey and up to 30 m (1 pixel) from registration error. The liberal assessment looked at a radius of 1 adjacent pixel around each checkpoint. If any of these pixels were correctly classified vis-à-vis the checkpoint, then the checkpoint was considered correctly classified. For both of the assessments an error matrix showing the users and consumers accuracy and absolute accuracy was created. Finally, a KHAT statistic for each assessment was calculated and an error analysis of each classification completed.

RESULTS

The Agroecosystem

Tat is one of ten hamlets in Tan Minh village, Da Bac District, Hoa Binh Province, and is positioned astride the Muong River, a tributary of the Da River, at an altitude of about 360 m asl. The Muong valley varies between a few hundred m to about 2 km wide and is bounded by peaks reaching 800 to 1000 m. The highest peak is Mt. Hen at 1176 m. The valley walls are extremely steep with slopes often exceeding 60% dissected by many small streams flowing into the Muong River. The valley is shaped like an amphitheater with buttress forming ridges extending up to the surrounding peaks.

The people of Ban Tat are mostly members of the Tay ethnic minority.[vi] They speak a language belonging to the T'ai family. Local oral history says that migrants from Son La settled the hamlet a little over one hundred years ago. Mobility appears to have been relatively high with several waves of immigration and emigration. According to one elderly informant there were only 7 households or approximately 50 people living in the village in 1954. Today the hamlet has grown to 69 households with a population of 389 people. This represents a population growth rate of roughly 4.9% annually—probably the result of a natural growth rate of 3.0% to 3.5% and in-migration. During this period population density grew from approximately 10 to 75 people per km^2. This is approximately twice the average population density for the Da River watershed.

For as far back as any informants can remember, the Tay of Tat Hamlet have been "composite swiddeners" (Rambo 1996). The defining characteristic of composite swiddening is that households simultaneously manage both permanent wet rice fields in the valley bottoms, shifting swidden fields on the hillslopes, and exploit wild resources of the forest. Similar composite systems are found among the Shan of Burma and northern Thailand (Schmidt-Vogt 1998), the Hani of Xishuangbanna

Prefecture in Southwestern China (Pei Shengji 1985), and the Ifugao (Dove 1983) of the Cordillera in the Philippines.

The distinctive characteristic of this system is that swiddening comprises an integral component of the total system. It is not a gradually vanishing survival of an earlier, more primitive pure swiddening adaptation that is in the process of being replaced by more advanced irrigated farming. Neither is swiddening present as a recent response to rapid population growth that has exceeded the carrying capacity of the wet rice fields and forced people to expand their farming onto the forested slopes.

Indeed, the household resource system of the Tay is notable for its incorporation of a wide range of subsystems. A typical Tay household manages a complex agroecosystem (Figure 2). Key subsystems include wet rice fields, home garden, fishpond, livestock, tree gardens, rice swiddens, and cassava swiddens. Fallow swiddens and secondary forest are also exploited, but management of these land units is the responsibility of the cooperative. Hence the landscape of Ban Tat is a mosaic of cultivated and fallowed fields, interspersed with protected forest areas. There does not seem to be any regular pattern to this mosaic. Swidden fields can be found anywhere on the slope from the toe of the hill to the top. A village people's committee that is responsible for allocating land and enforcing regulations regarding forest use administers the hamlet. This committee allocated land to villagers from two large areas near Ban Tat where swiddening is permitted—Suoi Co San and Suoi Muong. In principle, villagers should alternate their fields between the two areas every several years leaving one of the areas to fallow and regeneration. In practice, because land is scarce, both parcels are used at the same time.

Villagers' Perceptions of Land Cover

Farmers and resource users in Ban Tat suggested in the field interviews that there are seven distinct types of land cover found within the community's boundaries. These seven classes include 1) rice paddy; 2) active (planted) swidden; 3) grass; 4) mixed grass/bamboo/woody shrubs; 5) mixed bamboo/small trees/woody shrubs; 6) mixed tall bamboo/medium to large trees; and 7) medium to tall trees (no bamboo). Rice paddies are fixed on the landscape and do not change year to year. The class of medium to tall trees is also relatively fixed on the landscape—except if it is cleared for swidden purposes, it does not evolve into anything else. The other five classes, however, are actually stages of secondary forest regeneration. Active swidden fields are cut and planted in upland rice, corn, or cannas for 1 to 3 years after which they are usually left fallow. During the first 2 years of fallow, grass dominates the vegetation. By the third year the grass evolves into small bamboo and small shrubs. After 5 years the grass is shaded out, the bamboo grows taller, and the shrubs become small trees. At this stage, the field is usually cut again for swidden and the cycle starts again. If the field is not cut, however, then about 8 years after being left fallow the trees grow larger and, if the bamboo is of the type locally called *giang*, it will grow taller and the vegetation will be recognized as being in the sixth stage of regeneration. Finally, after 20+ years, the trees manage to out compete the bamboo and the vegetation may advance to the final stage of secondary forest.[7]

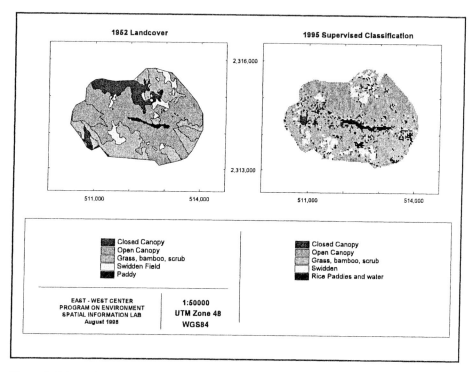

Figure 2. Schematic diagram of upland production systems

Image Processing Results

Classification results supported both our assessment that land cover of the area is complex and fragmented into small patches of different land covers, and the villagers' classification of 7 distinct types of land cover (Table 2 -1995 data).

Table 2. Land cover and fragmentation in Bat Tat in 1952 and 1995.

Land Cover/ Fragments	1952 Ha	1952 %	1995 Ha	1995 %	1952 # of fragments	1952 Mean size (ha)	1995 # of fragments	1995 Mean size (ha)
Secondary Forest	681	92	616	84	18	37	292	2
Closed-Canopy	81	11	19	3	4	20	19	1
Open-Canopy	400	54	110	15	5	78	40	3
Grass, bamboo, scrub	200	27	487	67	9	22	233	2
Swidden	52	7	73	10	35	1	75	1
Paddy	7	1	43	6	1	8	45	1
Total	740	100	740	100	54	13	412	2

For purposes of simplification we reduced this to 5 classes: 1) rice paddy; 2) active swidden; 3) grass, bamboo, and scrub (grass, and mixed grass/bamboo/woody shrubs), 4) open-canopy secondary forest (mixed bamboo/small trees/woody shrubs); and 5) closed-canopy secondary forest (mixed tall bamboo/medium to large trees, and medium to tall trees).

According to this classification the largest land cover class is grass, bamboo, and scrub (487 ha). This corresponded with the farmers' and resource managers' preference for cutting this type of vegetation to create active swidden fields and suggests that the community manages the landscape for this type of land cover. The classification also shows that the areas covered with closed-canopy tree cover (19 ha) open-canopy tree cover (110 ha) are limited. After the classification was completed, conservative and liberal accuracy assessments (as described above) were conducted. Tables 3 and 4 show the error matrix created for each accuracy assessment. These assessments show KHAT values of 93% and 65% respectively.

Land Cover Change Under Shifting Cultivation

Historical changes in land cover and fragmentation in Ban Tat between 1952 and 1995 are also summarized in Table 2 and shown in Figure 3. During this period 43-year period, the area covered by secondary forests or successional vegetation decreased from 92% to 84% of the landscape. The area covered by closed-canopy forest decreased from 11% to 3% of the landscape, the area covered by open-canopy forest decreased by 290 ha (39% of the landscape), and the area covered by grass, bamboo, and scrub increased by 287 ha (39%). While the total amount of land under secondary forests remained relatively constant, closed- and open-canopy forests were degraded to earlier stages of succession. Significant changes have also occurred in the spatial distribution of that land cover. Between 1952 and 1995 the number of secondary forest fragments grew from 18 to 292 while the mean size decreased from 37 ha to 2 ha.

Table 3. Liberal accuracy assessment land cover classification of 1995 image.

Ground cover as classified on image Class	\multicolumn{7}{c}{Land cover for ground-truth points as determined from field visits}	Total	User Accuracy
	1	2	3	4	5	6	7		
1	43	1	1		1			46	93.48%
2		40			1			41	97.56%
3		6	90	2	1			99	90.91%
4			1	25				26	96.15%
5				1	14			15	93.33%
6						8		8	100.00%
7							23	23	100.00%
Total	43	47	92	28	17	8	23	258	
Producer Accuracy	100.00	85.11	97.83	89.29	82.35	100.00	100.00		

Overall accuracy 94.19%

KHAT = 92.55%

1 = active swidden; 2 = grass; 3 = grass, bamboo, woody shrubs; 4 = bamboo and small trees; 5 = medium to large trees and *giang* bamboo; 6 = trees and no bamboo; 7 = rice paddy.

Table 4. Conservative accuracy assessment of land cover classification of 1995 image.

Ground cover as classified on image Class	\multicolumn{7}{c}{Land cover for ground-truth points as determined from field visits}	Total	User Accuracy
	1	2	3	4	5	6	7		
1	39	5	8		1		1	54	0.722222
2	3	32	3	1	1	1	1	42	0.761905
3	1	7	72	13	3	1	3	100	0.72
4			2	13	5			20	0.65
5			2	1	7			10	0.7
6		1	2			6		9	0.666667
7		2	3				18	23	0.826087
Total	43	47	92	28	17	8	23	258	
Producer Accuracy	90.70	68.09	78.26	46.43	.41	75.00	78.26		

Overall accuracy 72.48

Notes for Table 2.
KHAT = 64.59%
1 = active swidden; 2 = grass; 3 = grass, bamboo, woody shrubs; 4 = bamboo and small trees; 5 = medium to large trees and *giang* bamboo; 6 = trees and no bamboo; 7 = rice paddy

DISCUSSION AND CONCLUSIONS

The FAO (Rao 1989:6) distinguishes between deforestation and forest degradation as follows: 'Deforestation refers to the transfer of forest land to non-forest uses and includes all land where the forest cover has been stripped and the land converted to such uses as: permanent cultivation, shifting cultivation, human settlements, mining, building of dams, etc.' Degradation, on the other hand, 'refers to a reduction in the extent and quality of the forest cover due to such factors as: indiscriminate logging; inappropriate road-making methods; forest fires, etc.' It is notable that FAO defines deforestation as both a change in land cover, (i.e. loss of forest cover) and a change in land use (i.e. converted to other permanent uses). Based on this definition, FAO estimates that between 1990 and 1995 total forest cover in Vietnam declined by 6%.

This study shows that traditional swiddening does not entail permanent conversion but only temporary use of forestland. Hence we do not include the secondary forest that occurs after shifting cultivation as deforestation. We report only a small amount of deforestation (the increase in paddy from 1% to 6% of the landscape) for the study area over the 43-year period between 1952 and 1995. In Ban Tat, as in many swidden agricultural landscapes in Southeast Asia, rather than "deforestation" we observe a change in land cover from a fairly homogeneous forest cover (closed- and open-canopy) to a highly heterogeneous and fragmented cover of vegetation.

Other researchers have noted that estimated rates of tropical deforestation vary for a number of reasons including ambiguities surrounding the future of cutover land (Williams 1990, Myers 1991). If a substantial portion of cutover land is regenerating into forest, the rate of felling of primary forest is an overestimate of the overall net rate of change in forested area. This is illustrated most clearly in literature from South America where to date most of the research in forest patches and vegetation recovery has been done (Uhl, et al. 1988, Turner *et al.* 1993, Moran et al. 1994). Our work also suggests that the major land cover change in northern Vietnam is the degree to which forest fragmentation has occurred. Fox et al. (1995) found similar results in northern Thailand.

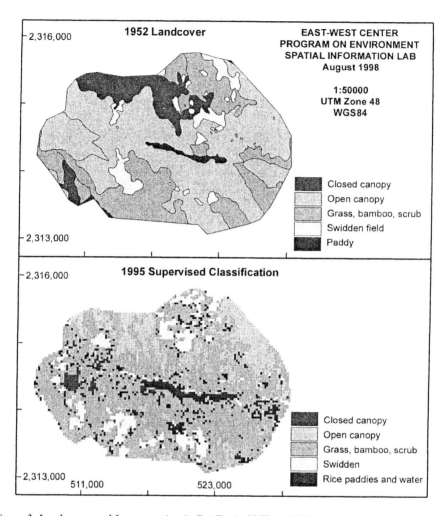

Figure 3. Land cover and fragmentation in Bat Tat in 1952 and 1995.

Our findings are at variance with the conventional wisdom that shifting cultivation, under conditions of increasing population density, inevitably results in extensive deforestation. This variance is due partially to how we define deforestation. Two other factors are of equal if not more importance. The first and perhaps most significant is the high sustainability of the composite swiddening system employed by the Tay which has not previously been appreciated by the *Kinh* or by Western observers. The second factor, also related to the Tay land management system, is the importance of secondary forests and the Tay's active efforts to maintain a mosaic of land cover across the full spectrum of the various stages of forest regeneration. This complex indigenous land use system thus both maximizes the stability of food production and the percent of the landscape dominated by secondary forests. Our research suggests that perhaps too much emphasis has been placed on changes in land cover, and insufficient attention paid to the stability of swidden agriculture as the main land use system in this region.

Contradictory evidence from a single case study cannot, of course, entirely refute an orthodoxy. The fact that forest loss or range of forest degradation may have been misinterpreted in one place does not necessarily mean they are being misinterpreted everywhere. But the fact that contradictory cases exist certainly casts doubt over the general applicability of received views, and calls for a more critical examination of the evidence that supports other cases.

This case study demonstrates the importance of using historical and 'time-series' data sets of various types to study the processes of landscape change more directly, documenting history rather than inferring it. This case study also demonstrates the logic and rationality of 'indigenous' knowledge and organization in natural resource management. By contrast, received wisdom would have much of the blame for the vegetation change perceived by outsiders as environmental 'degradation' rest with local land use practices, whether labeling them as ignorant and indiscriminate or, more commonly, as ill-adapted to contemporary social-economic and demographic pressures.

It is interesting to ask what would happen if policy makers were successful and farmers did switch from swidden agriculture to some form of permanent agriculture? In other words, what would happen if we finally did have a change in land use and not just a change in land cover? Permanent agriculture could result in a tree dominated land cover, (e.g. rubber, palm oil, cardamom, or tea), or it could result in a land cover composed of annuals (e.g. maize, cassava, and upland rice). Current trends towards tree gardens indicate the hamlet may be able to maintain a high percentage of tree cover. In either case, biodiversity, as measured by the number of species found on the landscape, would presumably plummet (Lawrence et al. 1998). Hydrological impacts could be considerable more severe than that experienced under secondary forests of the traditional swidden system (Zinke et al. 1978, Alford 1992, Forsyth 1994; 1996). The economic returns of converting from secondary forest to a market crop would probably be higher. In fact, as people in the northern Vietnam uplands continue to participate more deeply in a market economy, as the road system is expanded and improved, and as farmers become more or less secure in their rights to use hill lands, this is exactly the type of change we can expect to see over the next few decades. These changes will lead to further and perhaps drastic changes in land use practices and land cover.

Ban Tat has been fortunate in being able to survive the changes of the last 43 years with minor variations in the total amount of secondary forest. Government officials and planners need to recognize that perhaps the biggest effect of tropical "deforestation" so far has been due to a change from a relatively homogeneous forest to a highly heterogeneous and fragmented cover of secondary forest. The land cover may be "degraded" in terms of merchantable timber species. But it could well be that this "degraded" secondary forest, a product of the swidden land use system, may be the most species rich and hydrologically favorable (i.e. water- and soil-holding) land cover available. In addition, swiddening is the land use system most eminently suitable for meeting the needs of the local community.

Failure to see the benefits as well as the costs of secondary forest, and the swidden agricultural system, has led to government policies for settling swidden farmers—many of which have been failures. A more efficient, as well as humane, policy would be to invest in research on methods of maintaining the biodiversity associated with swidden fallows while increasing their productivity and soil sustaining properties. Failure to recognize secondary regeneration in swidden fallows has also led scientists to overestimate the extent of "deforestation" that has occurred in Southeast Asia. Finally, models of global climatic change have been based on an extreme scenario of forest conversion to degraded pasture or impoverished grassland (Giambelluca 1996). Failure to account for the effects of landscape heterogeneity may mean that significant effects of land cover change are not being recognized.

ACKNOWLEDGEMENTS

We would like to acknowledge financial support for our fieldwork in Vietnam from the United States National Science Foundation, the Japanese Ministry of Foreign Affairs, the Rockefeller Brothers Fund, the John D. and Catherine T. MacArthur Foundation, the Ford Foundation, and the Global Environment Forum. Analysis of remote sensing data on land cover change in Tat hamlet was supported by a separate grant to CRES from the Rockefeller Brothers Fund.

NOTES

[1] We distinguish shifting cultivation from other systems of "slash-and-burn" encompassing a wide range of land uses that rely on fire simply to clear forests for cultivation or further development.

[2] Spraying of chemical defoliants by the U.S. airforce is the second most frequently invoked official explanation for the decline in the area under forest cover between 1943 and 1995.

There is no question that herbicides destroyed large areas of forest in the southern part of the country in the 1960s and early 1970s. But the rate of deforestation has been even higher in areas, such as the Northwestern Mountain Region that were never the target of defoliant spraying. Unrestricted logging by state forest enterprises and clearance of land for resettlement of migrants from the lowlands have been much more important causes of deforestation. Even in the southern region, when speaking off the record, knowledgeable Vietnamese environmental scientists consider defoliation to be a relatively minor factor in explaining the current forest crisis.

[3] The aerial photographs were taken as part of a comprehensive mapping of Vietnam undertaken by the French military in 1952. We are grateful to Dr. Jean-Francois Dupon, Senior Scientist of ORSTOM for assistance in obtaining access to these photographs. We would also like to acknowledge the staff of the Institut Geographique National for their prompt and courteous response to our request for the photographs of Ban Tat.

[4] In remote areas of Vietnam it is not possible to set up a base station on a known, benchmark location. There were no benchmarks in the area where the fieldwork took place. Also, the nature of our GPS units (e.g. limited memory capacity and power source) meant that we had to download them frequently. This made it essential that the base station be close to the field site. Hence, we established a base station at a secure location in the community where the fieldwork was taking place and left on for 54 hours. Different studies suggest that the averaging of this amount of information can provide locational accuracy of +/- 1 to 2 m for the base station.

[5] As suggested by Congalton and Green (1999), most ground reference points were collected in the land cover categories thought to be the most spectrally similar (classes 1, 2, 3, and 4). This was done in order to maximize the use of the limited time and resources available for ground truthing and to maximize our ability to assess how well the classification differentiated between these similar classes.

[6] Although officially classified as "Tay", the people of Ban Tat are culturally quite distinctive. Some Vietnamese ethnologists argue that they are actually a branch of the White Thai. More probably, they are simply a unique local population, one of many variant groups that have evolved under conditions of relative isolation in the northwestern mountains.

[7] Village informants suggest this cycle is rarely completed as the fields are usually cut again before they advance into the fifth or sixth stage.

REFERENCES

Alford D. Streamflow and sediment transport from mountain watersheds of the Chao Phraya Basin, northern Thailand: A reconnaissance study. Mountain Research and Development. 1992; 12: 257-268.

Chu Huu Quy. Overview of Highland Development in Vietnam: General Characteristics, Socioeconomic Situation, and Development Challenges." In *The Challenges of Highland Development in Vietnam.* AT Rambo, RR Reed, Le Trong Cuc, MR DiGregoiro eds. Honolulu (HI): East-West Center, 1995.

Congalton, R., Green K. *Assessing the Accuracy of Remotely Sensed Data: Principles and Practices.* New York: Lewis Publishers, 1999.

Do Van Sam. *Shifting cultivation in Vietnam: Its social, economic, and environmental values relative to alternative land use.* London (UK): IIED Forestry and Land Use Series No. 3. London: International Institute for Environment and Development, 1994.

Dove MR. Ethnographic Atlas of the Ifugao: Implications of Theories of Agricultural Evolution in Southeast Asia. Current Anthropology. 1983; 24. 516-520.

Dove MR. Government versus peasant beliefs concerning imperata and eupatorium: A structural analysis of knowledge, myth, and agricultural ecology in Indonesia. Honolulu (HI): East-West Center Working Paper. No. 54. East-West Center, 1984.

Fairhead J, Leach M. *Misreading the African Landscape: Society and Ecology in a Forest—Savanna Mosaic.* Cambridge: Cambridge University Press, 1996.

Forsyth T. The use of cesium-137 measurements of soil erosion and farmers' perceptions to indicate land degradation amongst shifting cultivators in northern Thailand. Mountain Research and Development. 1994; 14: 229-244.

Forsyth T. Science, myth and knowledge: Testing Himalayan environmental degradation in northern Thailand. Geoforum. 1996; 27: 375-392.

Fox J, Krummel J, Yarnasarn S, Ekasingh M, Podger N. Land use and landscape dynamics in northern Thailand: Assessing change in three upland watersheds. Ambio.1995; 24: 328-334.

GAME-Tropics. Draft Implementation Plan of GAME-Tropics, web site: http://climate.gsfc.nasa.gov/~taikan/GAMET/Imple2e/imple2e.html#3221, 1996.

Giambelluca TW. "Tropical land cover change: Characterizing the post-forest land surface." In *Climate Change: Developing Southern Hemisphere Perspectives*, TW Giambelluca, A Henderson-Sellers Eds. Chichester UK: John Wiley and Sons, 1996.

Jamieson NL. Culture and Development in Vietnam. Honolulu (HI): East-West Center Working Papers Indochina Series No. 1. East-West Center, 1991.

Kummer DM. Turner II BL. The human causes of deforestation in Southeast Asia. BioScience. 1994; 44: 323-328.

Lawrence D, Peart D. Leighton M. The impact of shifting cultivation on a rainforest landscape in West Kalimantan: Spatial and temporal dynamics. Landscape Ecology. 1998; 13: 135-148.

Leach M, Mearns R. eds. *The Lie of the Land: Challenging Received Wisdom on the African Environment.* London: The International African Institute, 1996.

Le Duy Hung. Some Issues of Fixed Cultivation and Sedentarization of Ethnic Minority People in Mountainous Areas of Vietnam." In *The Challenges of Highland Development in Vietnam.* AT Rambo, RR Reed, Le Trong Cuc, MR DiGregoiro eds. Honolulu (HI): East-West Center, 1995.

Le Trong Cuc. 1996. Swidden Agriculture in Vietnam." In *Montane Mainland Southeast Asia in Transition.* Chiang Mai (Thailand): Chiang Mai University Consortium, 1996.

Lillesand T, Kiefer R. *Remote Sensing and Image Interpretation.* New York: John Wiley and Sons, Inc, 1994.

Moran EF, Brondizio E, Mausel P, Wu Y. Integrating Amazonian vegetation, land use, and satellite data. BioScience.1994; 44: 329-338.

Myers N. Tropical forests: Present status and future outlook. Climatic Change. 1991; 19: 3-32.

Myers N. Tropical forests: The main deforestation fronts. Environmental Conservation. 1993; 20: 9-16.

Pei Shengji. "Some effects of the Dai people's cultural beliefs and practices on the plant environment of Xishuangbanna, Yunnan Province, Southwest China." In *Cultural Values and Human Ecology in Southeast Asia*, K Hutterer, AT Rambo, G Lovelace eds. Ann Arbor (MI): Michigan Papers on South and Southeast Asia, No. 27. University of Michigan, 1985.

Poffenberger M. ed. *Keepers of the Forest: Land Management Alternatives in Southeast Asia*. West Hartford, Connecticut: Kumarian, 1990.

Potter L, Brookfield H, Byron Y. "The Sundaland region of Southeast Asia." In *Regions at Risk: Comparisons on Threatened Environments*. JX Kasperson, RE Kasperson, BL Turner II eds. Tokyo: United Nations University Press, 1994.

Rambo AT. "Defining Highland Development Challenges in Vietnam: Some Themes and Issues Emerging from the Conference." In *The Challenges of Highland Development in Vietnam*. AT Rambo, RR Reed, Le Trong Cuc, MR DiGregoiro eds. Honolulu (HI): East-West Center, 1995.

Rambo AT. "The Composite Swiddening Agroecosystem of the Tay Ethnic Minority of the Northwestern Mountains of Vietnam." In *Montane Mainland Southeast Asia in Transition*, Chiang Mai: Chiang Mai University Consortium, 1996.

Rao Y. Forest resources of tropical Asia. Environment and Agriculture — Environmental Problems Affecting Agriculture in the Asia-Pacific Region. World Food Day Symposium, 11 October 1989. Bangkok: Food and Agriculture Organization, 1989.

Schmidt-Vogt D. Defining degradation: The impacts of swidden on forests in northern Thailand. Mountain Research and Development> 1998; 18: 135-149.

Turner II BL, Moss RH, Skole DL. 1993. Relating land use and global land cover change: A proposal for an IGBP-HDP core project. International Geosphere-Biosphere Programme: A study of global change and the human dimensions of global environmental change, IGBP Report No. 24.

Uhl C, Buschbacher R, Serrão EAS. Abandoned pastures in eastern Amazonia. I. Patterns of plant succession. Journal of Ecology. 1988; 76: 663-681.

Williams M. "Forests." In *The Earth as Transformed by Human Action,* BL Turner II, WC Clark, RW Kates, JF Richards, JT Mathews, WB Meyer eds. Cambridge: Cambridge University Press, 1990.

Zinke P, Sabharsi S, Kunstadter P. "Soil fertility aspects of the Lua' forest fallow system of shifting cultivation." In *Farmers in the Forest* P Kunstadter, E Chapman, S Sabharsi eds. Honolulu (HI): University Press of Hawaii, 1978.

Chapter 18

LINKING BIOGEOGRAPHY AND ENVIRONMENTAL MANAGEMENT IN THE WETLAND LANDSCAPE OF COASTAL NORTH CAROLINA
The difference between nationwide and individual wetland permits

Nina M. Kelly
Ecosystem Sciences Division, Department of Environmental Sciences, Policy and Management
University of California – Berkeley, 151 Hilgard Hall #3110, Berkeley, CA 94720-3110, USA
mkelly@nature.berkeley.edu

Keywords: Landscape metrics, wetland permits, North Carolina, wetland management.

Abstract A set of individual wetland permit sites was compared with a set of nationwide wetland permit sites to determine the difference between their respective pattern on the landscape. Five landscape metrics (distance to water, permit site clustering, distance to other sites, amount of wetland found surrounding the permit site, and the variety of land cover found surrounding the permit site) were calculated for each permit site, and the statistical separability of the two permit datasets for each of these metrics was tested. The results indicate that in the study area of coastal North Carolina during the early 1990s, individual wetland permits and nationwide wetland permits were statistically different with respect to both their proximity to water, and to the amount of wetland vegetation found in the area surrounding the permit site. However, the two types of permits were not statistically different with respect to their spatial dispersion across the study area, and with respect to the amount of land cover heterogeneity found in the area surrounding each permit site. These results have management implications in light of recent changes to the wetland permit process in the United States.

INTRODUCTION

The spatial analysis of landscape-scale pattern has provided a rich literature in the last decade describing the links between landscape pattern and natural and anthropogenic processes (Ryskowski and Kedziora 1987, Turner 1990, Ales et al. 1992, Franklin 1995, Hulshoff 1995, Kamada and Nakagoshi 1996, Forman 1997, Ribe et al. 1998, Walsh et al. 1998). Such methods have also proved useful in analyzing patterns resulting from human-induced disturbances propagating through landscapes (Aspinall 1993, Kelly 1996, Kienast 1993, Li and Reynolds 1993). This chapter extends these biogeographical methods by applying them to a specific case of environmental management. The management of natural environments is an inherently spatial endeavor - requiring knowledge of context, proximity and change. Despite this, some forms of environmental management do not utilize techniques of spatial analysis to guide decisions (Allen 1994). This chapter reviews one application of spatial analysis of landscape pattern to understand a valuable, highly visible, and extensively managed habitat type - the wetland communities of coastal North Carolina.

These wetlands are ideal for spatial analysis, exhibiting measureable landscape structure (in discrete and continuous wetland communities) (Lyon and McCarthy 1995), function, and change (through increasing wetland development) (Turner 1990). Despite displaying all the classic characteristics necessary for landscape ecological analysis, rarely are they managed with a landscape approach (Bedford 1996, Kelly 1996). Indeed, a site-specific focus to wetland management in the United States has led to dramatic changes in the wetland ecosystem at landscape-scales, resulting in a re-configuration of the kinds and spatial distribution of wetland ecosystems over large areas, and a homogenization of wetland landscapes (Bedford 1996). These changes have occurred through the federal wetland management process that is organized under a permit system. There are two types of these permits: individual permits and nationwide permits. The legal definition clearly distinguishes between individual permits and nationwide permits, but the logic of the separation has been questioned. Landscape-scale pattern analysis, which has been used extensively in biogeography, but less often in environmental management applications, provides a novel approach to investigate the differences on the ground between the two permit types. In this work set in coastal North Carolina, a set of individual permit sites were compared with a set of nationwide permit sites to determine the difference between their respective pattern on the landscape.

The goals of this work are: (1) to test the differences on the ground between individual and nationwide permits; (2) to bring spatial analysis of landscape pattern to environmental management by examining a particular environmental law and its possible spatial context; and (3) to contribute to the body of work that insists that wetlands be managed with consideration of the spatial context of a wetland in a larger matrix of upland and wetland land cover.

WETLAND RESOURCES AND MANAGEMENT IN NORTH CAROLINA

North Carolina's coastal plain contains most of the state's wetlands (United States Geological Survey, 1996) (Figure 1). Estuarine emergent marsh (Cowardin et al. 1979) is the most common wetland type, consisting of

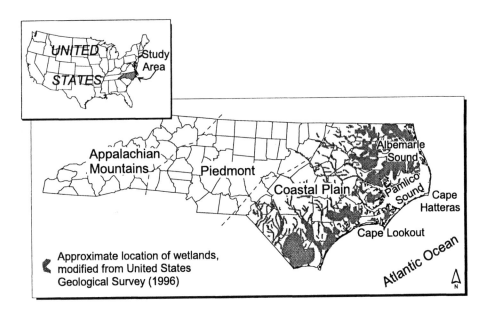

Figure 1. Wetlands of North Carolina and the physiography of the state. Most of the wetland area in the state is located on the Coastal Plain, redrawn from United States Geological Survey (1996).

salt marsh communities dominated by smooth cordgrass (*Spartina alterniflora*), and grading into rushes (*Juncus roemerianus*) in brackish areas (Mitsch and Gosselink 1993). These wetlands provide habitat for birds, spawning fish and other marine and coastal mammals, provide shoreline stabilization, and nutrient cycling (Thayer et al. 1978, Mitsch and Gosselink 1993). Coastal North Carolina also has extensive inland palustrine wetlands. Wooded marsh, in particular the famous pine pocosins and Carolina Bay wetlands, are found in the coastal plain area. These wetlands also provide habitat for bird and mammals, and perform important hydrological functions such as flood control. Wetlands here vary in size from isolated patches under 0.5 ha to large continuous stretches of 100s of hectares. Coastal urban development has been changing the amount and distribution of estuarine and palustrine wetlands. Since 1980, the number of applications for federal permits to alter wetlands, and number of permits granted, have increased annually in coastal areas (Kelly 1996).

Wetland Management and the Permit System

Wetlands are managed in the United States under the Clean Water Act through a permit system that requires the filing of a permit with the Army Corps of Engineers, the review of the permit by the public and by relevant state and federal agency, and finally the granting of that permit by the Army Corps of Engineers before dredge or fill material can be placed in a wetland (Dennison and Berry 1993, Want 1990). This permit system is designed to provide a regulatory burden on those wishing to alter wetlands (Dennison and Berry 1993). There are two types of permits issued for alterations to wetlands. The first of these are individual permits, which require extensive public and environmental agency review, and in many cases will be allowed only if compensatory mitigation (i.e. the creation or restoration of a wetland of equal or larger size) is performed (Berry and Dennison 1993, Zedler 1996, Brown and Lant 1999). The second class of permits is termed nationwide permits; these are issued for activities with no significant environmental impact. These activities require no review, and therefore they require no mitigation (These nationwide permits are actually a subgroup of a class of permits called "general" permits. In the study area in North Carolina, as in many areas of the United States, individual and nationwide permits are the two types of permits issued. For the sake of simplicity, this paper includes only the two types of permits found in the area: individual and nationwide permits). Of note is nationwide permit number 26, a focus for much of the controversy surrounding wetland management in the U. S. (Montana Audubon Society 1993, Race and Fonseca 1996, Race 1985). Nationwide-26 allows for alterations in isolated and headwater wetlands if the alterations are less than 4 hectares in size. The nationwide permit system in general, and the nationwide-26 permit in particular, has been heavily criticized. The nationwide permit system allows losses of small, but important wetlands (Semlitsch and Bodie 1998), and it virtually guarantees incremental, cumulative losses of wetlands (Race and Fonseca 1996). It also has been loosely applied, and with limited oversight, so that wetland alterations can proceed under the nationwide program that should require an individual permit (Montana Audubon Society 1993). Because of the first criticism, the federal law regarding permits is changing, and alterations to isolated and headwater wetlands (and now other non-tidal wetlands) can only proceed without review if they are up to 1.2 hectares in size (Deneen 1998). But the second criticism is harder to answer, and it introduces a final point about wetland management.

The two types of permits might be more similar than the law specifies. One way to investigate this question is to examine the information contained in each permit file to determine the amount and type of habitat lost with each permit, and make comparisons between nationwide and individual permits. Unfortunately, the permit files themselves can be hard to locate and the information contained within the files is often inconsistent, with variable accuracy (Kelly 1996, Kelly in review). A novel method to determine the separability of these two permit types is through the analysis of landscape-scale pattern. Since landscape pattern can indicate aspects of underlying physical process (Franklin 1995, Frohn 1998), landscape metrics might be a way to test the separability of these two permit types.

Study Area

The study area in eastern North Carolina covers a portion of the Neuse River watershed and a portion of the White Oak River watershed (Figure 2). This area is a nearly level plain of peninsulas dissected by bays and rivers. Elevations do not exceed 8 m above mean sea level; soils are unconsolidated sediments, and support estuarine, riverine, palustrine and lacustrine wetlands. The landscape is predominantly rural, with agriculture, forest, wetlands and shrub communities comprising nearly 90 percent of the area (Holman and Childres 1995). All wetland types have experienced significant alterations in the last 15 years through conversion to farmland and urban development (Holman and Childres 1995, United States Geological Survey 1996).

METHODS

Testing Permit Separability with Landscape Metrics

To test the separability of the two permit types, a series of ecologically and management relevant landscape metrics were calculated for two datasets of permit sites in coastal North Carolina (a set of the locations of 116 individual permit sites, and a set of the locations of 100 nationwide permit sites – these are described in detail later), and the statistical separability of each of the five metrics in each of the two datasets was tested. The landscape metrics measured (and a brief description) are:

- Distance from each permit site to the closest water body. This metric indicates the hydrological function of the wetland that was altered in the permit process - whether the wetland was estuarine or palustrine. The metric also suggests whether or not the proposed activity is "water-based," as in a dock, or not. Water-based activities usually require an individual permit (North Carolina Coastal Federation 1988).
- Spatial clustering of permit sites in the study area. This metric yields an overall statistic measuring the distribution of all permit sites in the area. A more dispersed distribution of permit sites would indicate permitted actions in areas where permitting is infrequent. Conversely, a highly clustered distribution would suggest permitting is taking place in highly localized areas, and might suggest that some areas of wetlands bear higher concentration of development than others.
- Proximity to other permits. This metric is different from (2) above. The closeness of permit sites to each other can be an indication of increased localized wetland development. This can take the form of a clustered development activity, for example, in a new marina development, several entities will apply for permits to build docks and alter wetland. This metric also gives an indication of the permitting frequency in an area. When activities are permitted increasingly close to each other, cumulative loss of wetland vegetation might be occurring.
- The amount of wetland in the area surrounding the permit site. This metric will yield information pertaining to the amount of wetland vegetation in the area surrounding the permit, and to the degree of isolation of the wetland being altered. Isolated wetlands, though often small in size, are extremely valuable for maintaining biodiversity and connectivity among species populations (Semlitsch and Bodie 1998). Removal of isolated wetlands can be detrimental to landscape-scale biodiversity provision and habitat. Small isolated wetlands are also altered frequently through the nationwide-26 permit program (Race and Fonseca, 1996), while applications to alter a portion of a larger wetland require the filing of an individual permit.
- The variety of land cover types in the area surrounding the permit site. The heterogeneity of land cover near the permit site is important for understanding the spatial context of permit activities. In newly urbanizing areas, where wetland permitting activity is frequent, the spatial context surrounding permit sites is varied, with high land cover heterogeneity (Kelly, 1996). In many cases, this landscape complexity is a result of conversion of large homogeneous areas of estuarine emergent wetland to different types of land cover - bare, and urban. Thus, in the

study area, landscape complexity is often associated with increased wetland alteration.

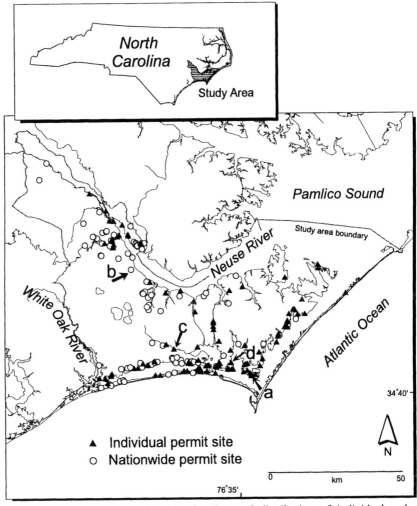

Figure 2. The study area in coastal North Carolina and distribution of individual and nationwide permit sites used in this study. Locations of four permits (a), (b), (c), and (d) used in Figure 3 are also shown.

Hypothesis Development

A null hypothesis was developed for each of the five landscape metrics to determine if the two samples (i.e. individual or nationwide permit sites) came from the same population, or from two different populations. (The nearest neighbor statistic was not tested for difference, as one only value was calculated for each point distribution. Instead R for individual permits and for nationwide permits is compared). Expectations were as follows, and are summarized in Table 1.

> (1) The distance to water from individual permits was expected to be greater than that for nationwide permits. Nationwide permits contain headwater alterations, and non-water-based alterations.
> (2 and 3) The distance between individual permit sites was expected to be smaller than that of nationwide permit sites, and a smaller nearest neighbor statistic for individual permits than for nationwide permits was expected. There were more individual permits in the study area, and their focus was water-based activity. In addition, nationwide permits were allowed in the upland areas, which would result in a more even spread of permit sites throughout the study area.
> (4) It was expected that individual permit sites would have more wetland in the area surrounding the permits than in equivalent areas surrounding nationwide permits, because individual permits are required when an alteration to a wetland is large enough to have a significant impact.
> (5) Finally, land cover heterogeneity was expected to be higher in the area surrounding individual permit sites as these are commonly in rapidly developing urban and newly urban areas with high land cover heterogeneity. Nationwide permits occur in upland areas, which are predominantly wooded, and permit activities in isolated wetlands.

Data and Preprocessing

Permit Sites

Wetland permit datasets were difficult to gather and ground-truth in North Carolina (Kelly in review). In other areas of the country by contrast, a long history of development of scarce wetlands has lead to an open, public and highly visible wetland management system (such as the case in California). In these areas wetland permit files are often published on the Internet, and public review is encouraged (i.e. California Resource Agency's California Wetlands Information System: http://ceres.ca.gov/wetlands/). Perhaps it is the extensive wetland resource base in North Carolina, combined with recent acceleration of permit applications that has lead to a permit process where the accuracy of the spatial information found in the permit files is often questionable (Kelly in review). Whatever the cause of the spatial errors, the site locations in the permit files required some form of ground verification.

Table 1. Development of null hypotheses to test the separability of individual and nationwide permits. Landscape metrics listed are: (1) Distance to water (m); (2) Clustering of permit sites; (3) Proximity to other permits (distance (m) to closest permit); (4) Wetland in area surrounding permit site (m^2); (5) Land cover heterogeneity in area surrounding permit site.

Landscape metric	Expectation for Individual permits	Expectation for Nationwide permits	H_0, H_a, test statistic, and decision rule
1.	Close: water-based development requires review.	Far: headwater and isolated wetlands, activity requires no review.	H_0: I ≥ N H_a: I < N t-test, $\alpha = 0.005$
2.	More clustered: same as above	Less clustered: same as above	Not tested, but compared.
3.	Close: increased development and focus along water	Far: activity throughout the study area	H_0: I ≥ N H_a: I < N t-test, $\alpha = 0.005$
4.	Large: large alterations, water-based activity	Small: isolated wetlands, small alterations, insignificant impact	H_0: I ≤ N H_a: I > N t-test, $\alpha = 0.005$
5.	High: urbanizing area	Low: isolated sites, in less complex upland	H_0: I = N H_a: I ≠ N Chi-square $\alpha = 0.005$

For this study, a collection of 116 individual permit sites which were field-verified using a Global Positioning System (GPS) were compared with a collection of 100 nationwide permit sites which were located on 1:24,000-scale topographic quadrangle maps. The method for finding the permit site, either in the field or on the map was the same - each permit file contained maps describing the location of the permit site. These maps were used to either locate the permit site in the field, where a GPS location was taken (Kelly 1996) or the map was used to locate the permit site on 1:24,000-scale maps and the location was transferred to a list. Because of the difference in ground verification methods, the two datasets have different measures of spatial accuracy and precision. The GPS location method is summarized in Kelly (1996) and provided spatial accuracy and precision of approximately 5 m. Locating the permit sites from the 1:24,000-scale maps allowed spatial accuracy and precision of approximately 20 m, which is the width of an arrow locating the permit site on the map. (The precision and accuracy of both datasets is acceptable for the analysis, which used a 100 m search area around the permit site to measure land cover variables). All of the permit sites were located within the study area. The individual permits cover permitted activities from 1989 through 1992, and the nationwide permits cover permitted activities from 1994 through 1996. The discrepancy in the dates is unfortunate, but these two datasets represent the only known compiled data for the area. Thus an assumption being made in this work is that the patterns expressed in each dataset were representative of overall changes, and that patterns in individual permitting will not have changed significantly after 1992.

The locations of the individual permit sites were transferred from the GPS data logger to Arc/Info point coverage format. The nationwide permit site database was copied from the North Carolina Division of Coastal Management in Arc/Info point coverage format. Both coverages were converted to the Universal Transverse Mercator projection system, using the NAD 83 datum. The permit sites are mapped in Figure 2.

Hydrography

A 100,000-scale digital hydrography coverage was used to determine the distance from each permit site to the closest water body. The hydrography layer was downloaded from the GeoData site provided by the United States Geological Survey (http://edcwww.cr.usgs.gov), imported to Arc/Info coverage format and then transformed to ArcView shapefile format. The data were re-projected to UTM (NAD83). The scale of this coverage provides sufficient detail to locate all the main waterways in the area. A coverage of finer scale data, provided at 1:24,000-scale has inconsistencies throughout the study area, and was not used. All hydrography data were provided and analyzed in line format.

Land Cover

Classified Thematic Mapper (TM) data were used to determine the amount of wetland in the area of the permit, and the variety of vegetation types surrounding the permits. A TM scene from September 1992 was processed according to NOAA's Coastal Change Analysis Program protocol (Dobson et al. 1995, Jensen et al. 1993) to produce a classified layer of eight land cover classes - water, urban/developed, grassland, cultivated, bare ground, sand, woody wetland, and emergent wetland. The final dataset had a spatial resolution of 30 m. The imagery was acquired in a morning pass, which minimized cloud cover and glare. The preprocessing routines used in this work and those covered by the C-CAP protocol include: geometric correction to within 0.5 pixel using GPS-gathered ground control points; radiometric correction using the black body method (in this case, black bodies included deep water, spectrally stable fallow agricultural fields, and an airstrip); and accuracy assessment yielding total accuracy of over 90% (Kelly 1996). The processed image was converted to Arc/Info's Grid format for further analysis.

Spatial Analysis

Distance to water was determined by a straight-line distance using the Pythagorean theorem embedded in a proximity routine in ArcView. The distance to water for each site was recorded in the point attribute table, and the data were exported to SAS for statistical testing. The proximity of each permit to its closest neighbor, and a measure of nearest neighbor statistic R were determined using a nearest neighbor routine provided by ESRI from their ArcScripts webpage (http://andes.esri.com/arcscripts/scripts.cfm), and modified. The nearest neighbor metric requires measuring the statistic R, which is tested against a random distribution to determine clustering. In the method, the expected average distance between points in a random point pattern is calculated: $R_{random} = 1 / 2(n/A)^{1/2}$, where n is the number of points in a pattern and A is the area of the study area; and the observed average distance between points ($R_{observed}$) is calculated as the mean of the distance from each point to its nearest neighbor. The nearest neighbor statistic R is calculated such that: $R = R_{observed} / R_{random}$. R is tested for significance by using a Z-score with 0.005 significance level. Where there is a clustered point pattern, $R < 1$, and $R = 0$ for a single point. For perfectly regular point patterns, R attains the maximum value of 2.149 (Barber 1988). The measure is highly susceptible to the size of the study area chosen to bound the point distribution, and so is not recommended as a comparison between different areas (Dale 1999). Since this work compares point distributions within a single study area, the results from the two permit datasets were comparable.

The amount of wetland in the area surrounding each permit site, and the variety of land cover types found in the vicinity of each permit site were determined using the focal and zonal functions in Arc/Info's GRID module. Focal functions calculate the characteristics of a user-defined area surrounding each cell in a raster array. Zonal functions determine the characteristics of a zone (a uniquely coded raster cell or group of cells) overlayed on a value grid. In this work the zone grid was a raster depiction of each permit site. A focalsum function was performed on the classified TM image to determine the amount of wetland within a circle of radius = 100 m. Determining the precise shape of a permit site is not possible, and the uniform shape of circle was used throughout this work to approximate the shape of the permitted alteration (Kelly, 1996). The size of the circle (area = 31,400 m^2, or 3.14 ha) was chosen to encompass most of the permitted actions found in the individual permit files. Similarly, a focalvariety function was performed to determine the variety of land cover classes found within a circle of radius = 100 m. The point datasets were each converted to raster format with cell size of 30 m. This process created two raster datasets, each with single pixels representing each permit site. The value of each cell in the raster grid was either zero, for background, or a unique number that distinguished it from its neighbor site. These two raster datasets were used as "zones" in the zonal functions that followed. A zonalstats function was performed for each permit dataset. In this operation, each zone grid (each permit site) was queried for its corresponding value in the "value" grid – the focalsum result and the focal variety result. The output were two grids for each permit type, each was a raster array whose value corresponds to either the sum of wetland in the vicinity of each permit site, or the land cover heterogeneity within each permit site. The two point datasets were then used to query each resulting zonalstat grid, and an item was added to the point dataset containing the necessary information on wetland amount or variety in the area surrounding the permit sites.

Hypothesis Testing

A decision rule and test statistic were decided for each metric. These are summarized in Table 1. The value for the test statistic for each metric was calculated in Statistical Analysis Software (SAS), and the significance of the test was determined. An unbalanced t-test was used to test the difference between nationwide permits and individual permits with respect to distance to water, distance between sites, and wetland in the area surrounding the permit site. A Chi-square test was used to test the difference between land cover heterogeneity in the area surrounding individual and nationwide permit sites.

RESULTS

Landscape Metrics and Differences Between Permits

The results of the spatial analysis are listed in Table 2. The individual permit sites were closer to water than were the nationwide permits. The individual permits ranged from 10 m to 1,000 m from water, and the nationwide permits ranged from 10 m to 5,600 m from water. The respective mean and maximum values confirmed the difference, and while the average distance of the nationwide permit site from water (718 m) was larger than that of an individual permit site (124 m), there were nationwide permits near to water. Nationwide permits were found much further away from water than Individual permits, with some permit sites over 5 km away from water. The nearest neighbor statistic R showed a similar clustering of individual and nationwide permit sites. $R_{expected}$ for individual permits was slightly smaller, due to a larger sample size (n = 116). The mean values of the closest neighbor were similar for the two permit types: 1,585 m (individual permits) and 1,805 m (nationwide permits). Both permit types displayed clustering (R= 0.5125 for individual permits and R = 0.5282 for nationwide permits), that was significant at the $\alpha = 0.005$ level. R for individual permit sites was slightly smaller than that for the nationwide permit sites, indicating a slightly greater degree of clustering.

Table 2. Results of spatial analysis. Landscape metrics listed are: (1) Distance to water (m); (2) Clustering of permit sites (nearest neighbor statistic R); (3) Proximity to other permits (distance (m) to closest permit); (4) Wetland in area surrounding permit site (m^2); (5) Land cover heterogeneity in area surrounding permit site (number of permit sites displaying various classes of land cover heterogeneity).

Landscape Metric	Measure		Individual permit	Nationwide permit
1.	Minimum		10.70	10.05
	Maximum		1,012.18	5,621.88
	Mean		124.66	718.57
2.	$R_{expected}$		3,093.8	3,418.5
	$R_{observed}$		1,585.6	1,805.6
	R		0.5125	0.5282
3.	Minimum		47.1	52.6
	Maximum		12,139.69	14,635.78
	Mean		1,585.6	1,805.6
4.	Minimum		900.00	0.00
	Maximum		31,023.52	2,281.84
	Mean		861.42	89.05
5.	Number of different land cover classes:	1	0	9
		2	6	10
		3	10	10
		4	23	26
		5	35	19
		6	21	18
		7	20	8
		8	2	0

The distance from each individual permit site to its nearest neighbor was almost identical to that of the nationwide permit sites. Both types of permits had examples of closely spaced permits (within 50 m of other sites), and both types of permits have isolated cases that were over 10 km from other permit sites. The maximum value might be unreliable for both types of permits, as isolated permit sites might be located close to permitted activity found over the study area border. In this case however, there were no problems with the study area border. The maximum value for the individual permits represents a permit site on the Outer Banks, in the southeast of the study area, and the maximum value for the nationwide permits represents a permit in the upland area in the western portion of the study area. The spatial distribution of permit sites is displayed in Figure 2.

Metrics 4 and 5 deal with land cover in the area surrounding the permit site. An example of each metric is shown for both individual and nationwide permits in Figure 3. Typically, individual permits displayed more wetland in

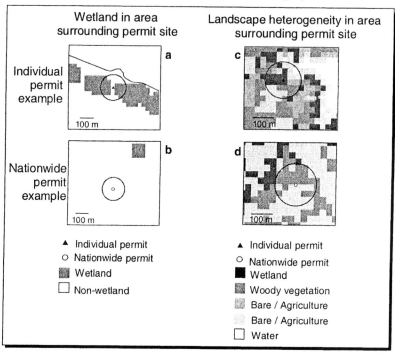

Figure 3. Examples of two individual permit sites and two nationwide permit sites showing land cover variation in the area surrounding each permit site. A 100 m buffer used to quantify the land cover surrounding each permit site is shown. Figures (a) and (b) show the amount of wetland in the area surrounding each permit site, and figures (c) and (d) show the amount of land cover heterogeneity in the area surrounding each permit site.

the area surrounding each permit (minimum = 1 pixel, or 900 m^2, and maximum = 31,023 m^2. The maximum figure approximates a circle filled with wetland habitat. Examples in Figure 3 include (a), which shows a typical individual permit site along a continuous strip of wetland habitat, and (b), which shows an individual permit without wetland habitat in its vicinity. Land cover complexity was similar between the two types of permits. The number of permit sites with different land cover classes is listed in Table 2. Both individual and nationwide permits showed a normal distribution with respect to the number of permits displaying a number of different land cover classes. Most permits in both classes were located in moderately complex land cover (4 - 5 different land cover classes in the vicinity of the permit). Figures 3 (c) and (d) show typical individual and nationwide permit sites.

Statistical Separability

Tests for statistical separability showed expected differences between permits, and revealed interesting similarities between the two types of permits. The results of the statistical tests are listed in Table 3. Individual permits were closer to water than were nationwide permits (t = 3.289) The difference was highly significant (α = 0.0001), and the null hypothesis was

Table 3. Results of Hypothesis testing. Landscape metrics listed are: (1) Distance to water (m); (2) Clustering of permit sites (nearest neighbor statistic R); (3) Proximity to other permits (distance (m) to closest permit); (4) Wetland in area surrounding permit site (m^2); (5) Land cover heterogeneity in area surrounding permit site (number of permit sites displaying various classes of land cover heterogeneity).

Landscape Metric	H_0, H_a, test statistic, and decision rule	Calculated statistic & level of significance	Result
1.	H_0: I $\geq$ N H_a: I < N t-test, α = 0.005	t = 3.289 α=0.0001	Reject H_0 I < N (permits were different)
2.	Not tested, but compared.		
3.	H_0: I $\geq$ N H_a: I < N t-test, α = 0.005	t = 0.076	Accept H_0 I $\geq$ N (permits were not different)
4.	H_0: I $\leq$ N H_a: I > N t-test, α = 0.005	t = 3.531 α=0.0005	Reject H_0 I > N (permits were different)
5.	H_0: I = N H_a: I $\neq$ N Chi-square, α = 0.005	Chi-square = 21.09	Accept H_0 I = N (permits were not different)

rejected. The difference between R for individual and nationwide permits was not tested for significance. The difference between the distance from individual permit sites to their closest neighbor, and those of the nationwide permit sites were not statistically separable (t = 0.076) and the null hypothesis of similarity was accepted. More wetland was found in the area surrounding individual permit sites than that surrounding nationwide permit sites (t = 3.531). This was a significant difference (α = 0.0005), and the null hypothesis of similarity was rejected. Land cover variety in the area surrounding individual and nationwide permit sites was not statistically different (Chi-square = 21.09), and the null hypothesis of similarity was accepted.

DISCUSSION

Significance of Separability

The results of separability for each metric have management significance. The degree of water-based development of a permit was measured as distance from a permit site to water. Development that is based on water usually requires extensive review (North Carolina Coastal Federation 1988, Beatley et al. 1994), and it is expected that development near water would require that an individual permit be applied for, and granted before activity began. This metric was the most obvious difference between the two types of permits, and results showed a statistical difference between the permit classes based on the proximity to water. It is also noteworthy that while many nationwide permits were located in an area of pine pocosin vegetation (with small, isolated wetlands in upland wooded complexes) on the peninsula between the Neuse and the White Oak, there were nationwide permits - and thus development activity - that occurred close to water, and presumably without review.

Individual and nationwide permits had an unexpectedly similar spatial distribution pattern. Clustering of permit sites indicates one of two things. First, clustering might indicate areas of intense local development, like a new housing and marina development. In such a development, several neighboring landowners would be expected to apply for permits to develop docks or dredge channels that connected them to the main water thoroughfare. These types of activities – if they alter wetland vegetation - require individual permits. Clustering of permits might also indicate that there is less wetland vegetation available for permitting. As development progresses, habitat "nibbling" occurs, and sites are inevitably closer together. With either explanation, nationwide permits were expected to be more evenly distributed across the study area. Instead, the distance between an individual permit site and its closest neighbor, and the distance between a nationwide permit site and its closest neighbor were not statistically separable. The nearest neighbor statistic confirms this similar clustering of both nationwide and individual wetland activities.

There was a difference between nationwide and individual permits with respect to the amount of wetland in the area surrounding each permit site. Individual permits were located in areas with more wetland vegetation than were the nationwide permit sites. This underlines an essential difference between nationwide and individual permits highlighted at the beginning:

nationwide-26 permits cover small, isolated wetlands that might not be visible in a classification of TM imagery, and individual permits are often applied for in areas of homogeneous estuarine emergent wetland, because such proposed activity would necessarily have a significant environmental impact.

In developing areas such as coastal North Carolina, wetland development occurs in areas with high land cover complexity (Kelly 1996). Since the upland area as depicted by classified TM is less complex than the urban area, this similarity between nationwide and individual permits means either that nationwide permits were occurring in rare upland areas that showed more complexity than expected, or that nationwide permits occurred in more complex urban areas.

Ramifications for Future Management

The nationwide permit system is changing in the late 1990s, and the U. S. Army Corps seeks to phase out the Nationwide-26 permit. The new rule will require that activities in all non-tidal wetlands under 1.2 hectares in size proceed under the nationwide program, and larger activities will require an individual permit and all the compensatory mitigation that type of permit requires. There are two aspects of this proposed change to consider here. First, the change will require that more activities be managed under the individual permit system. This work has shown that there are already important similarities between individual and nationwide permits, and suggests that such change is appropriate, and perhaps overdue.

The second aspect concerns compensatory mitigation. The wetland mitigation mandate that is associated with the Clean Water Act has been criticized for the small amount of created or restored wetland commonly required of mitigation projects, and for the lack of success of compensatory mitigation projects (Brown and Lant 1999). According to many, the mitigation system is not working to limit net losses of wetlands (Montana Audubon Society, 1993, Bedford 1996, Race and Fonseca 1996, Brown and Lant 1999). In light of this, all permitted wetland alteration should be limited. Changes to the wetland permit system that allow for more permitted actions to go forward, as long as they require mitigation might, because compensatory mitigation takes time to succeed or fail, be ensuring losses to wetlands that are not discernable until their impact is considerable and widespread. Clearly, the wetland management system still needs considerable review.

CONCLUSIONS

This work demonstrates an application of spatial analysis of landscape pattern to environmental management. By examining the spatial pattern of two types of permits jointly managed under the Clean Water Act's wetland permit system, it is clear that there are similarities on the ground that are not suggested in a reading of the legal definition which clearly separates the two types of permits. In the study area of coastal North Carolina during the early 1990s, individual wetland permits and nationwide wetland permits were statistically different with respect to their proximity to water, and the amount of wetland vegetation found in the area surrounding the permit site. The two types of permits were not statistically different with respect to their spatial dispersion across the study area, and the amount of land cover heterogeneity found in the area surrounding each permit site. These results were representative of the wetland management regime before the 1998 changes to the Army Corps changes to the nationwide permit system that tended to conflate the two permit systems.

Often wetlands are not managed at a scale appropriate for their function. Site-specific management of wetlands has lead to ecosystem-scale change to wetlands. This research contributes to the body of work that insists that wetlands be managed with consideration of the spatial context of a wetland in a larger matrix of upland and wetland land cover (Bedford and Preston 1988, Childers and Gosselink 1990, Johnston et al. 1990, Detenbeck et al. 1993, Schwarz et al. 1996, Semlitsch and Bodie 1998). Indeed, wetland legislation must focus not only on the size of wetlands, but on such spatial variables familiar to the landscape ecological lexicon: shape, context, and local and regional distribution patterns.

ACKNOWLEDGEMENTS

The author would like to thank several people who assisted in this research: Lori Sutter, formerly of the North Carolina Division of Coastal Management, for the nationwide permit data; Arlene Boscie of the National Marine Fisheries Institute Habitat Conservation Office, and Phyllis Ramel of the U.S. Army Corps of Engineers for assistance in collecting the individual permit files; Drs. John Helly and Dave Colby for statistical analysis.

REFERENCES

Ales RF, Martin A, Ortega F, Ales EE. Recent changes in landscape structure and function in a Mediterranean region in SW Spain (1950-1984). Landscape Ecology 1992; 7: 3-18.

Allen CD. "Ecological perspective: linking ecology, GIS and remote sensing to ecosystem management." In *Remote Sensing and GIS in Environmental Management*. VA. Sample, ed. Washington DC: Island Press 1994

Aspinall R. "Use of geographic information systems for interpreting land-use policy and modelling effects of land-use change." In *Landscape Ecology and GIS*. R Haines-Young, DR Green, SH Cousins eds. New York: Taylor and Francis 1994.

Barber GM. *Elementary Statistics in Geography*. New York: Guildford Press 1998.

Beatley T. Brower DJ, Schwab AK. 1994. An Introduction to Coastal Zone Management. Washington DC: Island Press 1994

Bedford BL. The need to define hydrologic equivalence at the landscape scale for freshwater wetland mitigation. Ecological Applications 1996; 6:57 - 68.

Bedford BL, Preston, EM. Developing the scientific basis for assessing cumulative effects of wetland loss and degradation on landscape functions: status, perspectives, and prospects. Environmental Management 1988, 12:751-771.

Berry JF, Dennison MS. "Wetland mitigation." In *Wetlands: Guide to Science, Law, and Technology*, MS Dennison, JF Berry eds. Park Ridge, NJ: Noyes Publications, 1993

Brown PH, Lant CL. The effect of wetland mitigation banking on the achievement of No-Net-Loss. Environmental Management 1999; 23:333-345.

Childers DL, Gosselink JG. Assessment of cumulative impacts to water quality in a forested wetland landscape. Journal of Environmental Quality 1990; 19:455-464.

Cowardin LM, Carter V, Golet FC, LaRoe ET. *A Classification of Wetlands and Deepwater Habitats of the United States,* Washington DC: Office of Biological Services, Fish and Wildlife Service, U. S. Department of the Interior, 1979.

Dale MRT. *Spatial Pattern Analysis in Plant Ecology*. Cambridge UK: Cambridge University Press, Cambridge, 1999.

Deneen S. Paradise lost: America's disappearing wetlands. E Magazine, November/December 1998; 36-41.

Dennison MS, Berry JF. *Wetlands: Guide to Science, Law, and Technology*. Park Ridge, NJ: Noyes Publications, 1993.

Detenbeck NE,. Johnston C, Niemi GJ. Wetland effects on lake quality in the Minneapolis/St. Paul metropolitan area. Landscape Ecology 1993; 8:39-61.

Dobson JE, Bright EA, Ferguson RL, Field DW, Wood LL, Haddad KS, Iredale H, Jensen JR, Klemas VV, Orth RJ, Thomas JP. NOAA Coastal Change Analysis Program (C-CAP): Guidance for Regional Implementation, Coastwatch Change Analysis Project, Coastal Ocean Program, NOAA, U.S. Department of Commerce, 1995.

Forman RTT. *Land Mosaics: the Ecology of Landscapes and Regions*. Cambridge UK: Cambridge University Press, 1997.

Franklin J. Predictive vegetation mapping: geographic modelling of biospatial patterns in relation to environmental gradients. Progress in Physical Geography 1995; 19:474 - 499.

Frohn RC. *Remote Sensing for Landscape Ecology*. Washington DC: Lewis Publishers, 1998.

Holman RE, Childres WS. Wetland Restoration and Creation: Development of a Handbook Covering Six Coastal Wetland Types. WRRI Project No. 50186. UNC-WRRI-95-3, North Carolina State University Water Resources Research Institute, 1995.

Hulshoff RM. Landscape indices describing a Dutch landscape. Landscape Ecology 1995; 10:101-111.

Jense, JR, Cowen DJ, Althausen JD, Narumalani S, Weatherbee,O. An evaluation of the CoastWatch change detection protocol in South Carolina. Photogrammetric Engineering and Remote Sensing 1993, 59:1039-1046.

Johnston CJ, Detenbeck NE, Niemi GJ. The cumulative effect of wetland on stream water quality and quantity: a landscape approach. Biogeochemistry 1990; 10:105-141.

Kamada M, Nakagoshi,N. Landscape structure and the disturbance regime at three rural regions in Hiroshima Prefecture, Japan. Landscape Ecology 1996; 11:15-25.

Kelly NM. An Assessment of the Spatial Changes to Estuarine Emergent Wetland in Coastal North Carolina under Section 404 of the Clean Water Act, Dissertation thesis, Department of Geography, University of Colorado, 1996.

Kelly NM. Spatial accuracy assessment of wetland permits in North Carolina. Cartography and Geographic Information Systems, in review.

Kienast F. Analysis of historic landscape pattern with a geographical information system - a methodological outline. Landscape Ecology 1993, 8:103-118.

Li H, Reynolds JF. A new contagion index to quantify spatial patterns of landscapes. Landscape Ecology 1993, 8:155-162.

Lyon JG, McCarthy J. *Wetland and Environmental Applications of GIS.* Boca Raton, FL: Lewis Publishers, Boca Raton, 1995.

Mitsch WJ, Gosselink JG. *Wetlands.* New York: Van Nostrand Reinhold, 1993.

Montana Audubon Society. *Protecting Montana's wetlands: an overview of Montana's section 404 program.* Helena MT: Montana Audubon Council, 1993.

North Carolina Coastal Federation. A Citizen's Guide to Coastal Water Resource Management, University of North Carolina Sea Grant College Program, 1988.

Race MS, Fonseca MS. Fixing compensatory mitigation: what will it take? Ecological Applications 1996, 6:94-101.

Race MS. Critique of present wetland mitigation policies in the United States based on an analysis of past restoration projects in the San Francisco Bay. Environmental Management 1985; 9:71-82.

Ribe R, Moranganti R, Hulse D, Shull R. A management driven investigation of landscape patterns of northern spotted owl nesting territories in the High Cascades of Oregon. Landscape Ecology 1998, 13:1-13.

Ryskowski L, Kedziora A. Impact of agricultural landscape structure on energy flow and water cycling. Landscape Ecology, 1 1987:85-94.

Schwarz WL, Malanson GP, Weirich FH. Effect of landscape position on the sediment chemistry of abandoned-channel wetlands. Landscape Ecology 1996, 11:27-38.

Semlitsch RD, Bodie JR. Are small isolated wetlands expendable? Conservation Biology 1998; 12:1129-1133.

Thayer GW, Stuart HH, Kenworthy WJ, Ustach JF, Hall AB. "Habitat values of salt marshes, mangroves and seagrasses for aquatic organisms." In *Wetland Function and Values: The State of our Understanding*, JRC Greeson, PE Clark, JE Clark eds. Minneapolis: American Water Resources Association, 1978.

Turner MG. Spatial and temporal analysis of landscape pattern. Landscape Ecology, 1990; 4:21-30.

United States Geological Survey. *National Water Summary on Wetland Resources.* Water Supply Paper 2425, U. S. Geological Survey, 1996.

Walsh SJ, Butler DR, Malanson, GP. An overview of scale, pattern, process relationships in geomorphology: a remote sensing and GIS perspective. Geomorphology, 1998; 21:183-205.

Want WL. Law of wetlands regulation. New York: Clark Broadman Co. Ltd., 1990.

Zedler JB. Ecological issues in wetland mitigation: an introduction to the forum. Ecological Applications 1996, 6:33-37.

Index

accuracy assessment of land cover mapping, 32, 294-296, 280
ADAR, 25-26
aerial photography, 73
   processing, 74
afforestation, 273
agroecosystem, 290-91, 296-297
airborne multispectral survey, 24-25
airborne data acquisition and registration 5500 (see ADAR)
area-perimeter index, 259, 261
ATE-BGC, 127-128, 130
AVHRR, 50-52, 54, 56, 59-60

β diversity, 110
barriers, 110, 113-121
   simulation modelling, 112-121
   trait analysis of *Ephigger ephigger*, 197-198, 201
biological inventory, 230

biophysical endowments, 102-103
Brauilo-Carrillion National Park (Costa Rica) (see also La Selva-Brauilo complex), 73

carbon balance, 131
   hypothesis, 125, 127
   krummholz, 127
   trees, 125
cellular automata, 109, 217
ceptometer, 50
confusion, in land cover mapping, 32-33
covariance, 203-204

database (MS Access), 217-218
deforestation, 70-71, 79-82, 92-93, 95, 97, 273, 301-304
digital elevation data, 25
digital elevation model, 172, 199
domains of scale, 54-55

ecological and elevation modelling, 29
ecotone, 123, 128
   biological spatial processes, 124
   location of, 125
edge density, 81
   habitat, 108
   total length, 100-101
end members, unmixing of, 16-17
environmental gradients, 241-242
Ephigger ephigger, 194-195
   contemporary selection and drift hypotheses, 196
   interpretation of traits across barriers, 198-200
   pattern analysis of traits, 203-205
error matrices (in land cover analysis and mapping), 32-33
error propagation, 20

feedback mechanisms, 132-135
forage,
   distance, 219-220
   geometry, 220
   quality, 220
   quantity, 220
   visibility, 221
foraging behaviour, 219
   landscape heterogeneity, 219-221
   inter-specific interactions with, 316
   intra-specific interactions with, 221
FOREST-BGC, 127, 130
FORSKA, 127, 131

forest mangement in Belize, 165
forest pattern analysis, 83-84
fractal dimension, 119-120, 260
   De Cola's 260, 266-268
   Gaustead's, 260, 264-268
   landscape, 113
fragmentation of,
   habitat, 81, 84, 109
   land, 92
   land cover, 297-300
   landscape, 76, 92-93, 110-113
   spatial distributions, 259
FRAGSTATS, 76, 100, 281

GIScience and effects of selection populations, 193
GPS, 75, 214, 292, 294, 317-318
   kinematic, 24-30
gradient directed ecological survey, 231
gradient directed sampling, 231
gradsect survey, 231
   efficiency of, 231
   representativeness of, 231
growth limitations, 134

habitat, 24-25
   Belding's Savannah sparrow, 34-42
   light-footed clapper rail, 28-34
   fragmentation, 80, 84, 109
   mapping, 32
household interviews, 292-293

image analysis, 294-296
   classification, 30, 76, 95-96

ISODATA, 30, 75
  unsupervised classification, 30-32
image, spatial variance in, 49, 55
image variance
  SAVI, 55, 57-60
  seasonal variations in, 57-60
implementation of GIS design, 224-226
inter-annual change in vegetation biomass, 93
interspersion/juxtaposition index, 99-101
inverse distance weighting interpolator (IDW), 222-223

James-Stein estimator, 198
Jostedalsbreen (Norway), 256-258

kriging, 20, 167, 171
krummholz, 9, 124-125, 127, 129-130

LAI, 7-8, 9, 11-12, 50-51, 131-132
  relationship with NDVI, 15-16
  relationship with tree cover, 10
  relationships with vegetation index, 8, 17-19
  tree, 16
land cover, 292-293, 297-301
  village perception of, 297
land cover change,
    physical environmental factors, 86
    proximity to intensive and recent land use, 86
    road networks, 82, 86
    shifting cultivation, 299-300
land cover maps, 77
  time series of, 77
landscape diversity, 81
landscape metrics, 81, 281-282
  wetland permit analysis, 313-315, 319-323
    amount of wetland in permit site, 313-315, 320-323
    distance to permit site, 313-315, 320-323
    land cover types near permit, 313-315, 321-323
    proximity of other permits, 314-315, 321-323
    spatial clustering of permits, 313-315, 320-323
land use and land cover change, 70, 78-79, 92, 97-106
  plant biomass, 96
  social processes, 103
La Selva-Braulio complex (see also Brauilo-Carrillion National Park), 79-80, 84-85
linear mixture modelling 13-14
litterfall, 50
LUCC (see land use and land cover change)
LULC (see land use and land cover change)

metapopulation dynamics, 139

MLIV (mean local image variance), 60
  seasonal vegetation dynamics, 61-63
Moran's I, 259, 262-265
Mojave Desert Ecoregion, 230
movement parameters of wildlife, 222-223
  moving direction, 222-223
  moving distance, 223
  moving sinuosity, 223
  moving velocity, 223
  temporal scale of, 223-224

NDVI, 8, 19, 93, 96-97
  LULC change, 97, 102
nearest neighbour analysis, 280-282
NetSURF, 200-201
non-linear species response, 133
nutrient averaging hypothesis, 125
  negative feedback, 127

object-orientated vector GIS, 218-219

parcel-level analysis, 279-282
Passerculus sandwichensis beldingi (Belding's Savannah sparrow), 24
patch mean size, 81-82, 100-102
  number of, 81-82, 100-102
pattern metrics and landscape structure, 100-101
phenology, of trees, 52, 55

population-environment interactions, 102-105
positive feedback mechanisms, 123, 134
post-classification sorting, 31
plant canopy analyser (LICOR LAI 2000), 11
plant-soil relationships, 165, 168-169
pre-settlement, 274-275
principal components analysis, 204-206

radiance, 12
radiometric calibration, 12
Rallus longirostris levipes (Light-footed clapper rail), 24, 28
reforestation, 99, 103
reflectance, 7
regional-level analysis, 279
regression modelling, 17
resource averaging hypothesis
  vegetation patterns, 133-134
RMS error, 13-14, 294-296

sampling, 258-259
salt marsh, 24, 28
Savanna aborizada, **50**
  image variance in, 60-61
SAVI (soil adjusted vegetation index), 56, 60-64
Seasonally-inundated Tropical Forest, 52
  image variance in, 63-64
Seasonal Semi-deciduous Tropical Forest, 52
  image variance in, 63-64
Senecio vulgaris, 179-180

differentiation of radiate and
non-radiate morphs, 186-187
non-radiate morph, 180
radiate morph, 180-181
evolution of, 181-182
inheritance of, 180-181
Senecio vulgaris var.
denticulatus, 180
Senecio vulgaris var. hibernicus, 180, 182
spread of radiate form, 184
S. vulgaris and *S. squalidus*,
spatial patterns of
hybridisation and
introgression between, 189
Seneco squalidus, 181-182
spread of, 183-184
rail network, 184-185
Shannon's diversity index, 81
shape index, 100-101
shifting cultivation (see swidden cultivation)
simulation modelling, 111, 127, 130
soil depth, 170-171
soil fertility and tropical forests, 164
soil moisture,
spatial variation of, 169
soil properties and tree species,
spatial variations in, 172-174
soil spatial variability, 164
species pattern
simulation modelling, 111
species richness, 120
succession of,
animal dispersed plants, 262
wind dispersed plants, 262
Sweetwater Marsh National
Wildlife Refuge (California), 25-26

swidden cultivation, 289-290, 299-304

Thematic Mapper (Landsat)
imagery, 12, 70, 72, 74-76, 95-96, 279-280, 292, 318
tidal influence on salt marshes, 35
topographic relief, 276-277
treeline ecotone, alpine, 8-9, 11, 13, 124, 126-127, 130-131

variance analysis of remotely
sensed, imagery, 56-57
variogram, 170-171
analysis, 169-171
vegetation,
greening-up and dieback, 48
phenological triggers, 48-52
vegetation index (VI), 14-15
VI-LAI relationships, 15-16
(see also SAVI)
vegetation pattern, 125-127

wetlands,
coastal, 24, 311-312
urban development, 312
wetland management permits, 311-312
individual permits, 310, 312
nationwide permits, 310, 312
wildlife movement, 213
GIS representation of, 214